FIBER-TO-THE-HOME TECHNOLOGIES

Fiber-to-the-Home Technologies

by

Josep Prat

Pere E. Balaguer

Joan M. Gené

Oscar Díaz

and

Sergi Figuerola

Universitat Politècnica de Catalunya

SPRINGER SCIENCE+BUSINESS MEDIA, LLC

A C.I.P. Catalogue record for this book is available from the Library of Congress.

ISBN 978-1-4419-5297-4 ISBN 978-1-4757-5219-9 (eBook)
DOI 10.1007/978-1-4757-5219-9

Printed on acid-free paper

TABLE OF CONTENTS

CHAPTER 3

Protocols and standardisation 117

CHAPTER 4

Projects & field trials 149

CHAPTER 5

Components 189

CHAPTER 6

Transmission impairments and monitoring 303

CHAPTER 7

PREFACE

In this book, an analysis of optical technologies involved in Optical Access Networks has been carried out. Current research status constitutes the main focus, although commercial level has also been addressed. These optical technological issues cover from network architectures to component development and integration in our aim to obtain a broad analysis of all technical elements and concepts that may be involved in near-future deployment. This step forward in access media implies a strategical revolution that depends on these techno-economical issues.

Optical fiber is foreseen as the future technology for the Access Networks (AN) in a remarkable common agreement among most of the telecom players: known as "Fiber-to-the-Home" (FTTH) approach. Only the optical fiber technology provides an unlimited capacity and performances needed for the near future broadband services. For example, high quality video services require about 10 Mbit/s per channel, and there is an increasing interest for personal bi-directional video communications, like video-conferencing between individuals. These and many other services can only be efficiently supported by the optical media, with all services (Video-over-IP, Voice-over-IP, etc.) integrated into a unique IP data flow (IP-over-optics).

This motivation leads the fiber to be considered as the right choice in new access network deployment in the near future and even today for new operators. However, because of the presence of the old copper pair/cable to the home, operators are trying to stress its capacity by using complex modulation techniques, like in ADSL, HFC, . . . , probably awaiting for the maturity of the FTTH technology, and minimizing the investment risks in the meantime; however, the copper cable is reaching its fundamental physical limits (bandwidth × distance) and is also a source of problems like crosstalk and electro-magnetic interferences especially in the return path. In a short period of time, it will constitute the bottleneck of the network for the new broadband demanded applications without any further extension possibility. Therefore, in this temporal impasse, there is a strategic urgency for the FTTH investigation and maturity.

This text is a state-of-art document that aims to cover all the related technological issues of Optical fiber Access Networks that have been studied, experimented, developed or tested up to this moment: architectures, modulation/codification techniques, electro-optic devices, optical fiber technology advances, FTTH field trials, international recommendations and consortiums, etc. In the book, both commercial and research perspectives are covered. They are at different levels, but the coverage may offer a deeper knowledge of the current state and future tendencies of FTTH technology. The main bibliographical sources are the relevant international journals, letters and conferences, standards of ITU-T, Eurescom, IETF, OIF, consortium projects of ACTS, IST, etc. Other references and Internet sites of interest have been also examined. Our aim is to cover all quality publications of the last decade up to this edition.

This book is geared to a wide range of interests and backgrounds. Best useful for engineers in electronics or telecommunications with fiber technology interests, an important part of the book is self-content for any person interested in the FTTH field. Since it covers all topic related in optical fiber access, we recommend it as a textbook in a master course in this field.

After the introductory chapter, in chapter 2, a review of the different network architectures is made. The PON (Passive Optical Network) architecture seems to be the best to offer bandwidth services to the customers. Different network parameters as scalability, up-gradability, cost, security, reliability, power consumption and others have been considered in the study of several optical access networks. The modulation formats have also been reviewed for their possible applicability to short and medium haul links. Finally, multiplexing techniques (TDM, WDM, OCDMA . . .) are also analysed in access.

In chapter 3, FTTH standards are reviewed. They are still in the development stage, but with a resolute way to progress forward. Roughly, half of PON products now in the market support the Full Service Access Networks specifications (G.983-FSAN) released recently by the International Telecommunication Union (ITU). More recently, a new series of specs is being developed by the First Mile Study Group of the Institute of Electrical and Electronics Engineers (IEEE) with a transport based on the IP Ethernet protocol.

In chapter 4, a relevant number of projects deal with optical access networks and their components. The aim of the projects is mainly to achieve a technological trade-of between cost and performances for the new broadband demanded services.

The different physical components used in optical access networks are analyzed in chapter 5, like fibers, transceivers, amplifiers, connectors, couplers, multiplexers, etc. Key points here are, for example, the fiber type and the integration of the electronic, optic and mechanical elements in the transceivers.

Other important factors in the performance of future access networks are the functions of maintenance and supervision of the operation and quality of the transmission. The different techniques proposed are analysed in the chapter 6, as well as a study of the different optical degrading effects present in the system.

The last chapter deals with the economical aspects related to the deployment of an optical access network. It has been illustrated by solving a typical case study. Point-to-Point and Point-to-Multipoint PON solutions are studied in order to figure out the main advantages and drawbacks in a practical deployment case.

The techno-economical advances in all-optical access networks, both in research and commercial levels, presented in this text, reveal that Fiber-to-the-Home can feasibly progress from the research stage to the massive commercial field in a hopefully short term period.

Acknowledgements

The work of this book comes from the collective effort of a group of the UPC in collaboration with external individuals and companies that we want to sincerely thank. Their professional opinions and technical information have undoubtedly contributed considerably to the enhancement of this undertaking.

Above all, we wish to refer to an important support from Nortel Networks that has given us valuable assistance in many ways. Our special thanks to Robin Rickard, Stewart Bruyn, Vincent Handerek, Richard Epworth and Peter Dyke. Their help has encouraged us to endeavor vigorously the development of this work.

We would also want to point out here that the general opinions or technical approaches expressed in this book are solely and exclusively the responsibility of the authors. It is therefore and obviously not the view, as they appear here, of Nortel or any other institution. The concentrated content of the book is basically an extended analysis of all substantial and valuable work in FTTH found worldwide, as well as our experiences as the Group of Optical Communications of the *Universitat Politècnica de Catalunya (UPC)*, that have been acquired in the last years of research and consultancy on FTTH and fiber communications in general.

Also, our profound thanks and acknowledgement go to Ricardo Tobar, Manuel Borras, Jordi Compte and the rest of the staff of the local operator flash10.com, for these years of close collaboration and efforts that we consider priceless.

Close to us, of course, we would like to kindly appreciate the cordial and unrelenting support of our colleagues Jaume Comellas, Sam Buxton, Gabriel Junyent and all those who have assisted in one way or another that are not mentioned here.

Chapter 1

INTRODUCTION

We are nowadays witnessing an extraordinary increase of transmission capacity in the core transport networks, thanks to the recent opto-electronic technology advances. These are often complex and expensive technologies to ensure the capacity and the resilience needed for the huge amount of transported information from thousands of individual users.

On the other hand, at the user end, the drastic increase in performances of the personal computers and devices, has allowed an expanding demand of Internet and multimedia services. IP data traffic is today surpassing the voice traffic in an exponential growth.

Between both the user end and the core network, there is a clear bottleneck, which limits the possibilities of the individuals to make use of this huge capacity available in the core network. The way to approach this capacity to the user in an efficient way does not have a clear solution currently. This is the goal of the Optical Access Networks (AN). The AN links every user to the local office or node network, which concentrates the traffic into huge digital pipes to the core network via the metro network.

Telecommunication and electronic media have commenced a process of transformation from "narrow-band" towards "broadband". The main driving forces that are responsible for this are the new broadband services and a need to integrate services into the house. The content of these services will play an important role in the process.

The evolution of a network is related to the evolution of the services. The congestion and bottlenecks due to new services can basically occur in the access network, the transmission network or in the servers. As it has been commented at the beginning, the access network is an important bottleneck

J. Prat et al., Fiber-to-the-Home Technologies, 1–10.

for most users. As new technologies and network deployments looking for more bandwidth are introduced in the access network, the bottlenecks may shift to the servers and it is likely that a cyclic problems and solutions will develop.

The new services that are now emerging in the market are mainly: Video of Demand, High Speed Internet, Videoconferencing, Telemedicine, Gaming, Telelearning, . . . All of them need a great bandwidth in order to offer a good Quality of Service (QoS); then, the access networks have to provide more bandwidth to run them and have to be designed to deliver these services with an admissible QoS.

According to the nature of the services received by the users, they can be classified into:

- *Audio Broadcast or direct video*: This kind of applications is similar to the ordinary radio and TV broadcast. The broadcasting can be done with an unicast transmission to each user or, more efficiently, with a multicast transmission.
- *Audio on demand*: Audio on demand applications allow customers to achieve comprised audio files which are located in servers. These files can be comprised of several types of content, from lectures or music to radio programmes. Wherever and whenever, a customer can demand an audio file from a server. Most audio on demand products for Internet market, after seconds delay, begin to reproduce the audio in the customer's device while the file continues being transmitted from the server. Several products offer also the possibility to interactive with the user (when the transmission begins, pauses, . . .).
- *Video on demand (interactive television)*: This type of application is analogous to the audio on demand applications, but video information is added in the data flow.
- *Telephony over IP*: The user can communicate with other users in real time through the network like the PSTN. These applications can offer both local and long distance telephone services with a lower price. Also it is possible to give group communication in real time, directory services, call filters, . . .
- *Videoconference in real time*: This service is analogous to the IP telephony, but the users are also connected in visual way. During a session, an user can open windows in order to see all the users connected (if they are placed in more than one place).

Services can also be classified according to the recover procedure of the information through the network:

- *Conversation applications*: The IP telephony and the videoconference in real time are examples of this kind of applications. The user can talk to and move whenever they want. The delay, from the user speaks or moves to the action is seen in the receiver, must be less than hundreds of milliseconds. For voice, the delay must be less than 400 ms (the top delay). Other values are not admissible.

- *Interactive applications with stored data*: Typical applications are both audio and video on demand. This kind of applications use stored data, because both the video and the audio have been loaded and stored into a disc before they are injected through the network. They are interactive because the user can demand to the server to start the transmission, to return to the beginning, to do pauses, . . . The delay, from a petition and the result in the user's host, is between 1 to 10 seconds to get an acceptable quality (time used en buffering). Jitter and delay requirements are not as restrictive as both telephony and video applications in real time.
- *No interactive applications*: For example, audio and video broadcast. The emitter generates the content without interruptions, and the user must only tune the broadcast like to it happens in traditional radio or television. Several tens of seconds can be allowed to get an acceptable quality service.
- *Mail applications*: The data traffic of these applications need neither temporal nor bandwidth requirements. They are, for example, the electronic mail and the electronic multimedia mail.

The next tables (Figure 1.1 and Figure 1.2) show a comparison of recommendations for both video coding and audio coding and give the bit rate necessary for each transmission

In addition to the services explained above, there are a lot of emergent services that begin to penetrate in the society, such as teleworking, telelearning, entertainment or videogamming, telemedicine . . . The Figure 1.3 depicts the relation between information content (in bits) and response time requirements for different services.

Since the necessary optical and IP technology is already feasible, the only current limitation is the cost although this is already comparable to the old technologies. Considering the expert's opinion, when the access fiber will be widely deployed is an issue limited to a short-range period between one to ten years. The AN has the highest impact on the global cost per user in the telecom network: there is a physical point-to-point connection to every user home (local loop), without concentration or bandwidth sharing. The high cost of the Core Network is divided by the high number of clients in the national area, resulting in a low impact per client. The investment is so high that it has to be carefully analyzed before in all aspects: technologies, infrastructure and services. However, it is assured that the investment will be amortized, since it is a durable solution where no time limit is foreseen in the capacity of the access fiber. Because of this there is an important effort for standardization. In some evolution studies, the broadband IP access is

Video compression system	Application/ target	Resolution (number of pixels × number of lines)	Bit rate	Application network	Related system recommendation	Features
H.261	Videophone, teleconferencing	180×144 – 360×288	56/64k– 2 Mbps	ISDN, digital leased line	H.320, 321, 322, 323	Basic video compression
MPEG1	Storage media	360×240/288	1.1 M– 1.5 Mbps	CD-ROM	–	Fast forward and reverse (back prediction) word
MPEG2/H.262	Storage, communication, broadcasting (digital broadcasting, etc.)	720×480 – HDTV	3M– 80 Mbps	ATM, satellite, radio, xDSL, DVD, etc.	H.310	Improvement and extension of H.261
H.263	Portable terminal, mobile unit Internet	128×96 – 1408×1152	10k– 2 Mbps	IP network, analog telephone line, radio	H.323	Improvement and extension of H.261
MPEG4	Portable terminal, mobil unit Internet	180×144 – 720×480	5k– 5 Mbps	IP network, analog telephone line, radio	MPEG4	Object base coding

Figure 1.1. Comparison of recommendation for video coding.

Audio compression system	Application/target	Audio band (khz)	Bit rate (kbps)	Features
G.726	Corporate communication (digital leased line), DCME	300–400	16/24/32/40	ADPCM, G3 FAX possible at 40 kbps
G.722	Voice in teleconferencing, AM broadcasting transmission	50–7000	48/56/64	SB-ADPCM
G.728	Corporate communication	300–3400	16	LD-CELP, delay time 2 ms or less
G.729	Portable terminal, corporate multimedia communication	300–3400	8	CS-ACELP, echo canceller necessary
G.723.1	Analog TV communication	300–3400	5.3/6.3	V-CELP
MPEG1	Storage and recording media	15/20	Layer 2, 128/ ch	Sub-band coding, psychological auditory model
MPEG2BC	Application expansion to broadcasting (surrounding)	7/20	Layer 2, 128/ ch	5.1 ch multi-channel and multi-lingual mode added to MPEG1
MPEG2AAC	Application expansion to broadcasting (surrounding)	4/40	128/ 2 ch	Bit rate improved in addition to MPEG2BC

Figure 1.2. Comparison of recommendation for video coding.

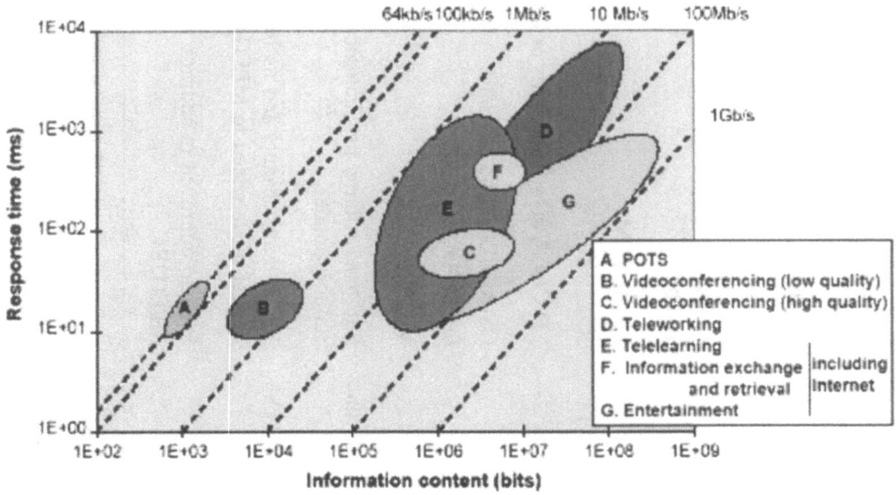

Figure 1.3. Main Access Services requirements (Source: HHI).

expected to almost reach the number of conventional CATV accesses by the year 2010 (see Figure 1.4).

In the access context, the optical fiber shows great advantages compared to conventional techniques, based on copper or radio. The main basic advantages of the fiber,

- the low loss and
- the high bandwidth

are important here but not critical as they surpass the necessities of the access. There are other advantages that are equally appreciated here:

- its immunity to the electromagnetic interferences (that constitute a real limiting problem with copper in dense urban areas);
- the low size and weight; in the AN, one or two fibers per home or business are to be installed, leading to a very high fiber and cable counts in the massive deployment. The ease for the installation of dense cables and the last drop of fibers to the home become differentiating advantages compared to the copper cable;
- other issues, like the fiber isolating material that avoids the necessity for grounding, are also important in this context.

All these points lead the fiber to be considered as the first choice in new access network deployment in the near future and even today. FTTH investigation has to deal with the possible problems of this technology. Its main classical drawbacks considered, the cabling easiness supporting small radius curves and the fragility of the connections, are being already solved by research

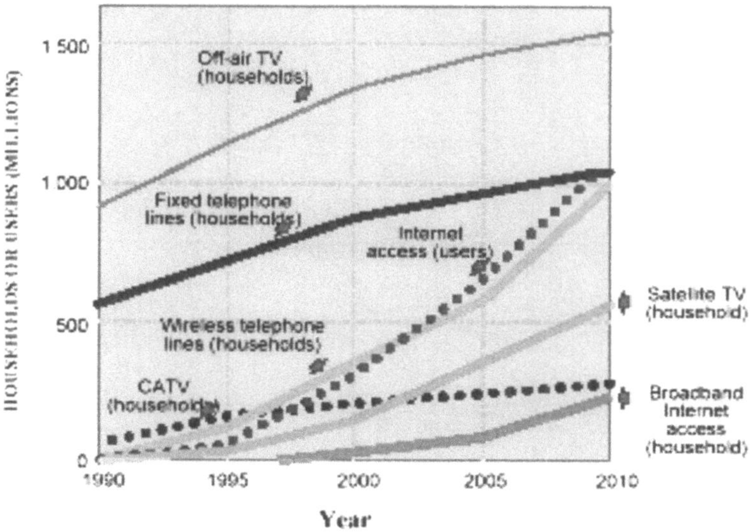

Figure 1.4. Evolution of access techniques (Source: CSELT).

labs and manufacturers. So, the principal limiting point is the cost: not the cost of the fiber, or of the connections, or the cabling (they are similar and becoming lower than the copper's), but the cost of the electro-optical converters and other specific elements. These are the main extra elements that are needed in FTTH, and those have a direct impact on the cost of the AN investment and deployment. Research & Development have an important open field here, aimed to optimize cost via electro-optical-mechanical integration and saving techniques, maintaining or even increasing the performances. The residual cost will be clearly compensated by the incredible broadband bidirectional applications and service quality reaching homes and businesses.

When talking of access, there are other techniques that combine fiber and copper for the last drop, getting closer to the user, that have to be mentioned (see Figure 1.5) for reference:

- FTTcab (fiber to the cabinet at the local area, and copper distribution from there to the homes); this is often used in xDSL.
- FTTCurb (to the last group of houses/buildings). This is typically used in HFC-CATV.
- FTTBuilding. The distribution inside the building to the apartments is made with copper (e.g. category 5 twisted pair), and an active element is installed at the basement or top. This is also sometimes referred to as FTTBusiness.
- FTTHome. The fiber reaches the inside of the user premises without any copper cable in between, having then a personal direct unlimited capacity link for its own future needs. This last is the target scope of this book.

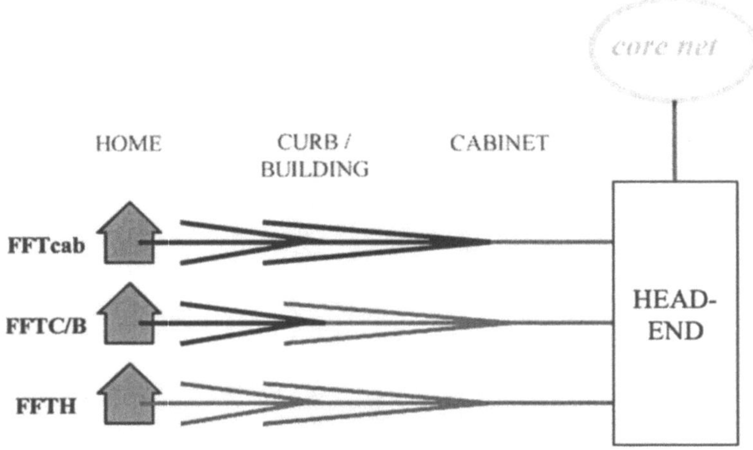

Figure 1.5. Fiber/Copper Access Techniques.

In all these techniques, the copper last drop will sooner or later become a bottleneck; the active intermediate element will have to be powered, re-installed in every up-grade, etc. An all-optical (passive) access network (FTTHome) definitively can solve this. FTTH is also often known as FTTU (user) or FTTD (desk), stressing the idea of reaching close to the end user without any other physical medium in between.

The technological solutions that have to be adopted in an optical access network are basically different compared to other optical networks, because its fundamental requirements are completely different. The basic requirements are defined by the scenario and telecom demands of the users. We can graphically represent them in the next two figures. In Figure 1.6, the different optical networks in terms of the distance span and the demanded bandwidth (bit rate) for the current and new services are presented.

An Optical Access Network covers a local area which commonly is in a distance of 10 Km or shorter from a central Head-End of the local telecom service center. This average distance is analysed in Figure 1.7, where the average local loop distance length is done for several countries. The information bandwidth that is expected to be transported in each fiber is between 10 MBit/s and 1 GBit/s expressed in bits/second since the new services are digital services, coming from the IP layer in the new integrated data services approach.

Neighboring the AN, there are the Local Area Networks (with a shorter reach) and the Metropolitan Networks (to about 100 Km and higher capacity). In between, we may have some smooth borders where the specification may be similar.

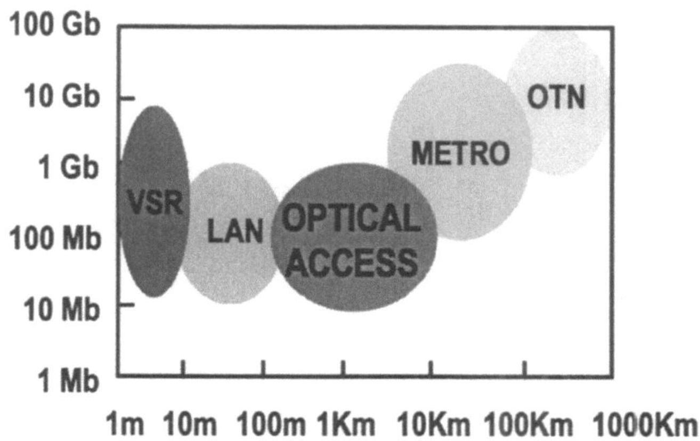

Figure 1.6. Optical Network scenarios regarding distance and bit-rate.

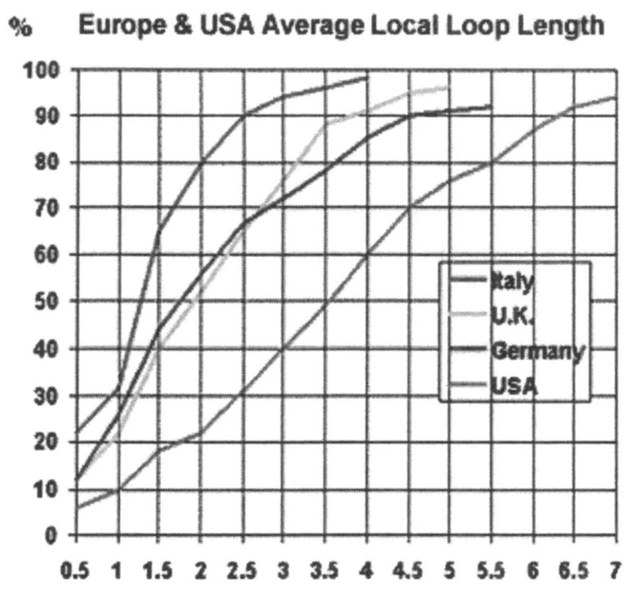

Figure 1.7. % Europe & USA Average local loop length.

In a second specification graph, the requirements are further differentiated: Figure 1.8 shows the number of terminals (or end users or nodes). In an AN, a very large number of homes (up to some thousands) can be serviced, while the metro networks have a very limited number of nodes.

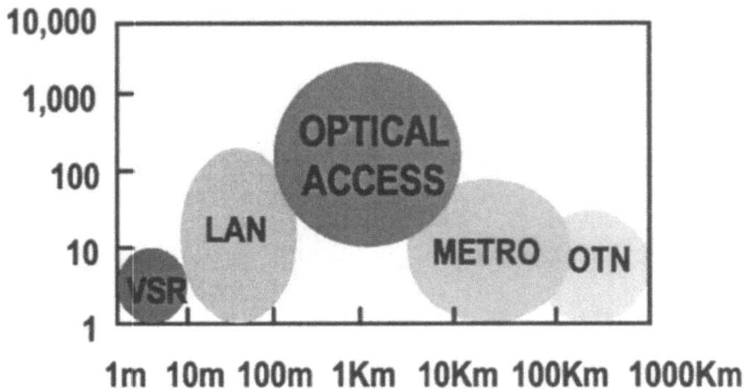

Figure 1.8. Optical Network scenarios regarding distance and number of E/O interfaces.

In conclusion, the design requirements of an Optical Access Network and its elements are completely different from the rest that are more mature optical networks. New elements and architectural concepts will have to arise.

DOCUMENT OVERVIEW

This report is a state-of-art document that aims to cover all the related technological issues of Optical fiber Access Networks that have been tested, experimented, studied or developed up to this moment: architectures, modulation/codification techniques, electro-optic devices, optical fiber technology advances, FTTH field trials, international recommendations and consortiums, etc.

Our main bibliographical sources have come from the principal international journals, letters and conferences, standards of ITU-T, Eurescom, ODSI, IETF, OIF, consortium projects of ACTS, IST, etc. Other references and Internet sites of interest have been also examined. Our aim is to cover all quality publications of the last decade up to the publication date.

Chapter 2

ACCESS TECHNIQUES

INTRODUCTION

In this section we will deal with several aspects concerning the different optical access techniques. Is difficult to summarize all the matters that must appear in a study like this but we have tried to put them into a logical order. The criterion chosen has been the layer depth, from physical level to MAC level.

In section 2.1 Network Architecture, a definition of the access network model has been given as a starting point. After that some network elements have been defined individually. Then a review about the different types of distribution networks has been performed giving place, this way, to a presentation of all possible optical access networks architectures. The most important one is the Passive Optical Network (PON), that will be taken as a reference in most cases under study in this work. Finally a comparison between passive solutions versus active solutions has been pointed out in order to check the advantages and drawbacks of both techniques.

Section 2.2 Modulation Techniques deals with several possible signaling systems to be used in a future optical access network. Probably the importance of this point would be greater in a metropolitan environment but due to the close relationship with the access loop, we considered an important issue to be taken into account. Basically there exist two kinds of modulation techniques: with direct detection and with coherent detection. The first ones are the most extended due to its simplicity and the big hurdles of the optical coherent technologies. Together with the classical NRZ and RZ modulation formats, multilevel signaling schemes have been taken into account being

J. Prat et al., Fiber-to-the-Home Technologies, 11–116.

the Optical Duobinary Modulation the most attractive one due to its high dispersion tolerance and bandwidth reduction.

In section 2.3 Multiplexation Techniques, a review of possible multiplexation schemes has been performed: Optical Time Division Multiplexing (OTDM), Sub-Carrier Multiplexing (SCM), Wavelength Division Multiplexing (WDM) and Optical Code Division Multiplexing (OCDMA). The last one has taken a lot of attention in last few years due to its asynchronous mechanism and its study is larger compared to the rest.

Finally, we discuss about the different solution to the Medium Access Control mechanism proposed in section 2.4 Multiple Access Techniques. The main classification can be as follows : using deterministic strategies, using random strategies or using Request-Grant strategies. In principle seems that the last ones will be the most suited by all standardization committees involved. In this sense a Dynamic Bandwidth Allocation (DBA) can be implemented.

2.1. NETWORK ARCHITECTURE

Introduction

In this chapter generic access network is defined and evaluated according to its architecture. The functions of each part of the network are studied. There is a deep study about PON (Passive Optical Network), thus it is the kind of technology that may be deployed in the near future. Its elements are studied and the best architecture solution is met. A comparison between active and passive systems has been done and also a comparison between unidirectional and bi-directional systems. Finally, a brief description about projects which are developing advanced architectures and a review about most important field trials are exposed.

2.1.1. Access network

An Access network is defined as the set of equipment and infrastructure needed to perform the connection between the user's equipment and the carrier premise. The carrier premise is the point at which service providers interconnect with the access network. The local access network is also known as "the last mile" and "the local loop".

Several types of access networks can be deployed, but this book is focused on optical ones, since they are considered the best way to build a flexible, future-proof and full-service network platform with potentially unlimited capacity. In the past, access networks were optimised for a specific service; nowadays FTTH can be designed from the beginning as general service platform.

An optical access network is generally integrated by an OLT (Optical Line Termination), ORNs (Optical Remotes Nodes) and ONUs (Optical network unit). The OLT is the key element to perform the union between the access network and the metropolitan network. The remote nodes are in charge for distributing the information from the hub to the ONUs. Then, each hub can be connected to several remotes nodes and each remote node can be connected to several ONUs. The network between the hub and the remote node is called *feeder network* and the network between the remote node and the ONUs is called *distribution network.*

In the telecommunication literature, the elements of an access network can have other names depending on the type of networks. The OLT is also called Hub, HE[1] (Head-End), CO[2] (Central Office) and LS (Local Switch). The ORN is sometimes called ODN (Optical Distribution Node). Finally, several names are used to design the ONU: NIU (Network Interface Unit), ONT (Optical Network Termination) and CPE (Customer Premises Equipment).

Before classifying optical access networks, it is important to observe the need to implement them in the near future due to the new services proliferation. For example, if the services shown in Figure 2.1.2. are considered as types of services it is possible to define several classes or profiles of users. The bandwidth (amount of information that flow from one site to another in a given amount of time) of each service corresponds to downstream direction. It is

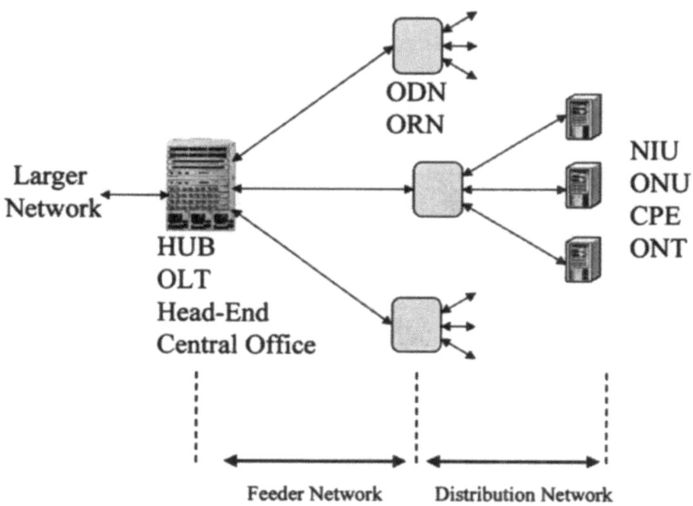

Figure 2.1.1. General structure of an access network.

[1] Usually in cable television networks
[2] Generally for telephone systems

Service	Banwidth	Profile 1	Profile 2	Profile 3	Profile 4
Telephone	0,064	X	X	X	X
Internet access	2	X	X	X	X
Video broadcast	4		X	X	X
Video on demand	4			X	X
Voeoconference	2				X
Total bandwidth (downstream)		2,064	2,064	10,064	12,064

Figure 2.1.2. Example of table of services.

important to take into account the current trend that each user can be content producers, and therefore the upstream direction would have to be similar capacity to the downstream direction. Residential communications could move from a broadcast model where few entities manage and produce information to a all-producers model where each costumer can generate content.

In this case, five services have been taken into account and four user profiles have been defined. More services and other combinations of them are able to originate other profiles. Also, only one channel or capacity for service has been considered, i.e., it is possible to have two TV receivers in home with two different channels and the capacity of video broadcast will increase until 8 Mbps. In all cases, these profiles are the basic ones or the minimum capacity needed for each profile. The total bandwidth values of each profile can be changed if new emerging services and more channels for user are considered.

In order to calculate the total bandwidth of each PON it is necessary to estimate a percentage of each profile in each PON and consider two simultaneous indices: the first one for simultaneous users and the second one for the simultaneous services.

Access networks can be classified according to the type of distribution network or the type of feeder network.

Types of distribution network

Distribution networks can be broadcast or switched. In the broadcast networks, the remote node sends all information (this information may be in different optical channels) that it receives from the OLT (without electronic process) to all ONUs which are in charge of discarding or processing the information frames. In the switched networks, the remote node processes the information and sends the data frames to the destination ONUs (independently whether the service is broadcast or not). The information process implicates an added delay in the transmission if electronic devices are used (and other disadvantages: power supply, environmental conditions, etc.); however, this process can be done optically if WDM technology is used with AWG devices. Obviously, switched networks have a high security level compared to broadcast networks where one user receives all the users' information.

Types of feeder networks
Feeder networks can either assign to each ONU its own bandwidth or share the total bandwidth among all the users. In this case, each ONU has all its bandwidth in short time slots and a medium access control protocol (MAC protocol) is necessary to coordinate ONUs access.

If traffic to/from ONUs is burst, it is more efficient to share the total bandwidth among all the users than to assign a fixed bandwidth to each ONU. However, with a fixed bandwidth to each ONU, it is easier to guarantee a Quality of Service and to have electronic/optical devices that work at less speed.

2.1.2. Functions to be done by access network elements

The main functions are the following:

Optical Line Terminator
- Optical reception and transmission.
- Control functions in the distribution network: control of the transmitted power and received power, FEC (Forward Error Correction) and interleaving.
- OLT is in charge of bandwidth arbitration for the upstream channel in the MAC protocol (if this protocol is necessary).
- Optional is to do functions of higher levels, as address resolution, address assignments, tunnelling, etc.

Optical Network Unit
- Optical reception and transmission.
- To change the optical signal into electrical signal and vice versa.
- Cooperation with the OLT to control the transmitted power from the user to the operator.
- FEC and interleaving.
- In cooperation with the OLT, the ONU is in charge of supervising the MAC protocol in the upstream channel.
- The ONU and the residential gateway are responsible for multiplexing/demultiplexing sessions from multiple sources (voice, data, video) in the optical link.

Because of the complicated nature of the interface functions, a component called the residential gateway or home network gateway will possibly decouple the purely optical functions of an ONU from high-layer protocol functions. The residential gateway is similar to a cable modem used in HFC networks or a DSL modem.

Some believe that an outdoor ONU (typical in the garage or basement) would increase location flexibility, facilitate maintenance access, and provide a cleaner demarcation for the network interface. Others believe that the FTTH system will benefit more from the cost reduction of an indoor ONU, thereby making it look more like a piece of consumer electronics.

The ONU function is to convert the optical signal into electrical signal and the reverse operation (electrical to optical). However, it is also possible for the residential gateway and the set-top-box to meet in the same device. Nevertheless, these set of functions can be put in separated devices (Figure 2.1.3.)

2.1.3. Power budget

A key factor in an optical system design is the power budget. Many other issues with respect to the final design such as operating wavelength, selection of transmitters, receivers, fibers, compatibility of components and cost versus

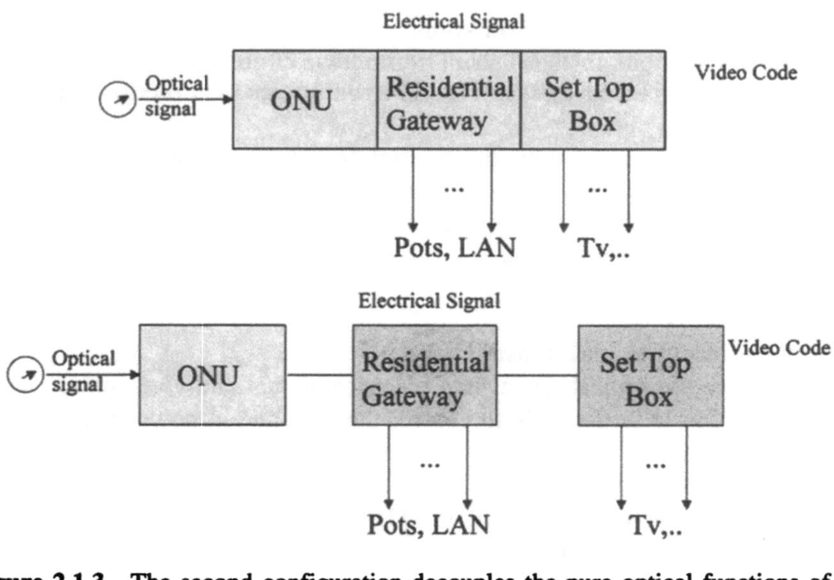

Figure 2.1.3. The second configuration decouples the pure optical functions of an ONU from high-layer protocol functions: ONU, residential gateway and set Top Box. The former configuration has all the functions in one device.

performance will be treated. The role of the power budget is essential for the correct performance of the network. The aim of the power budget is to ensure that enough power reaches the receiver to obtain and maintain good performance during the system lifetime.

The power budget takes a simple form when it is expressed in decibel units and the optical powers in dBm.

$$\overline{P}_{tr} = \overline{P}_{rec} + C_L + M_S$$

where
$\overline{P}_{tr}$ is the average launch power specified for each transmitter;
$\overline{P}_{rec}$ is the receiver sensitivity (minimum average power required by the receiver);
C_L is the total channel loss;
M_S is the called system margin (additional sources of power penalty that might develop during system lifetime due to component degradation or other events). A system margin of 6–8 dB is generally allocated during the design process.

The takes into account all possible sources of power loss, including connector, splice losses and device losses. Therefore, a possible expression for channel loss could be:

$$C_L = \alpha_f \times L + \alpha_{con} + \alpha_{spl} + \alpha_{dev}$$

where
α_f is the fiber loss in Db/Km;
L is the maximum fiber length;
α_{con} is the connector loss;
α_{spl} is the splice loss;
α_{dev} is the device loss.

Sometimes splice loss is included within the loss of the fiber cable. Also, the connector loss includes connectors at the transmitter and receiver ends and other connectors if used within the fiber link as devices connectors.

Figure 2.1.4. shows a practical example to calculate the receiver sensitivity of a system to correctly keep the power budget of the system.

A maximum distance of 20 km link between OLT and ONU is considered. The transmission is in the third window and the parameter of attenuation and splice is the 0,25 dB per Km. The optical link has an optical splitter which insert a 12 db loss to give the signal to 16 users. Each connector has 0,5 dB as insertion loss. The system margin considered is 6 dB. The power launched by the transmitter is –4 dBm.

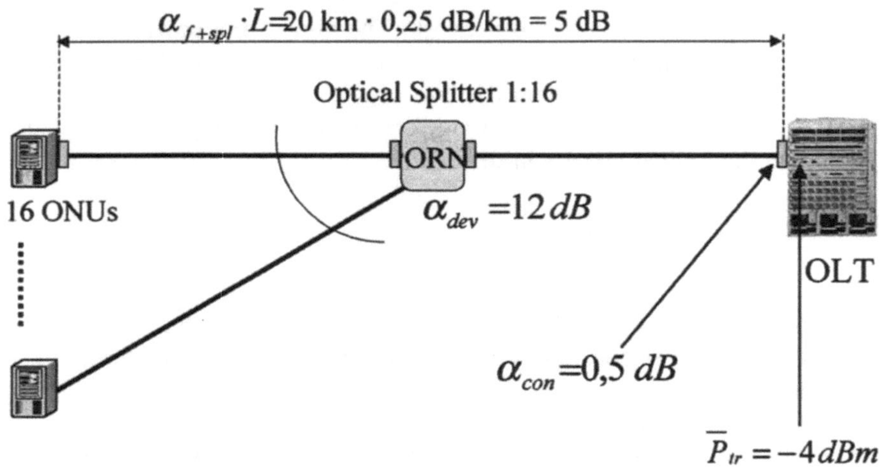

Figure 2.1.4. Power Budget example.

The contributions for attenuation parameter are as follow:

$$(\alpha_f \alpha_{spl})L = 0{,}25\,\frac{\text{dB}}{\text{Km}} \times 20\ \text{Km} = 5\ \text{dB}$$

$$\alpha_{con} = 0{,}5\,\frac{\text{dB}}{\text{connector}} \times 4\ \text{connector} = 2\ \text{dB}$$

$$\alpha_{dev} = 12\ \text{dB}$$

And C_L has the value of

$$C_L = 5\ \text{dB} + 2\ \text{dB} + 12\ \text{dB} = 19\ \text{dB} \rightarrow C_L = 19\ \text{dB}$$

Finally, the equation of power budget in dB is:

$$\overline{P}_{tr} = \overline{P}_{rec} + C_L + M_S \rightarrow -4\ \text{dBm} = \overline{P}_{rec} + 19\ \text{dB} + 6\ \text{dB}$$
$$\rightarrow \overline{P}_{rec} = -29\ \text{dBm}$$

In order to keep the power budget in the designed system, the receiver sensitivity has to be less than –29 dBm.

Finally, the Figure 2.1.5., the table depicts the optical power value in several points of the system.

At the H point the value of the optical power is –23 dBm. This value has to be decreased due to the margin system. The margin system is 6 dB. Then, the receiver sensitivity is –29 dBm as it has been calculated before.

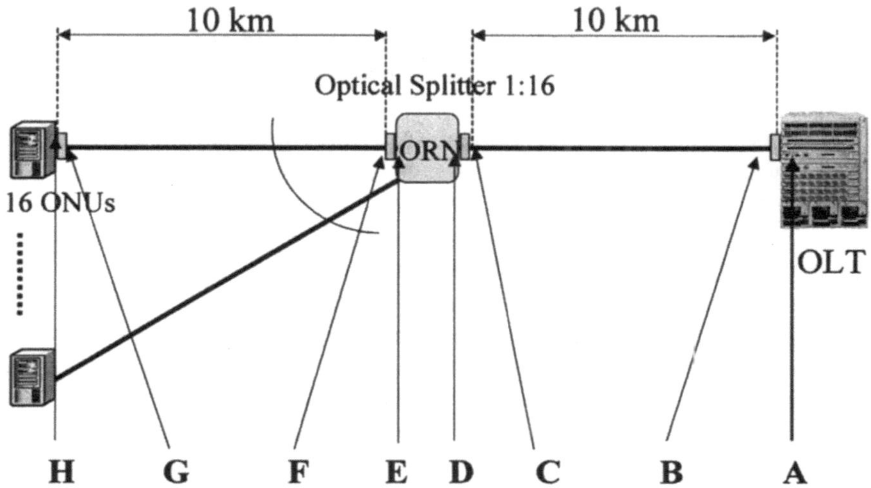

Point	*Optical Power*
A	-4 dBm
B	-4,5 dBm
C	-7 dBm
D	-7,5 dBm
E	-19,5 dBm
F	-20 dBm
G	-22,5 dBm
H	-23 dBm

Figure 2.1.5. Optical power value in several system points.

2.1.4. Active versus passive architectures

In this chapter, the features of using or not electronics in the remote nodes are analyzed. Several peculiarities from distribution networks and feeder networks are picked up and classified into two groups: active networks and passive networks.

In an active network, the information is processed and sent to the suitable output port/s;[3] i.e., the last link (RN-ONU) is a point-to-point link and each ONU only receives the information addressed to it whereas in a passive network the information is not processed electrically in the remote node (it could send or not all information depending on the designed architecture)

[3] It is like a switch or brigde in Ethernet words. It is not a simple repitter.

and the ONU is in charge of accepting or discarding the received information. For the active systems, the MAC protocol is not necessary and for the passive systems if the optical channel is shared among users, a MAC protocol will be mandatory.

As it is mentioned above, the main active networks advantage is that the link between the remote node and the ONU is dedicated and, therefore, it is possible to use electrical/optical devices with a less cost in the ONUs. In the passive networks, all the devices have to work with the total bandwidth of the optical channel, and the ONU cost is increased.

From the information privacy point of view, the active networks are better. Normally, strong encryption algorithms are taken for the passive networks.

However, for the active networks there are several issues to be taken into account: power supply, environmental conditions for the equipment, the maintenance of the network, cable structure, less reliability for using electronic in the remote node and the added cost for the use of these features. In several countries, back-up batteries are mandatory to solve power outages. These can be located either at the RN or at the Central Office. However, other countries, like Sweden, this requirement is not necessary.

Network scalability is more complicated and costly for active networks, since it is generally necessary to change RNs to adequate new capacities and modulation formats. A passive network is completely transparent to these changes and only ONUs need the new capacity features to be modified. In addition, passive networks can be extended to all available optical bands.

New PONs designs can change the features explained below. However, general and present-day considerations about both networks have been reflected.

The Figure 2.1.7. depicts both a passive and active architectures. Several issues mentioned before can be noticed in these configurations (see Figure 2.1.6.).

For the active systems, the theoretical guaranteed bandwidth per user (R_G), considering all the users connected and no bandwidth allocation policies applied in both downstream and upstream directions, is determined by the dedicated link between the ODN and the ONU, or the capacity of the feeder network divided among all the users. The more restrictive limitation is taken as R_G (see Figure 2.1.8.).

$$R_G = \begin{cases} R_b' & \text{if} \quad R_b' \leq \dfrac{R_b}{n} \\ \dfrac{R_b}{n} & \text{if} \quad R_b' \leq \dfrac{R_b}{n} \end{cases}$$

Depending on the protocols used in the network design the R_b is reduced due to the overhead, the guard time (in bits/seg) and OAM (Operation And

	Advantages	**Disadvantages**
Active	– Optics/electronics at less bandwidth – Dedicated link between RN and ONU: no MAC protocol – No encryption – Better QoS	– Not efficient for burst traffic – Must be powered and maintained – No transparent bit rates & modulation – Expensive solution
Passive	– Efficient for burst traffic – Reliability – Easy of maintenance – Not powered network – Transparent bit rates and modulation – Easy to upgrade	– MAC may be necessary in upstream channel – Optics/electronics at the total bandwidth – Encryption algorithm – Ranging protocol may be needed – QoS difficult to achieve at the current moment

Figure 2.1.6. Comparison between active systems and passive systems.

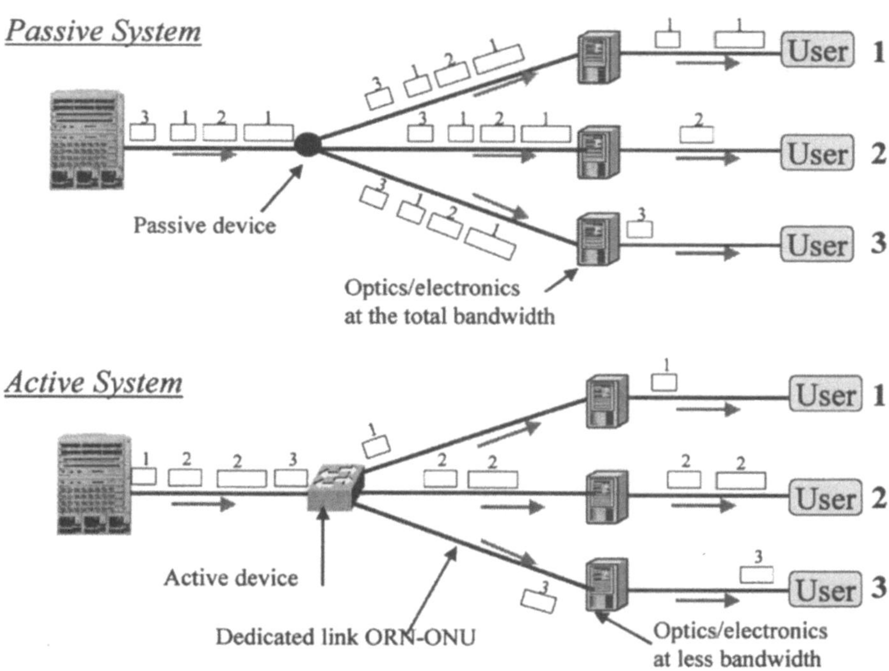

Figure 2.1.7. Active systems and passive systems structure.

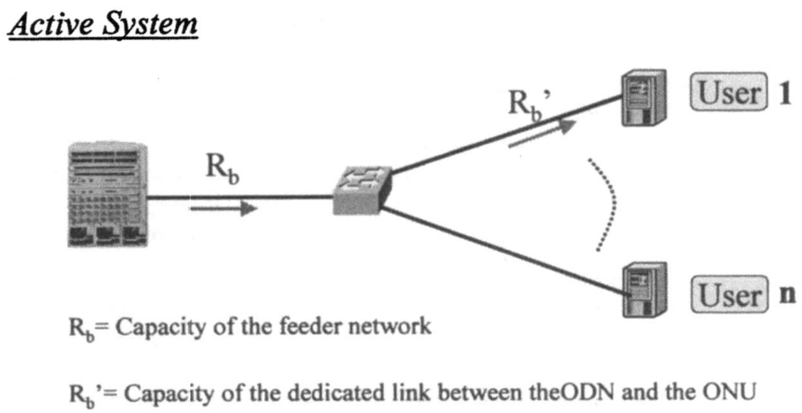

Figure 2.1.8. Active system bandswidth.

Maintenance) information. Then, the real bandwidth per user is reduced, and strictly, R_b should be modified by $R_{b,\text{ real load}}$:

$$R_{b,\text{ real load}} = R_b - \text{overhead} - \text{OAM} - \text{guard time}$$

Similar analysis could be done for PON. For this case, there is no dedicated link between ODN and ONU, then, $R_b = R_b'$ (see Figure 2.1.9).

The theoretical guaranteed bandwidth per user (RG) in the PON is given by the following equation:

$$R_G = \frac{R_{b,\text{ real load}}}{n} = \frac{R_b - \text{overhead} - \text{OAM} - \text{guard time}}{n}$$

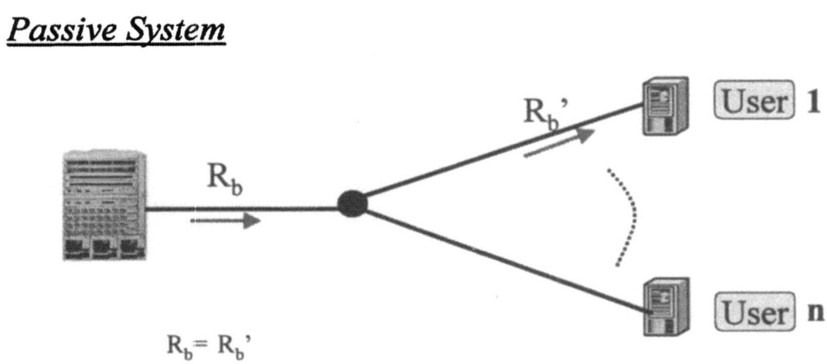

Figure 2.1.9. Passive system bandswidth.

2.1.5. Unidirectional versus bidirectional systems

An unidirectional system uses two fibers, one for each traffic direction. A bidirectional system only has one fiber for the two traffic directions. In the following both systems are compared in a WDM system with *W fixed optical channels per fiber*.

1. An unidirectional system has the capacity to handle W full-duplex channels over two fibers. A bidirectional system only has the capacity to support W/2 channels over one fiber. Therefore, a bidirectional system has the half capacity with respect to unidirectional system, but it allows a better gradual increase of the total system capacity.
2. With only one fiber, only bidirectional systems can be deployed
3. From the capacity efficiency point of view, considering that each network can support W wavelengths, the bidirectional system is less efficient with respect to the unidirectional one, i.e., the former is less scalable.
4. Bidirectional systems are more difficult to design than unidirectional systems, since more transmission impairments have to be taken into account.
5. Amplification can result more complicated in bidirectional systems, but in these systems, it is possible to deliver a high optical power output and flat gain with respect to unidirectional systems, since only a half of the channels is amplified in each direction.
6. Bidirectional systems need guard band between channles that run in opposite directions to avoid crosstalk problems.

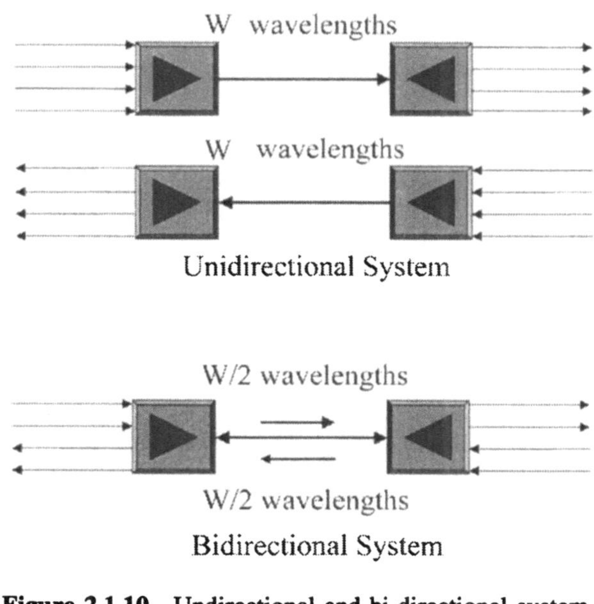

Figure 2.1.10. Undirectional and bi-directional system.

2.1.6. Optical access network architectures

Optical access network architectures must be simple, and the network must be easy to operate and service. This means that passive architectures, where the network itself does not have any switching in it and does not need to be controlled, are preferable to the active ones. The ONU must be kept very simple in order to reduce cost and improve reliability. This rules out using sophisticated lasers and other optical components within the ONU. Preferably, the components used in the ONU must be capable of operating without any temperature control. The hub equipment can be somewhat more sophisticated, since it resides in a controlled environment, and its cost can be amortized by the many subscribers served out of a single CO unless a point-to-point approach is used. In this case, the cost equipment has to be amortized by one user.

This kind of networks are normally called PONs. They use some form of passive component, such as an optical star coupler or static wavelength router, as the remote node if a point-to-multipoint network has to be designed; the physical connections between devices consist only of passive components: optical fibres, connectors, splitters, combiners and splice points. It is impor-

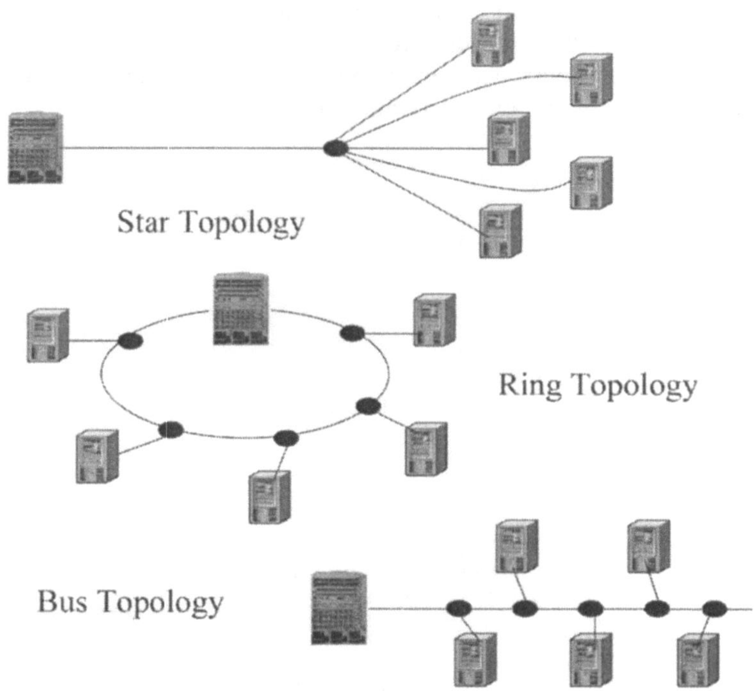

Figure 2.1.11. Physical topologies: star, ring and bus.

tant to note that each split consumes precious power. Splitters can be positioned to create PON point-to-point networks or point-to-multipoint networks. The latter group is comprised of star, ring, tree or hybrid configurations.

The optimum access architecture depends on the geographical conditions and user distribution in each country and each operating company. PON seems to be the most promising approach to establishing a cost-effective optical access approach; an excellent cost is achieved by sharing optical fiber and office equipment among users.

It is possible to join all these type of network topologies as access networks in the same metropolitan area network.

In the PON architecture, electronics can be placed inside the home and powered by the customer, then power supplies in the outside plant cabinets and pedestals are not needed. In the chapter about point-to-multipoint approach only the passive solutions are considered; point-to-multipoint active solutions are out of the scope.

2.1.6.1. *Point-to-point fiber approach*

This PON architecture uses a separate fiber-pair from the OLT to each ONU. The main problem with this approach is the cost of the hub equipment scales with the number of ONUs. Moreover, one needs to install and maintain all these fiber-pairs. This architecture is sometimes called Home Run and has the least degree of network sharing among the several FFTH architectures.

Instead of providing a fiber-pair to each ONU, a single fiber can be normally used with bi-directional transmission. However, the same wavelength is not usually used to transmit data simultaneously in both directions because of the uncontrolled reflections in the fiber. One way is to use time division multiplexing so that both ends do not transmit simultaneously. Another is to use different wavelengths (1310 and 1550 nm, for example) for the different directions, CWDM (coarse wavelength division multiplexing).

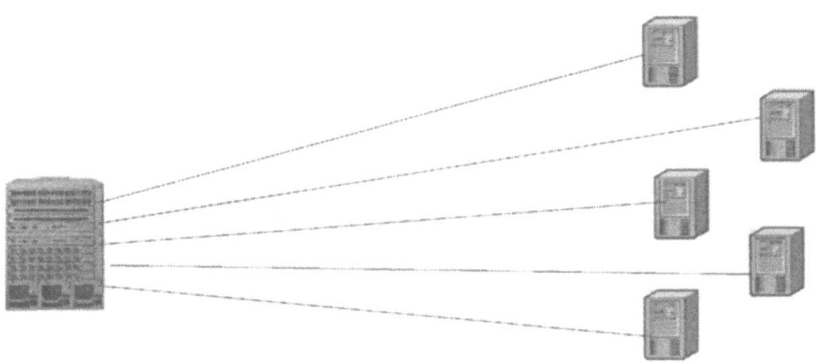

Figure 2.1.12. Point to point approach.

2.1.6.2. *Point-to-multipoint networks*

In this chapter, the typical point-to-multipoint configurations are treated and are analyzed from the reliability and restoration point of view. Many considerations are common in these structures, but every one has its own characteristics that make it ideal for a specific network deployment.

2.1.6.2.1. *Passive star coupler architecture*

This architecture is called in some telecommunication literature as Double Star. In this architecture, the downstream traffic is broadcast by a transmitter at the CO to all the ONUs by a passive star coupler. As it is mentioned in 2.1.3, passive splitting requires special actions for privacy and security. Though the architecture is a broadcast architecture, switched services can be supported by assigning specific time slots to individual ONUs based on their bandwidth demands. The ONUs either terminate or discard the cells based on the cell header addressing. Subscriber data security is ensured via a sophisticated encryption algorithm.. For the upstream channel, the ONUs share a channel that is combined using a coupler, again via fixed time division multiplexing, or some other multi-access protocol.

This architecture allows the CO equipment to be shared among all the ONUs and makes use of fairly mature low-cost optical components. In addition, the point-to-multipoint connectivity of a PON also reduces fiber congestion in

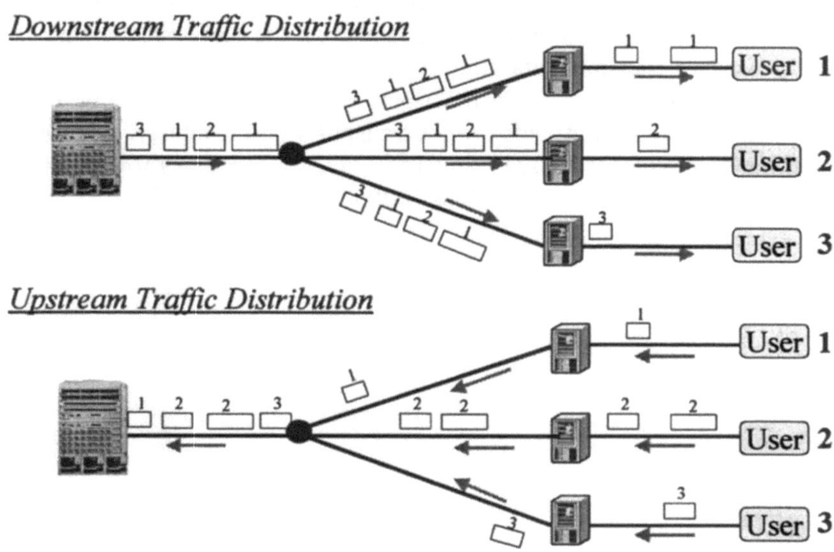

Figure 2.1.13. Downstream and upstream traffic distribution in a PON configuration based on TDMA.

the OLT site compared to point-to-point approach. The number of ONUs that can be supported is limited either by the splitting loss in the star coupler or the required bandwidth per user.

The star topology is attractive due to the fact that the transfer from a narrowband network to a broad-band customer access network is easy and effective. If the number of subscribers is increased, the star network can be easily broken up into several subnets, being a versatile and flexible architecture for network expansion.

Star architectures usually involve a passive optical broadcast star. The broadcast star architectures may have tunable transmitters and fixed receivers, tunable receivers and fixed transmitters, or tunable transmitters and receivers.

Stars present many weaknesses in terms of reliability. Failure of the central device can bring down the whole network, but total failure of a passive broadcast element is unlikely. However, many partial failure scenarios exist, including amplifier failures, port connection failures at the access nodes, transmitter or receiver failures at access nodes (e.g., because of laser failure or tuning skew), and cabling failures in the fiber. Such failures generally lead to failure of one or more arms of the star.

2.1.6.2.2. *Bus architecture*

In this architecture, the central node is connected to other nodes by a common link that is shared by all the nodes. Considerations about upstream and downstream are the same as the last network topology.

As it is explained in [Lumetta01] "bus schemes treat wavelengths as shared media between nodes, allowing nodes to put and retrieve traffic according to MAC protocol. In a folded bus, a single bus, originating at a head-end node, serves all nodes (ONUs). Generally nodes use the bus first as a collection bus, onto which they insert traffic, say in the left to right direction. The last node folds back the bus to make it travel in the right-to-left direction. In the right-to-left direction, nodes collect traffic placed onto the bus. The traffic may be read only or read and removed. In a dual bus architecture, two buses are used, each with its own head-end. In particular, they offer an effective way to share bandwidth between users".

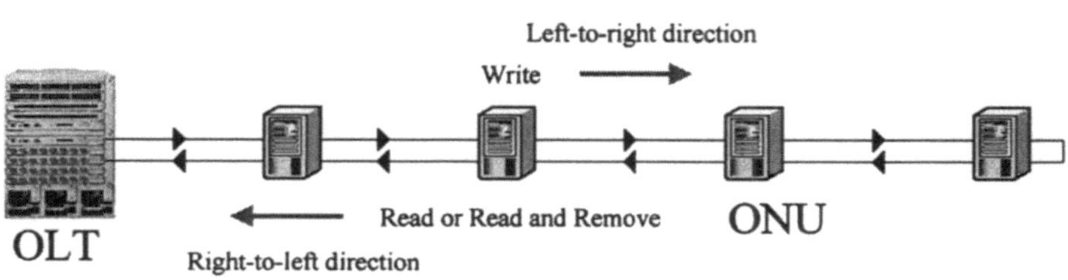

Figure 2.1.14. Typical mechanism of writing/reading in a single bus.

Also in [Lumetta01] the matters about folded and dual buses reliability are explained: "Figure 2.1.15. shows a folded bus and a dual bus after a failure. Partial recovery (a) can be affected by creating a bus on either side of the failure, several nodes will be out of service. For a dual bus architecture (b), the node immediately upstream of the failure needs to be able to fold the bus. In order to re-establish full connectivity after a failure, the end nodes of the original buses must be able to transmit traffic that was destined to traverse the cut". In this case, all the nodes will recover the communication.

Another approach that can be considered bus topology from the cable point-of-view is the solution proposed by Nortel (Gepon) where it is possible to use the same fiber for several customers as shown in Figure 2.1.16.

The fiber that is being used as feeder network in a PON can be used as distribution network in another PON. The design of this network allows to use one fibre in different PONs without connection between them. Figure 2.1.16 also depicts several scenarios to put the splitter (RN) in the FTTH technology.

2.1.6.2.3. *Ring architecture*

In this architecture, a common link is shared by all the nodes, but it forms a closed loop around itself. Considerations about upstream and downstream are the same as the other network topologies.

Ring topologies are attractive for robustness and are the building block behind many LAN and MAN architectures for reliable communications. Rings can recover from a failure using the path protection or link/node recovery.

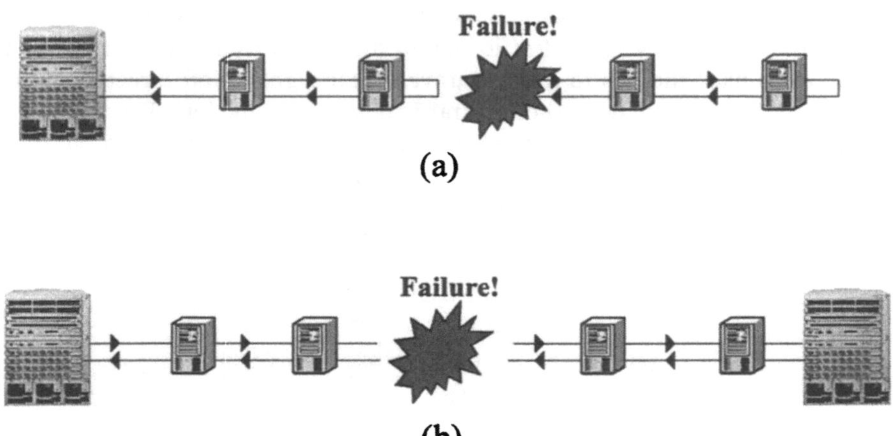

Figure 2.1.15. Folded bus and a dual bus after a failure. (a) Partial recovery; (b) Dual bus architecture [Lumetta01].

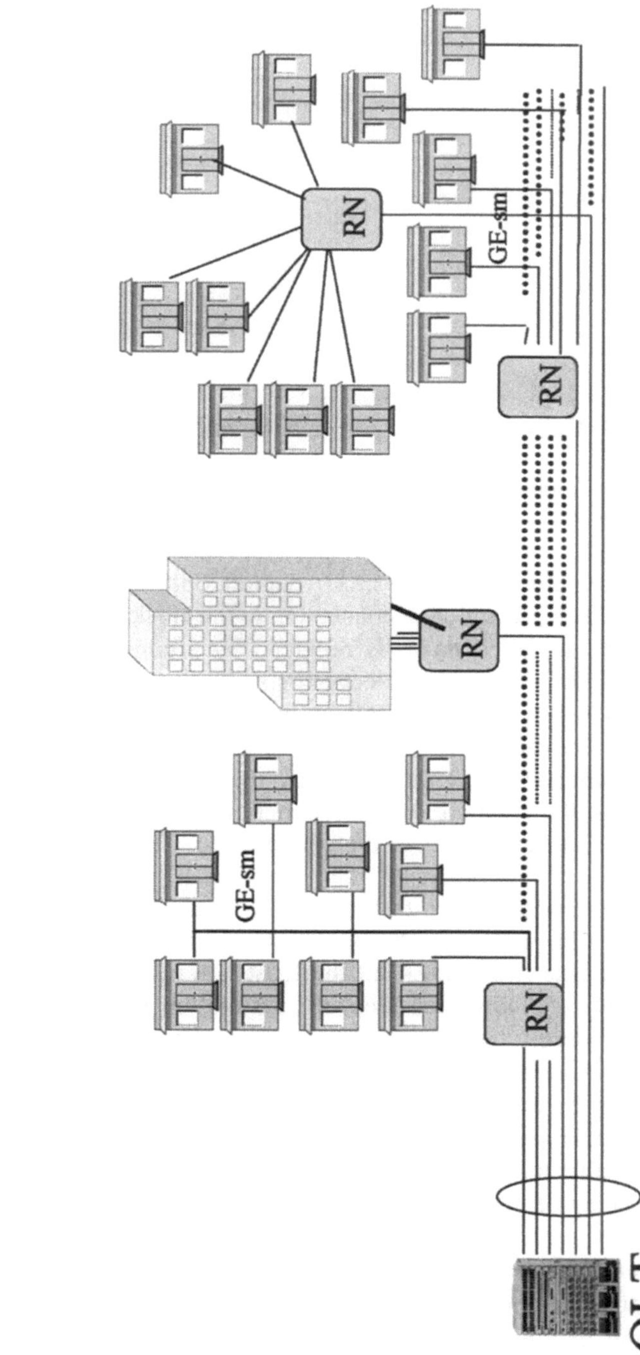

Figure 2.1.16. GePon: Nortel Solution.

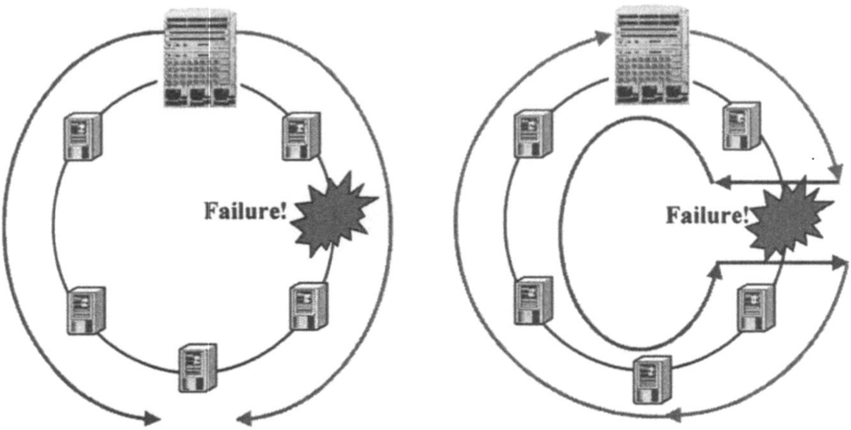

(a) Automatic Path Switching (b) Link/Node rerouting

Figure 2.1.17. Recover from a failure in a ring topology using (a) path protection or (b) link/node recovery.

As is shown in [Lumetta01] Figure 2.1.17 depicts both types of operation. (a) illustrates the path protection: a connection in the clockwise direction, severed by failure of the failed node, is replaced by a connection in the counterclockwise direction. The path protection is similar to the operation of unidirectional path-switched rings (UPSRs) in synchronous optical network (SONET). Link/node recovery, in (b), is similar to the operation of SONET bi-directional line-switched rings (BLSRs). In this case, the clockwise direction is changed to the counterclockwise direction by the node immediately upstream of the failure rather than the source node. This redirection is termed *loopback*.

In fiber-based recovery, the entire traffic carried by a fiber is backed by another fiber, as shown in Figure 2.1.18. The number of wavelengths carried by the fiber is in this case irrevelant – all the light is transferred from one fiber to a second one. In WDM-based recovery, restoration is performed on a wavelength-by-wavelength basis, and recovery capacity may or may not be available for all wavelengths. WDM-based recovery requires only two fibers, although it can be applied to any number. WDM-based recovery is shown in (b). If the primary fibers operate at half of the total capacity or less, only two fibers rather than four are needed to provide recovery. Moreover, certain wavelengths may be selectively given restoration capability.

2.1.6.2.4. *Hybrid solution*

In this structure, ONUs are connected by a mix of the below access networks. Considerations about upstream and downstream are the same as the individual ones.

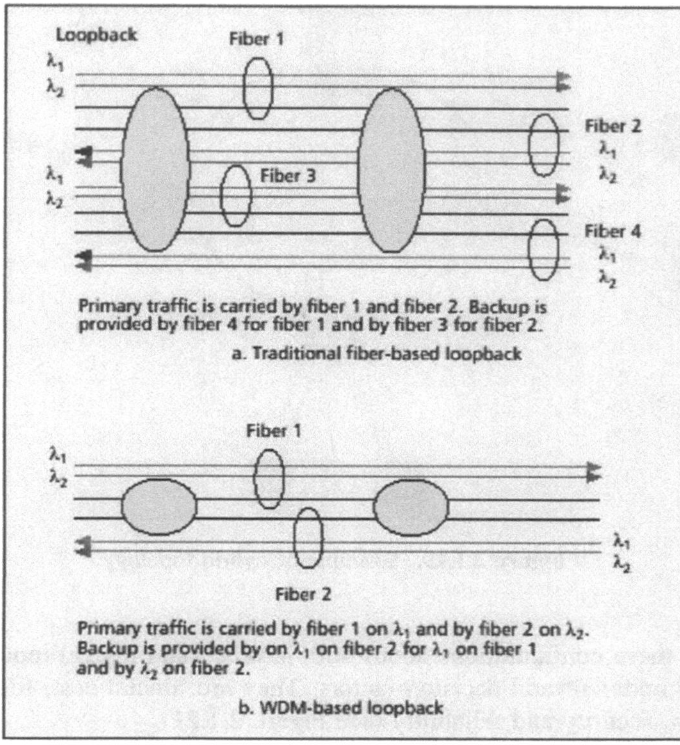

Figure 2.1.18. Loopback recover techniques (a) Traditional fiber-based loopback; (b) WDM-based loopback **[Lumetta01]**.

Figure 2.1.19 depicts a mix of topologies. Technical specifications have to be taken into account, such as power budget, reliability, protection, etc. This hybrid topology only wants to show the possibilities to combine topologies.

2.1.6.3. *Physical topology and logical topology*

Physical topology is referred to the network physical attributes. The typical topologies are: ring, bus and star. Hybrid combinations are also possible. Logical topology indicates how the users get the information. For example, if each user can only consent to private frequency band, wavelength, slot time . . . , it is said that the logical topology of the network is a logical star. On the contrary, if the resources are shared by the network users, it is said that the logical topology is a logic bus or ring.

Logical topologies and physical topologies can be combined (see Figure 2.1.20).

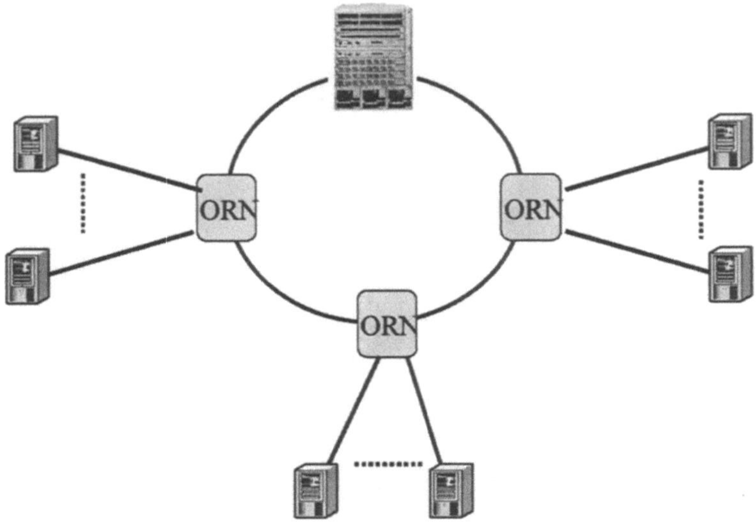

Figure 2.1.19. Example of hybrid topology.

Next, these configurations about both physical and logical topologies are analysed under several decision factors. They are: inicial cost, life, network evolution, security and reliability (see Figure 2.1.21).

Many companies have solutions that use active networks. Several users are connected to different switches (Level 2); they generally compose a ring (other configurations are possible) that is closed with the OLT. Each switch can have from 12 to 48 users; the number of switches in a ring can be limited by the protocol used or by the bandwidth capacity that want the operator to give to every user without congestion.

2.1.7. Main optical effects/issues

Several issues have to be taken into account in order to have a good transmission. The first three effects are important in TDMA (Time Domain Medium Access) technology; upstream optical reception and transmission cause these main difficulties that do not arise in point-to-point transmission:

Power deviation
A packet arriving from an ONU close to the OLT may have a signal level much higher (due to lower path length-dependent losses) than a following packet arriving from a distant ONU, and the receiver must adapt to the new level within the guard time between packets.

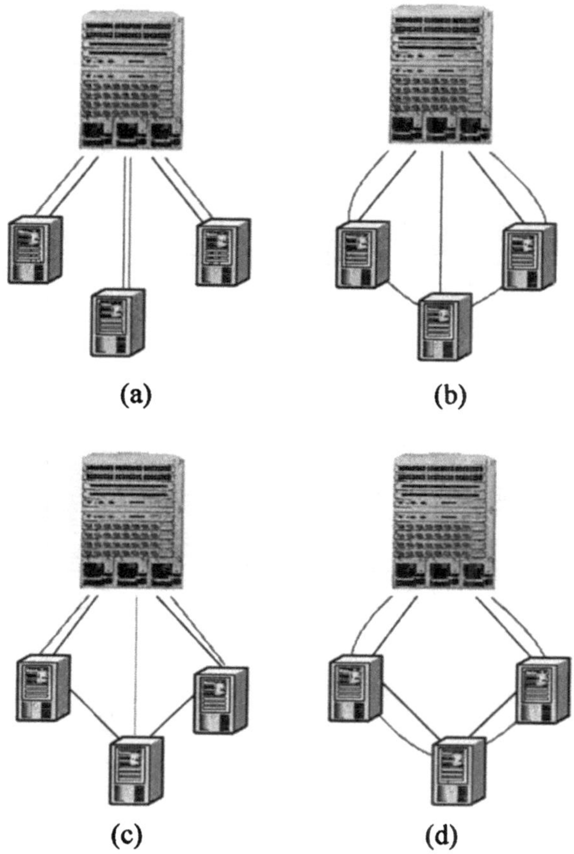

Figure 2.1.20. Logical/physical network configurations: (a) Logical star/Physical star; (b) Logical ring/Physical star; (c) Logical star/Physical ring; (d) Logical star/Physical ring.

Timing deviation
The timing of the upstream signals is complicated by a path delay that differs from ONU to ONU. In addition, the time-dependent variations in path length – for example, as a result of temperature changes – can also be very different from ONU to ONU. The timing of transmissions from each ONU must be adjusted for the packets to arrive at the OLT separated by the guard time.

Optical noise
The sum of the spontaneous light from several tens of (nontransmitting) lasers held at threshold is comparable to the signal level from the one transmitting laser, leading to a significant increase in the bit error rate in PONs. This effect is accentuated during the reception of packets from distant ONUs.

	Physical Star		Physical Ring	
	Logical Star	**Logical Ring**	**Logical Star**	**Logical Ring**
Initial Cost	More fiber Lower O/E cost More O/E	More Fiber Higher O/E cost Fewer O/E	Less Fiber O/E cost depends on implementation	Less Fiber Higher O/E cost Fewer O/E
Life	Easy bandwidth administration	Difficult bandwidth administration	Moderable bandwidth administration	Difficult bandwidth administration
Evolution	Excellent	May require overlay	Depends on implementation	May require overlay
Security	Excellent	Low	Depends on the technique used	Better for active systems than passive systems
Reliability	Excellent	Difficult to diagnose and isolate faulty nodes	Depends on technique used for the distribution	Better for active systems than passive systems

Figure 2.1.21. Configurations about logical and physical topologies of a network segment. (a) Physical star/Logical Star; (b) Physical Star/Logical Ring; (c) Physical Ring/Logical Star; (d) Physical Ring/Logical Ring.

OBI Effect

Another important issue to take into account is the Optical Beat Interference or also known as OBI effect. OBI is caused by light from multiple sources coherently interfering, creating noise at difference frequency. This impairment is specially important for subcarrier-multiplexing (SCM) systems. More information about this impairment is found in chapter 6.

2.1.8. Packet-switched approaches

There is a trend to make part of the signal processing with optical mechanisms. The access networks designers are aware of this and will try to take advantage of new steps in this field. In this section, functionalities are explained which operators want to insert in the optical field. Most optical functionalities must be studied whether being incorporated in the access market or not.

Packet-switched approaches to optical access look for, to develop in optics, a subset of the functionality provided by traditional optoelectronic packet-based networks. Attractive functionality includes:

- Header recognition
- Buffering
- Packet insertion
- Packet reading
- Packet retrieval
- Rate conversion

Supporting these operations in the optical domain is challenging; single for those operations that have been successfully demonstrated in laboratories, there is no one criterion that indicates the best implementation. Nevertheless, there are functions more difficult to treat than others. For example, operations that involve timing issues, or introduce loss and distorsion on the data stream, are among the most challenging. Replicating a stream, for instance, is a relatively straightforward operation requiring only passive optical splitters. Merging streams is very challenging: in contrast to the electronic domain, no practical method exists of delaying an optical signal for a dynamically chosen interval. Packet-level operations are among the most challenging, but can also be grouped by difficulty. Reading signals from an optical data stream is relatively easy since a small portion of an optical signal, taken of the main signal, is sufficient for such operations. Removing a packet (perhaps after reading it) is very difficult since it involves carrying out an operation in the whole stream, as well as timing, phase, and polarization issues. Operations such as packet switching are also challenging because of the issues of timing and optical switch speeds. Thus, *fully optical packet-switched systems replicating the entire operations of electronic systems are still distant.*

The overhead required for recovery in packet-based systems depends critically on which functionalities are implemented in the optical domain. Approaches that redirect traffic around a failure on a packet-by-packet basis are challenging in the optical domain since they require operating on the whole data stream and possibly separating packets from a stream. To avoid packet-level operations, flow switching on a stream-by-stream basis (say using multiprotocol label switching, MPLS) is a promising alternative. Stream-based operations are more amenable to optical processing, and stream-based processing reduces roughly to circuit-based recovery.

2.1.9. Advanced architectures

There are several projects which treat possible architecture configurations and efforts to achieve an efficient solution for optical access network. The Sonata project, the Harmonics, the Darpa project and the Boban project are examples of such projects. In this chapter, these projects are briefly explained and are currently considered as advanced architectural trends in the optical access market. Also, the main field trials and their topology are discussed. More detailed information is explained in the chapter dedicated to Projects and Field Trials.

SONATA

The SONATA project responds to 'Switchless Optical Network for Advanced transport Architecture'. The Switchless optical network is based on the use of both WDMA and TDMA technologies to give interconnexions between passive optical networks without the necessity to use electrical commutation nodes. Figure 2.1.22. depicts the structure of this architecture and its key elements.

The network design is thought of on the national basis. The challenge is to get 20 millions of users distributed in 40 PONs. This implies that the PWRN have to have 804 inputs and 804 outports; they are necessary to cover the 400 PONs, the backup of the channels and the wavelength converter array. The channel spacing have to be 0,05 nm (6,25 Ghz); the system is coherent. This requirement is also needed in the transmitters and the receivers. The formers are tunable lasers with tunable range of 40 nm; the wavelength precision has to be 100 Mhz and the tunable speed has to be less than 1 μs. The receivers have the same characteristics as the transmitters. The devices work in burst mode and the emission speed is 622 Mbit/s.

Network users are distributed in different PONs. Each PON is connected to a PWRN (Passive Wavelength Routing Node) port. Then, an user who wants to connect to another has to tune to his transmitter and receiver, in the assigned time slot, to a specific wavelength or optical channel. The PONs are connected by the PWRN and the information flow is achieved.

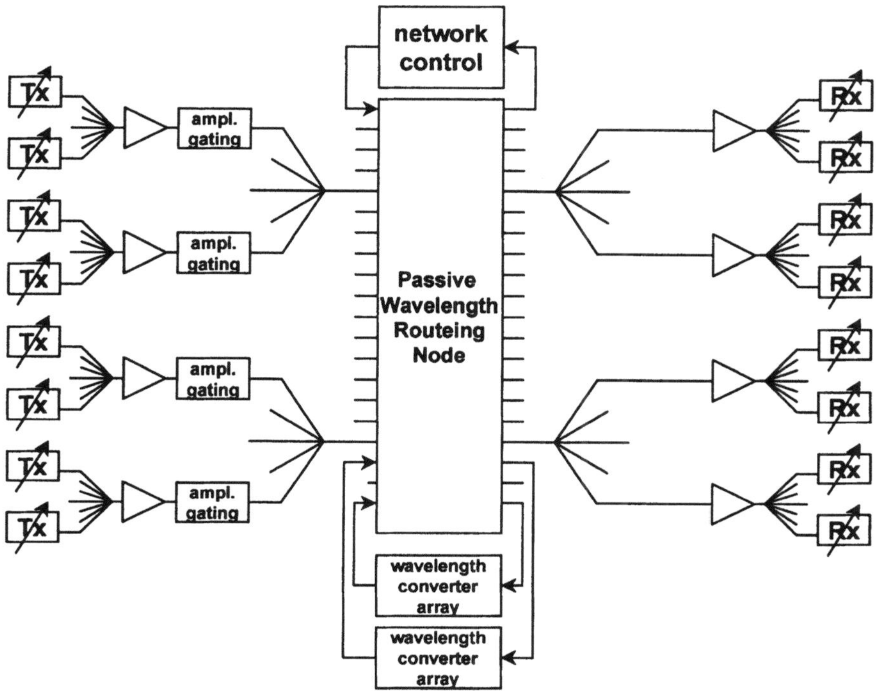

Figure 2.1.22. Network architecture for Sonata project.

Harmonics

Harmonics is an investigation project which began in May 2000 and may finish in April 2003. It belongs to the IST program of the European Union. The main challenges of Harmonics are the following:

1. Design an infrastructure of feeder network of optical fibre supporting a big number of access networks (UMTS, MMDS, MVDS, LMDS, HFC architecture, VDSL o FTTH technology.
2. MAC protocol based on WDMA/TDMA and packets format; capacity to offer a dynamic bandwidth allocation.
3. Control protocol to deliver several QoS based on IP services.
4. Definition of convergence strategies of hybrid access networks to achieve standarization.

Network architecture is comprised of a router which is an element of high speed transmission; an OLT which is the responsible for joining metropolitan network and access; the optical feeder network which is the main objective of Harmonics together with the ONUs; and the distribution network which can be developed under several technologies as is mentioned in the above paragraph.

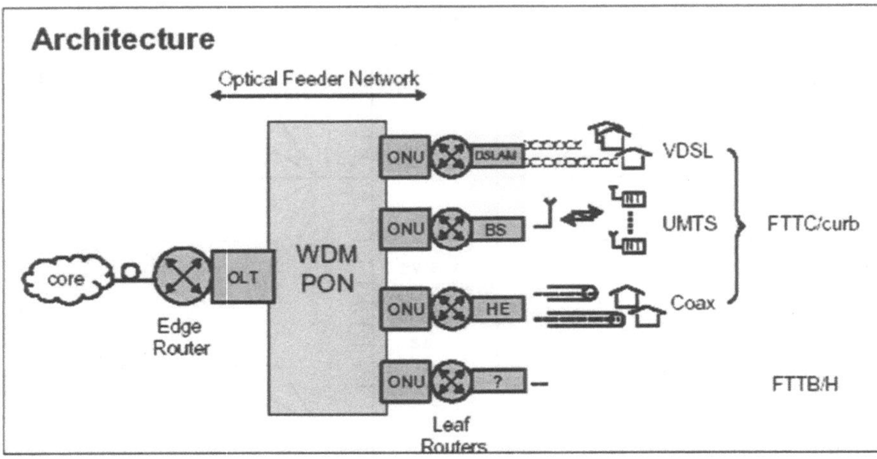

Figure 2.1.23. Harmonics Project: Architecture.

Darpa

In the Darpa Project, a metropolitan and access networks topologies are presented and the structure of the node (OLT) is shown. The content of this project follows the structure and explanation presented in [Kuznetsov00]. The challenge is to develop an optical regional access network which combines electronic IP routing with intelligent networking functionality of the WDM layer. The optical WDM layer provides such networking functions as network logical topology reconfiguration, optical flow switching to off load traffic and bypass IP routers, wavelength routing of signals, protection switching and restoration in the optical domain, and flexible network service provisioning by reconfigurable wavelength connectivity. Network scalability in terms of number of users and network capacity is also an important factor. Figure 2.1.24 illustrates the key physical architecture features of the desired metropolitan access network.

The access network is divided hierarchically into a feeder network and multiple distribution networks (according to the initial definition of feeder and distribution network[4] for a PON at the beginning of the chapter this distribution network is referred to both; the feeder network is the MAN architecture in this project). End users are locally connected to the distribution networks which in turn are connected to the access nodes in the feeder network. In addition to connecting one access node to another, the feeder network has one or more connections to the communication backbone via backbone access nodes. In the figure above, the feeder network has a ring

[4] In section 2.1.1.

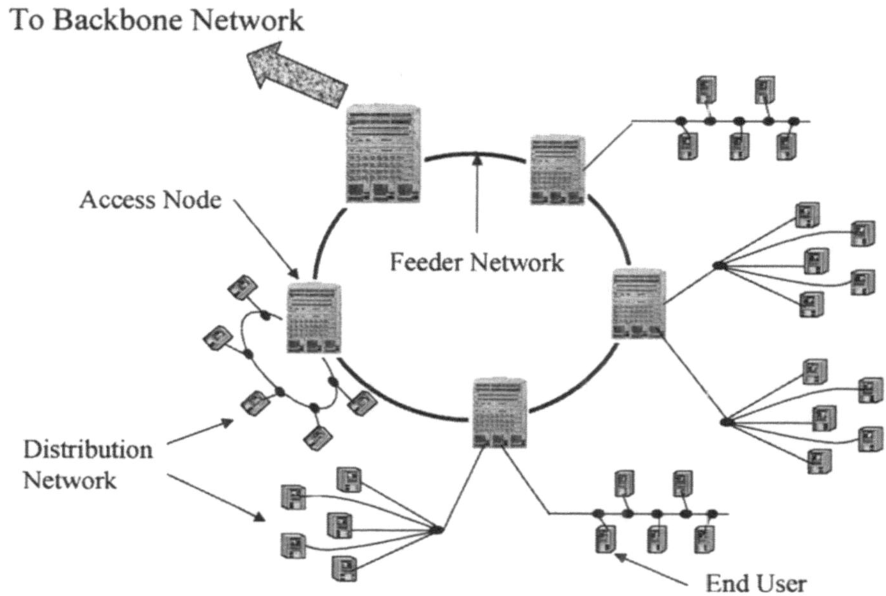

Figure 2.1.24. Physical architecture features of the desired metropolitan and access networks for Darpa Project [Kuznetsov00].

physical topology, although other topologies are also possible, such as a mesh. Feeder access nodes can be interconnected by multiple fibers for greater capacity and route diversity.

It is expected to have systems with 10 to 20 access nodes around a metropolitan area and 20 to 100 users in each distribution network. The feeder ring (for example) can carry 10 to 100 wavelength channels at data rates of 2.5 Gb/s, 10 Gb/s and potentially higher; transparent optical connections, for example with analog-modulated signals, are also possible. Passive distribution networks can use various physical topologies, such as a tree, ring or bus.

BOBAN

One of the most important aims of this project is about the introduction of DWDM technology in the access network. It evaluates the potential of the technology and how it can add value to the telecom business. Although current demand for broadband services does not necessarily justify the deployment of DWDM in the access network today, demand forecasts indicate the need for DWDM technology in the near future as it has been indicated in this chapter. Also technology maturity is another key factor of deployment. In fact, BOBAN studies shows that for the time being the immaturity of DWDM technology poses a serious limitation to the introduction of DWDM in the access network. Although DWDM offers huge transmission capacities and supports a wide variety of input signal formats, it still suffers from limited flexibility and networking functionality compared to its full potential.

The following key applications fields have been identified as driving forces for the WDM access networks:

- Wavelength leasing
- Wavelength on demand
- Intra-Enterprise Connectivity/Optical VPN

Field Trials

Nowadays, several field trials are being developed in the world (see Section 4.4) In the United States, Grant County (Washington) and Palo Alto (California) are serving as adopters for FTTH technology and are deploying FTTH networks. In Europe, FTTH or FTTx technologies are being also implemented step by step. It is important to point out that widespread deployment of FFTH networks will probably be motivated when telephone companies decide to compete in video market or when new applications requires higher capacities. Until now, these motivations are not strong in the current telecommunications market.

The project in Grant County is called ZIPP. Currently, the ZIPP network serves 100 homes and 30 business. The services that are delivered to the users are: Internet access, voice and video services. The network uses an Ethernet over Star architecture and runs the Gigabit Ethernet protocol. The Network equipment includes CISCO routers and CPE and switching equipment provided by World Wide Packets. Two strands of fiber connect each user to a community hub (unidirectional system). Currently, each hub serves 48 customers.

The Palo Alto project serves 70 homes. They started deploying the network in March 2001. The planned services include Internet access of 7 Mbps downstream and 4,5 Mbps upstream, telephone and cable TV service. The network is Ethernet over PON architecture using Deep FiberTM products from Marconi. The system is bi-directional.

In Europe there are also several field trials. One of the most important trials is in Sweden. The name is Bredbandbolag and is supported mainly by Telia and NTL. The idea is to get 350.000 apartments and the project has the aid from the Sweden Government. The main services are the Internet Access, Video on Demand and telephone. The network is a switched Ethernet. The fiber is deployed in the feeder network. The distribution network (from the swith to the user's terminal) is coaxial cable. The speed is 10 Mbps. This field trial is not strictly using fiber to the home technology, the correct technology name is fiber to the building.

2.2. MODULATION FORMATS

Introduction

In this section, a review of the existing modulation formats is developed in order to investigate their possible applications to the access networks. The main interest of the major techniques is probably in the metropolitan networks. But due to the close relationship between the metro and the access, it is necessary to see which modulation formats can give better performance. Furthermore, the use of multi-mode fibers with low bandwidth parameters makes these techniques interesting in the access environments.

The modulation formats are classified into: basic, advanced and electrical. The first ones are the most important at this moment because of the current and near future available technology. As direct detection techniques, binary signaling formats (basic) as well as multilevel signaling (advanced) formats have been studied. In recent publications, a lot of attention has been paid to the Duobinary modulation. This is also included in this work. Another modulation family included in the advanced formats is the coherent systems. The phase noise of the lasers and their instability make the heterodyne-homodyne detection unfeasible. Therefore the Coherent Systems can be seen as a far future solution. The last point deals with electrical modulations used in systems where fiber can be seen as a transparent channel. Some examples are the Hybrid Fiber Coaxial (HFC) systems or the x-type Digital Subscriber Line (x-DSL) systems.

2.2.1. Basic modulation formats

Basic Modulation techniques are those using the Intensity Modulation with Direct Detection (IM/DD) and binary signaling. There is no phase recovery at the receiver and the use of local oscillator is not mandatory. The basic binary signaling formats are the Non-Return to Zero (NRZ) and the Return to Zero (RZ).

2.2.1.1. *NRZ vs. RZ comparison*

An issue still remains as to determining whether NRZ or RZ is the more optimal modulation scheme for the use in optical systems. The data rates used in the Access Systems are in general very low compared to the Metropolitan Systems. Some special situations exist where the data rates can be hardly increased, for example, when using some Optical Code Division Multiplexing schemes (see section 2.3). With the exception of these special cases, the reasons to choose one modulation format or another are extremely different from the high capacity systems ones.

The following basic statements can be made: 1) due to higher peak power, RZ may suffer more non-linearity, and considering shorter pulse-width, RZ may suffer more dispersion and 2) due to a longer pulse time and longer interaction time between wavelengths and bits, NRZ may suffer more non-linearity.

In **[Hayee 99]**, a comparison between NRZ with RZ modulation for high capacity dispersion-managed systems (Standard Single Mode Fiber SMF alternating with dispersion compensating fiber DCF) operating at the data rates up to 40 Gb/s is made. In 10–40 Gb/s systems, it is found that NRZ is more adversely affected by non-linearity whereas RZ is more affected by dispersion. Typically, 10- and 20-Gb/s systems are limited mostly by non-linearity, whereas 40 Gb/s systems are limited mostly by dispersion and ASE noise. Taken together, these two statements suggest that: 1) 10- and 20-Gb/s systems, in general, operate better using RZ modulation because non-linearity dominates and 2) 40-Gb/s systems, being limited mostly by dispersion, favor the usage of RZ for few channels, but require NRZ as the number of channels increases due to the dispersion.

In Figure 2.2.1, a complete comparison between both modulation formats by means of simulations is presented. To compare NRZ and RZ systems, the average power per channel is first varied for a 16-channel system with and without including non-linear effects (non-linear effects are excluded from the simulation by making the non-linearity coefficient zero). The resulting eye closure penalty of the poorest-performing channel is shown in Figure 2.2.1 [left (a), (b) and (c), respectively, for 10-, 20-, and 40-Gb/s systems]. In 10- and 20-Gb/s systems, the eye closure penalty is large for lower channel powers because the EDFA noise is dominant [Figure 2.2.1 left (a) and (b)]. For higher channel powers, the eye closure penalty decreases since the signal power now overcomes the EDFA noise. However, at the highest channel powers being considered, the eye closure penalty increases as a function of power due to non-linearity. The penalty due to non-linearity is more severe in the NRZ as compared to the RZ. Since the RZ performs better than the NRZ in both 10 and 20 Gb/s systems. The RZ is less affected by non-linearity than the NRZ. This is because: 1) isolated RZ pulses take advantage of soliton-like pulse compression in Self-Phase Modulation SMF, whereas the variable number of adjacent "1's" in the NRZ are non-uniformly degraded by non-linearity, thereby causing the rail of "1's" to spread, and 2) long strings of "1"s in the NRZ as compared to the RZ have a much longer cross-wavelength interaction time, thereby producing more severe penalty from non-linearity.

In a 40-Gb/s system [Figure 2.2.1 left (c)], the narrower RZ pulses are more susceptible to dispersion; note that the dispersion-based penalties grow inversely as the square of the pulse-width. Moreover, the required larger channel spacing of 1.6 nm for 40-Gb/s transmissions produces more dispersion in the end channels since the total wavelength range is increased. An

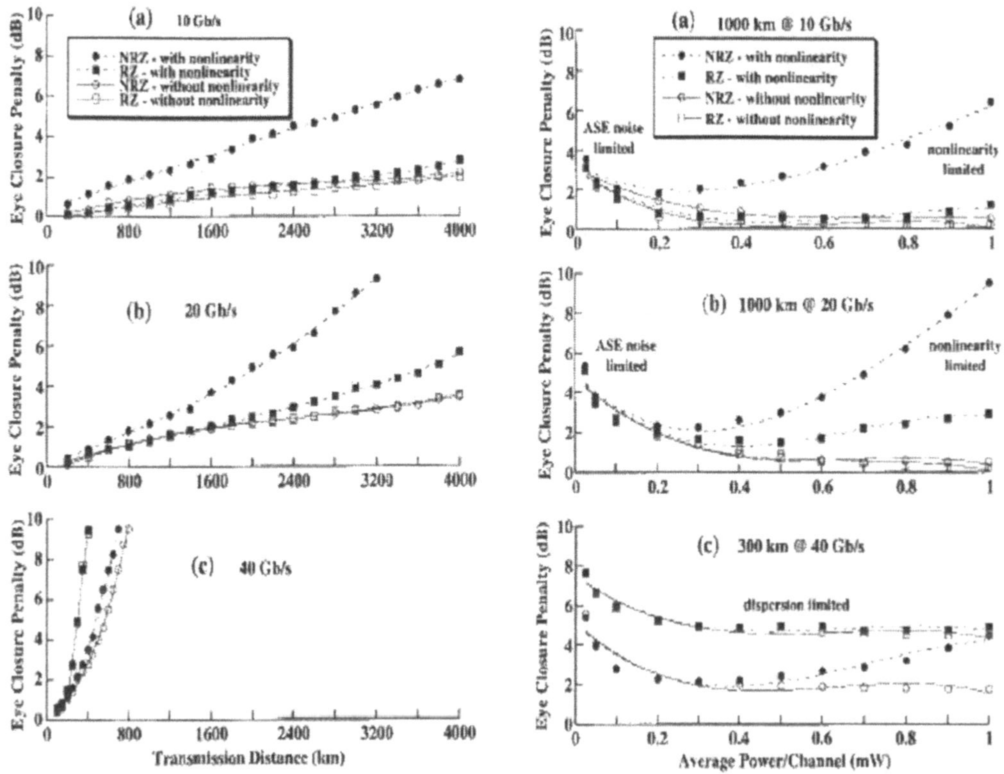

Figure 2.2.1. The Q factor of the worst of 16 channels versus average-power/channel for NRZ and RZ systems at right (1) 10 Gb/s, (b) 20 Gb/s, and (c) 40 Gb/s. The Q factor of the worst channel versus transmission distance for 16 channel NRZ and RZ systems at left (a) 10 Gb/s, (b) 20 Gb/s, and (c) 40 Gb/s. The average power/channel is 0.3 mW **[Hayee 99]**.

indication that RZ is more affected by dispersion can be derived from the fact that 40-Gb/s RZ systems have a larger eye closure penalty as compared to the NRZ systems for low and moderate channel powers. For higher channel powers, the penalty for the NRZ increases because of non-linearity, whereas no appreciable change in penalty is noticed for the RZ system.

In general, the non-linearity induced penalty for all the channels in 10- and 20-Gb/s NRZ systems increases with the transmission distance. The eye closure penalty as a function of distance for the poorest-performing channel in a 16-channel system is shown in Figure 2.2.1 [right (a)–(c), respectively for 10, 20, and 40 Gb/s]. The penalty is larger for the NRZ than for the RZ at 10 and 20 Gb/s when non-linearity is included. Furthermore, the penalty increases more rapidly with distance in the NRZ as compared to the RZ, thereby showing effectiveness of the RZ modulation format in combating non-linearities. However, the penalty increases for 40 Gb/s more quickly with distance in the RZ as compared to the NRZ. Moreover, there is a negligible change in the penalty for the RZ when non-linearity is included indicating that the penalty is mainly due to dispersion. For the 40-Gb/s the NRZ system, the penalty increases slightly because of non-linearities.

The transmission distances in the access networks are limited to a few tens of kilometers. In some cases, the metropolitan network is meshed together with the access one. It extends the distance up to 100 km making this study more interesting. But in most situations talking about access schemes, implies short distances up to tens of km, and slow data rates, up to 1 Gb/s. So the factors to decide which kind of signaling scheme to choose are extremely different: the multiplexation scheme used (see section 2.3), the synchronism mechanism or the source used.

When using Optical Time Division Multiplexing OTDM or Code Division Multiplexing (see section 2.3), the use of the RZ format is mandatory. About synchronism, the RZ format is much simpler to maintain than the NRZ format.

In summary, the choice of the RZ or the NRZ format depends on the environment and bit rate. In the Metropolitan or access networks, maybe the RZ format is more attractive due to the lower speeds used. Perhaps the biggest advantage of the NRZ format is the simplicity and the cost. To have a cost-effective solution is mandatory in Metro-Access networks.

2.2.2. Advanced modulation formats

Apart from the basic modulation formats, there exist some advanced modulation schemes that can be taken into account for special cases. We have included two advanced techniques: multilevel signaling schemes and coherent detection systems. The first have been strongly studied for high capacity systems in order to reduce the bandwidth and to increase the chromatic dispersion tolerance. In principle, bandwidth or chromatic dispersion are not limiting factors in the access environment but the use of multi-mode fibers implies a strict bandwidth limitation. The last are a well-known family solution in electrical communications. Their direct application to optical systems is not easy mainly because the local oscillator is not a trivial issue. The application to access systems seems to be a far future solution mainly because of cost.

2.2.2.1. *Multilevel signalling*

Recently multilevel signaling formats have taken a lot of interest due to its robustness to chromatic dispersion and spectral efficiency. In the access networks, dispersion is not a limiting factor because of the limited work distances (10 km). But when thinking about a metropolitan network (100 km), it can become an important hurdle. The spectral efficiency can be beneficial to use multi-mode fibers.

The most recent attempt in this direction is the Duobinary Modulation format where the use of phase changes gives an important increase both in dispersion tolerance and spectrum efficiency **[Pennickx 97]**. At this point, a review of different multilevel signaling techniques is presented and some example references are given to understand the argumentation.

Ask M-ary signals

This technology is well-known in electrical communications as a way to improve the spectrum efficiency while loosing signal to noise ratio SNR. The principle is very simple. The two levels signal is replaced by a M-levels signal where each symbol has a $\log_2 M$ bit information.

Where B i's the bit rate for a binary transmission, M the number of symbols and SNR the signal to noise ratio.

2^M bits per symbol

Bandwith Reduction $\quad BW = \dfrac{B}{\log_2 M}$

Sensibility Degradition $\quad SNR - \log_2 M$

Additional Power $\quad P_M = \dfrac{M-1}{\sqrt{\log_2 M}}$

In Figure 2.2.2, a generation scheme of M-ary signals is shown together with a receiver scheme. As can be seen, the complexity of the receiver is dramatically increased and the need of digital processing becomes necessary **[Walklin 99]**.

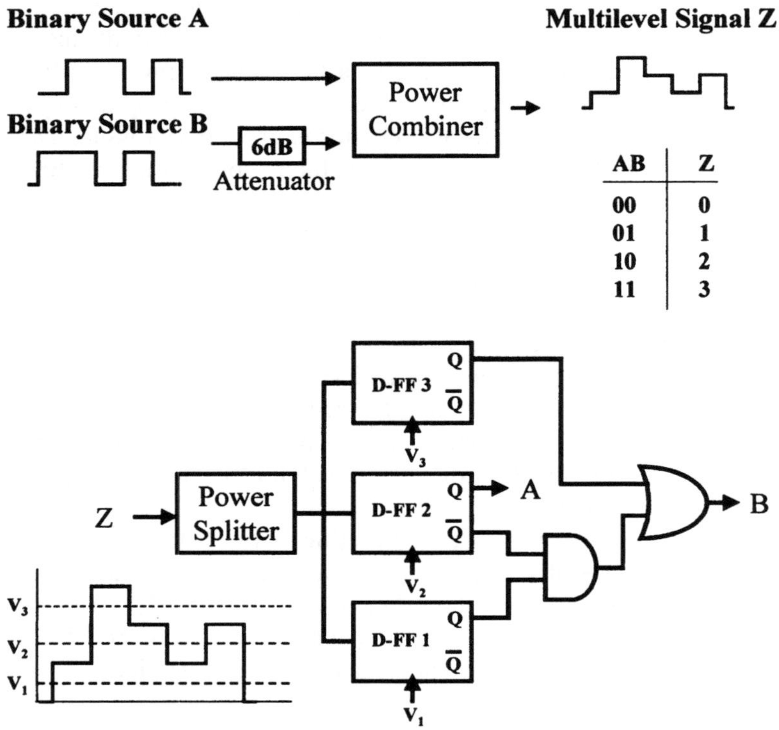

Figure 2.2.2. Multilevel signal generation and receiver **[Walklin 99]**.

$$\frac{B}{2f_c} = ld(m)$$

$$\frac{L_m}{L_2} = [ld(m)]^2$$

$$P_m/dB = -5\log\left[\frac{ld(m)}{(m-1)^2}\right]$$

m	B/2fc	Lm/L2	Pm/dB
2	1	1	0
4	2	4	3,3
8	3	9	6,1
16	4	16	8,8
32	5	25	11,4
64	6	36	14,1

Figure 2.2.3. Trade off between disperson tolerance and sensibility degradition in m-ary multilevel signals **[Wedding 99]**.

In **[Wedding 96]** a M-ary spectrum efficiency (dispersion tolerance) study has been performed to analyze the trade off with the signal sensibility degradation. In Figure 2.2.4, a comparative table is showed.

Where B is the binary bit rate, f_c is the cut-off frequency, L_m is the dispersion length for an M-ary signal and P_m is the sensibility penalty for a M-ary signal with respect the binary signal.

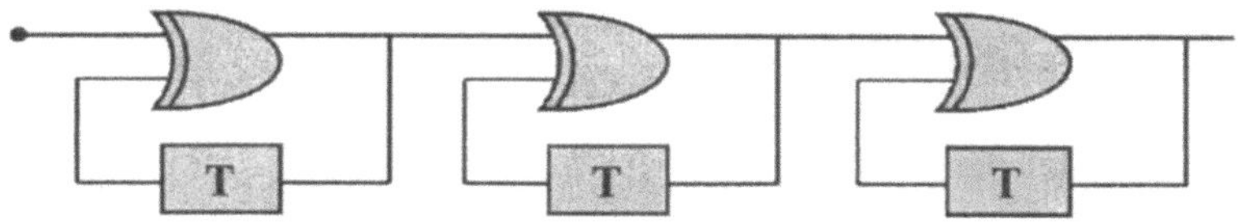

Figure 2.2.4. Pre-coder for a poly-binary signal.

To conclude for this family of multilevel signaling, a loss in sensibility is not an available choice in most cases. Furthermore, the complexity increasing mainly at the receiver avoids the M-ary signals to be a good solution.

Ask polybinary signals

Another well-known solution to increase the spectrum efficiency in classical communications is the use of delay and add filters to give some correlation to the symbols. This correlation reduces the signal bandwidth and hence improves the dispersion tolerance.

In Figure 2.2.4, a scheme of Polybinary signals generation is given. The number of delay and add filters applied to the binary signal gives the order of the polybinary signal. For instance, when only one filter is applied, an ASK-duobinary signal is obtained.

The bits correlation in the transmission is skipped by mean of a pre-coder that gives place to a pre-coded sequence. This sequence is filtered by the delay and add filter (differential encoder), and in the reception, a modulo-b b-levels receiver decodes the message.

In Figure 2.2.5 the pre-coder of this family of signals is shown. The number of stages is the same as the poly-binary signal order minus one. For example in duo-binary case, the number of stages is one.

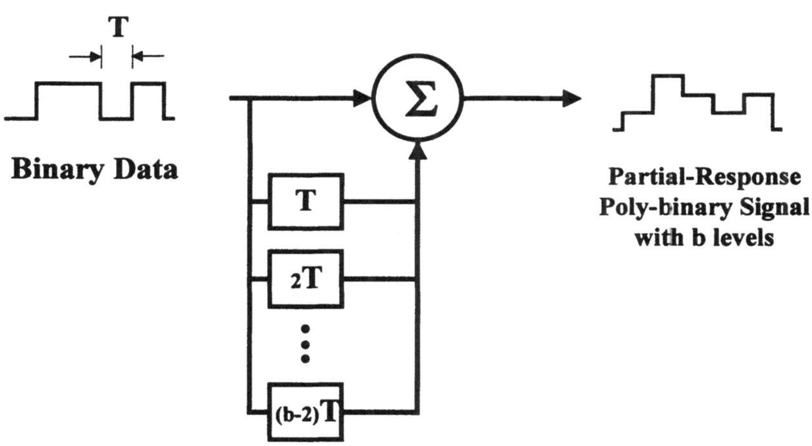

Figure 2.2.5. Poly-binary Signal generation [**Walklin 99**].

The mathematical expressions for the poly-binary generation can be written as follows **[Walklin 99]**:

a_k original data sequence
b_k precoded data sequence
c_k polybinary data sequence
$b_k = a_k \oplus b_{k-1} \oplus b_{k-2} \oplus \ldots \oplus b_{k-(b-2)}$
$c_k = b_k + b_{k-1} + \ldots + b_{k-(b-2)}$
$a_k = c_k \bmod 2$

The spectral efficiency of this kind on modulation format is also well-known. The principle is different from the M-ary case where the number of bits of information in a symbol is increased. The number of bits per symbol is one but the transitions are alleviated by the use of the delay and add filters.

The general spectrum expression for ASK-Poly-binary signals can be written as follows **[Walklin 99]**:

$$S(f) = \frac{(b-1)^2 A^2 T}{4} \sin c^2[(b-1)fT]$$

Where b is the number of levels, T is the bit time, A is the amplitude and f the frequency.

In Figure 2.2.6, the bandwidth reduction can be seen for a duobinary signal compared to the Non Return to Zero (NRZ) format. The energy is more

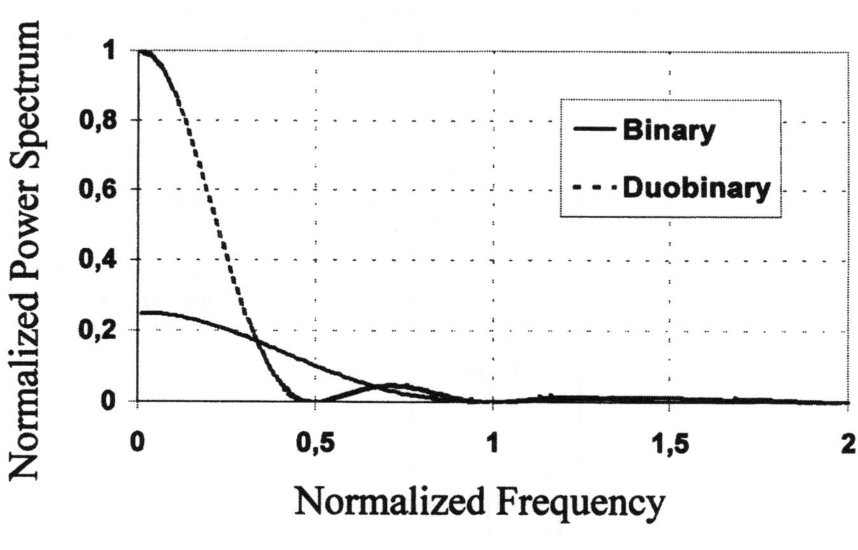

Figure 2.2.6. Spectrum efficiency of duobinary signals **[Shtaif 99]**.

concentrated around the carrier and hence can be filtered at half of the binary bandwidth. Not only it can be filtered but it must, because if not, the performance degrades severely **[Shtaif 99]**.

As in M-ary case the higher the number of levels the worse the sensibility of the modulation format. So the performance is not increased.

2^M bits per symbol	
Sensibility Degradition	$SNR - \log_2 M$
Additional Power	$P_M = \frac{M-1}{\sqrt{\log_2 M}}$

AM-PSK polybinary signals

This case is the same as the previous one but with a single modification: the application of the driving signal to the optical modulator is different. The signal is scaled and biased before being applied to the modulator in order to reach twice the electrical swing. This way some phase changes can be reached and used to increase the dispersion tolerance.

In Figure 2.2.7, the philosophy of those techniques is shown by mean of the AM-PSK Duobinary case. As can be seen, the voltage swing of the electrical signal is twice the binary case. The optical output signal has only two

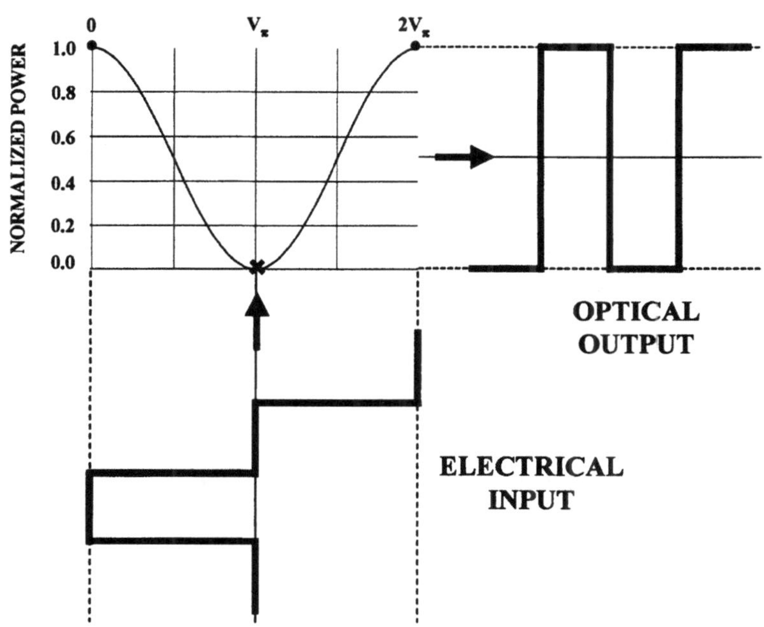

Figure 2.2.7. AM-PSK Duobinary modulation **[Walklin 99]**.

levels instead of three compared to the electrical one, but with 180 degrees phase changes in some "ones" (high level) **[Walklin 99]**. This means that there is no SNR loss because of the number of power levels that remains the same as the NRZ binary case.

Those phase changes, together with the bandwidth reduction, are the responsible for the increase in dispersion tolerance of this kind of modulation formats. When increasing the number of levels, the philosophy remains the same but the phase changes exist only when crossing the zero level and so the pulse expansion due to dispersion cannot be avoided in all the cases except the "101" transitions.

The number of output optical levels is (b + 1)/2 where b is the Polybinary order. In the case of AM-PSK Duobinary Modulation, the output signal has two levels and a conventional Intensity Modulation with Direct Detection (IM/DD) NRZ receiver can be used.

The delay and add filter is mandatory to be used together with a low pass filter in order to take advantage of those techniques **[Shtaif 99]**. There are a lot of possibilities of electrical filtering. Some extended studies have been performed in **[Penninckx 98]** and **[Walklin 97]**.

As seen in both figures, a new method to implement the three-level electrical signal uses an electrical filter at a quarter of the bit rate. This technique is well-known in classical communications. The delay and add filter, and the low-pass filter are replaced by a single filter with a cut-off frequency of only a quarter of the bit rate. This filter acts as an averager between the two last bits giving a Duobinary signal.

The AM-PSK Duobinary modulation using a Bessel electrical with a cut-off frequency of a quarter of the bit rate is named Phase-Shaped Binary Transmission (PSBT) **[PTL 9-2-259]**, **[PTL 10-4-612]**. The main idea of this new modulation format consists of having phase changes in all cases, even between zeroes (low level). In Figure 2.2.8, this philosophy is showed. The electrical input is a three levels signal but with a special characteristic. there are some oscillations between consecutive zeroes. Those forced oscillations, when translated to the optical domain, convert into small pulses with alternate phase 0–180 while maintaining the zero level. The number of consecutive small pulses equals the number of consecutive zeroes minus one.

The reduction of the bandwidth, both electrical and optical, is another important advantage of those techniques especially in 2.8 GHz Bessel filter case. The electrical requirements of the transmitter and the optical channel separation are extremely reduced. In Figure 2.2.9, the measured optical spectrum of the PSBT modulation is plotted together with the NRZ one. The high reduction of bandwidth can be appreciated.

In recent studies, some drawbacks have been found for the PSBT Modulation. In **[Royset 98]**, the study of the applied electrical signals symmetry requirements in Duobinary Modulation is performed while com-

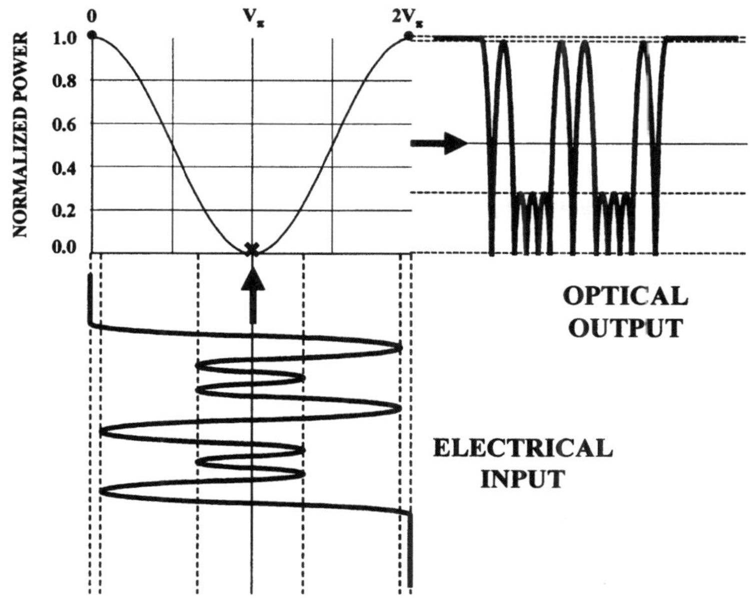

Figure 2.2.8. The Phase-Shaped Binary Transmission (PSBT).

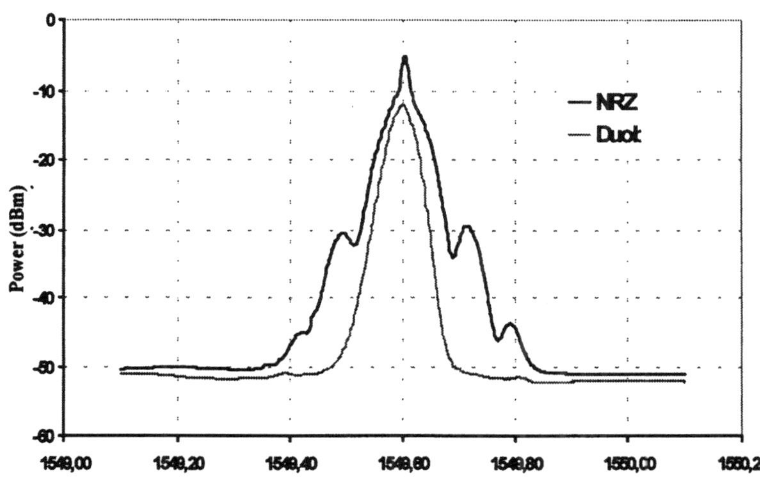

Figure 2.2.9. Spectrum of the PSBT modulation compared with the standard NRZ.

paring with the NRZ format. This modulation is very sensitive to asymmetries: bias asymmetries, amplitude asymmetries (chirp), delay asymmetries and amplitude swing deviations.

Another important drawback is that Duobinary Modulation is very sensitive to non-linear effects like Self Phase Modulation (SPM). In **[Kaiser 00]**, a study of the SPM effect in Duobinary transmissions is performed. The conclusion is that the SPM effects reduces the dispersion tolerance and must be taken into account when the power levels are relatively high and when the transmission distances are very long. The good performance of the Duobinary Modulation is completely lost when the SPM level is high.

More recent studies **[Pennincks 97]**, **[Antona 00]** and **[Bissessur 00]** have been developed the PSBT technique in order to optimize the performance. The main idea is to optimize the relationship between the maximum power level and the ripple power level.

Poly-M-ary signals

The last category of the multilevel signaling techniques mixes the two previous main ideas: multilevel signals (ASK M-ary) and correlation (Polybinary) **[Walklin 99]**. In Figure 2.2.10, the Poly-M-ary principle is showed.

The investigation on this technique is not too large so the application is far from being clear. The main attention has been focused in Duobinary Modulation.

For instance, the Duoquaternary Modulation can be taken as an example of those formats family, where a 4-level signal is filtered by one delay-and-add module. In this case, the involved sequences can be written as **[Walklin 99]**:

a_k original data sequence
b_k precoded data sequence
c_k polybinary data sequence
$b_k = a_k - b_{k-1} \bmod 4$
$c_k = b_k + b_{k-1}$
$a_k = c_k \bmod 4$

Performance comparison

Some of the reviewed multilevel signaling techniques have been compared by means of simulations in **[Walklin 99]**. Order Bessel filter (PSBT) has the best performance compared to binary modulations. When using multi-level modulations 4-ary ASK gives even a better performance due to its high bandwidth reduction. The drawback is the loss of sensitivity in the back-to-back configuration and the need of amplification to take advantage of so high transmission distances.

The best choice seems to be the AM-PSK Duobinary Modulation due to its simplicity and good performance. Not in access networks but in metropolitan networks with transmission distances over 100 km can be an interesting technique to be used. The dispersion tolerance and spectral efficiency are surely the most important advantages.

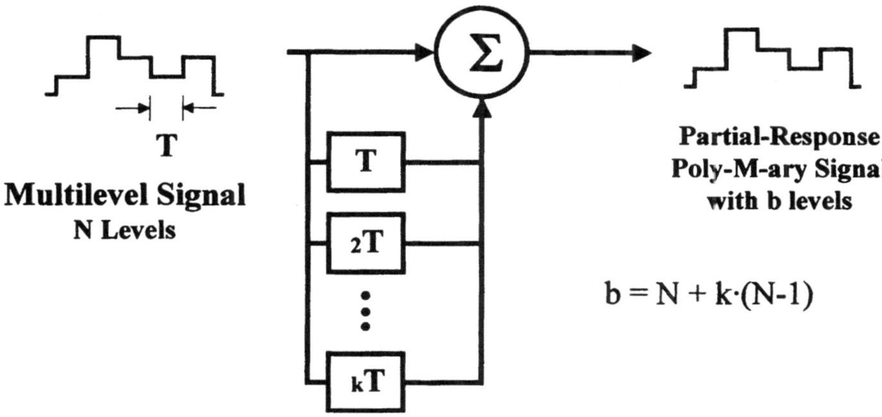

Figure 2.2.10. Poly-M-ary signals philosophy **[Walklin 99]**.

When using multi-mode fibers, the bandwidth reduction of those techniques is extremely interesting in the fiber access systems.

2.2.2.2. *Coherent systems*

Introduction
In this section, a completely different type of optical modulation and detection are presented. The coherent systems emulate the heterodyne RF receiver in the optical domain. They offer interesting performances compared to the conventional Intensity-Modulation Direct-Detection systems.

Basis
The coherent systems are characterized by a different kind of detection: the optical heterodynation. A local laser (LOL) in the receiver tunes the incoming channel wavelength. Both incoming lights are combined in a coupler and put into a photo-detector (single or balanced). The output current is an electrical intermediate frequency: the difference of both optical frequencies (see Figure 2.2.11).

Therefore, in the coherent systems, the transmitted signal is translated from the optical domain to the electrical domain without distortion keeping its shape in amplitude and also in phase, on the contrary to the conventional incoherent IM-DD (intensity modulation – direct detection) systems, where only the optical power is detected. In the coherent systems, any modulation technique can be applied (of phase, frequency, amplitude). It is demodulated after the intermediate frequency stages obtaining the base-band signal normally digital.

In electronics, the word coherent is reserved for systems where the detector

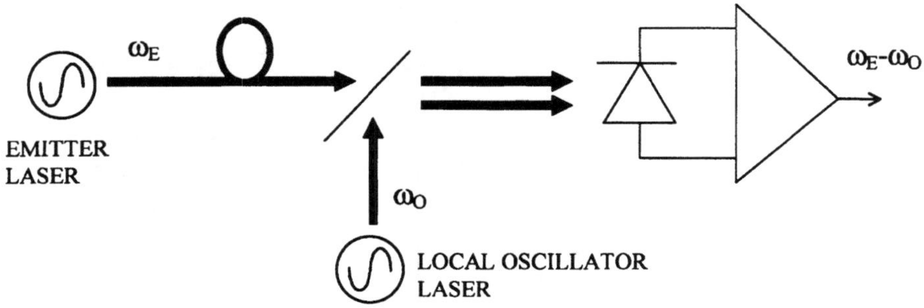

Figure 2.2.11. Optical Heterodynation.

"locks on" to the phase of the received signal. In optics, it makes reference to these systems with optical heterodynation where highly coherent laser sources are necessary.

Optical coherent detection has two major advantages over IM-DD detection:

1. Receivers can be much more sensitive than "direct-detection" ones (by 15 to 20 dB). This is because the local oscillator laser acts as a kind of optical pre-amplifier (the IF amplitude is proportional to the square root of the product of the signal by LOL powers). Also because of the more advanced modulations that can be used.
2. In the WDM systems, where there are many channels packed closely together, coherent detection allows a much better rejection of the adjacent channels as they are filtered in the electrical domain at the intermediate frequency stage. This allows the channels to be spaced more closely.

In addition, there are some other advantages, such as, better noise rejection and the use of the possible types of modulation (ASK, FSK, PSK, PolSK). They may offer additional advantages, as PMD rejection.

However, they require a laser linewidth significantly narrower than the bandwidth of the modulating signal. This means that the incoherent sources like the LED transmitters or unmodified Fabry-Perot lasers cannot be used. Also, they cannot operate in the presence of any significant level of frequency chirp. In fact, very good quality DFB or tunable lasers have to be used.

In the following Figure 2.2.12, the requirements and main characteristics of the most relevant coherent modulation formats are detailed:

- Modulation can be Direct to the laser or External to an optical modulator, depending on the modulation format (for example, ASK/OOK normally requires external modulation to avoid the effect of the transmitter chirp).
- Type of electrical demodulator; it is asynchronous because they do not require so narrow laser line-width.

Modulation Format	Laser/Exter Modulation	Demodulation	Sensib.: – quantum-limit (photon/bit)	Linwidth[1] ($\Delta v/Rb$)	BW IF[2] (BW/Rb)
PSK	External	Synchronous	18	0,07%	1.6
DPSK	External	Differential	20	0,34%	1.6
CPFSK	Direct	Differential	20	0,34–0,5%	1.1–1.6
FSK-DoubleFilter	Direct	Filter + envelop D.	40	5%	3
FSK-SingleFilter	Direct	Filter + envelop D.	80	5%	1.6
ASK	Dir/Ext	Synchronous	72	0,07%	1.6
ASK	Dir/Ext	Envelop D.	80	10%	1.6

Figure 2.2.12. Requirements and main characteristics of most relevant coherent modulation formats.

- Ideal receiver sensitivity, in photons per bit, for a bit error rate of 10^{-9}.
- Maximum laser line-width scaled to the bit rate. The higher the bit rate, the cheaper are the lasers.
- Minimum IF bandwidth normalized to the bit rate. This approximately corresponds with the optical signal spectrum bandwidth. The narrowest bandwidths are obtained with CPFSK modulation **[Tomesen 94, Prat 98]**.

Since channel selection is performed with an electrical filter, heterodyne detection can be seen as a perfect optical filter, thus very narrow channel spacing can be obtained.

The receiver can be tuned to any optical frequency (in a range) by varying the LO Laser frequency. It generates a wavelength very close to the optical frequency of the signal carrier. Of course, feedback in various forms is necessary to stabilize the intermediate frequency.

There are few significant drawbacks of the coherent reception, they (of which) can be solved, but at a high cost:

1. Optical coherent receivers are highly polarization sensitive and, a standard single-mode optical fiber does not respect polarization (any polarization fed to the fiber will evolve randomly so it will changed at the output). However, polarization changes slowly and can be compensated for with different mechanisms. In multichannel WDM systems, acquiring the polarization of a channel quickly after tuning is quite a challenge. Various solutions (including using diverse detectors with orthogonal polarization) have been tested **[Walker 90, Prat 95]**.

2. Stabilization of the frequencies of both transmitter and receiver within close tolerances is another crucial issue. This can be done with a tracking system that sense the IF center frequency and controls/adjusts the LOL wavelength.
3. IF frequencies are usually high (above the signal base-band spectrum) so the receiver has to include complex microwave electronics.

Coherent systems have been rarely applied to the access networks. Although they offer high density for user wavelengths, the cost is important. A relevant field trial was done in the RACE COBRA project **[Bachus 96]**. Four demonstrators where implemented within this project carrying SDH, uncompressed video, videoconference and ATM multimedia applications. The HDTV studio demonstrator was carried out locally in the BBC in the UK with a line rate of 2.5 GBit/s CPFSK modulated, 5 terminals, and 3 optical carriers [Figure 2.2.13 and Figure 2.2.14]. The node was a simple passive star allowing for more than 100 terminals. In this project, a fully integrated heterodyne receiver was developed: a four-section DBR local laser, a 3 dB coupler, balanced photodiodes, polarization hybrid and a preamplifier are integrated in a single InP chip, demonstrating that this is a very promising technology to achieve low cost integration.

Summary

Prototype coherent receivers have been available in laboratories since the middle of the 1980s and have been used in many experiments. However, the coherent receivers are not available as commercial systems, although they are theoretically the best detection technique. They may only be a far future solution, when electro-optical integration becomes more mature.

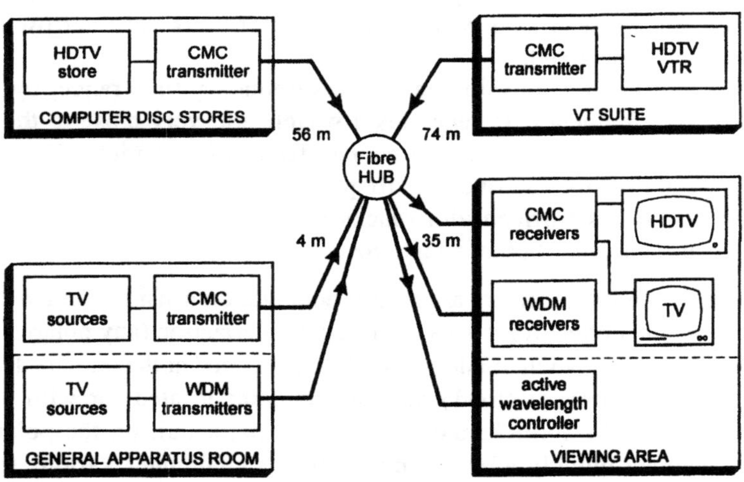

Figure 2.2.13. Video Studio coherent demonstrator.

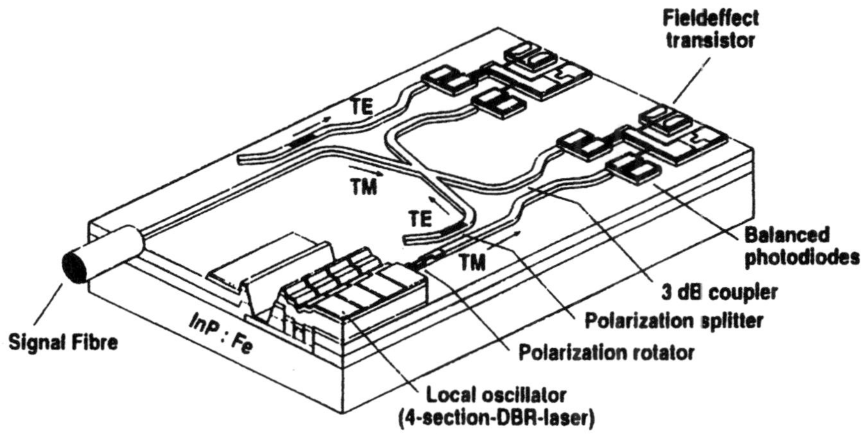

Figure 2.2.14. Single OEIC integrating several opto/electrical functions in a coherent receiver.

2.2.3. Electrical modulation

The modulations reviewed till now are those concerning optical modulations, but one alternative is to maintain the electrical systems using the fiber as a transparent channel. There are several cases where this solution is adopted, for instance, in the Hybrid Fiber Coaxial environment or the Fiber/Copper networks. In the first case, the idea is to use the extended Cable Television (CATV) networks to transport the information to the customer. The other case is to take advantege of the old copper telephony infrastructure with some complex modulations x-type Digital Subscriber Line (x-DSL).

The information can be either analogical or digital. For instance, the next standard in the television systems in Europe, Digital Video Broadcasting, includes the digital technology in television broadcasting instead of the classical analogical television PAL/SECAM. The modulation schemes used in one case or another is completely different, but the translation to the optical domain is basically the same. In Figure 2.2.15, a generic scheme of a electro-optical modulation is showed. The electrical signal goes through an optical channel in a transparent way.

Hybrid fiber coaxial (HFC) systems

As mentioned above, there is a lot of km of CATV networks already installed all around and those systems make use of the fiber as a transport network but completely transparent to the modulation format. These networks can provide the old analogical television and new digital television. The modulation of each one is completely different.

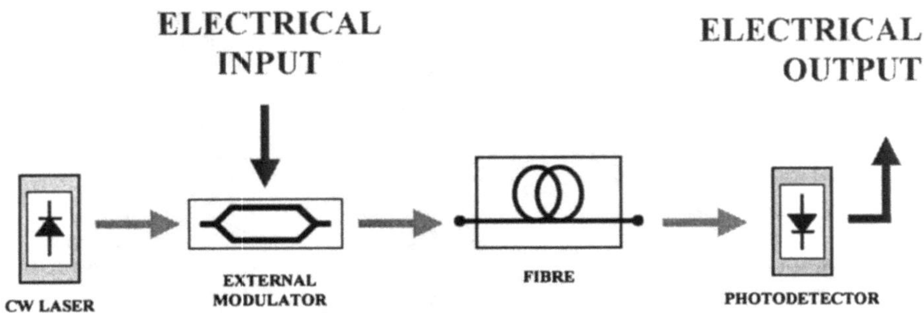

Figure 2.2.15. Electro-optical conversion.

Analogical TV. The old broadcast TV formats can be transported by optical fiber maintaining its analogical nature. There are several standards depending on the country but in Europe the most important is the Phase Alternation Line-rate (PAL) that makes use of the Vestigial Side Band (VSB) Amplitude Modulation (AM) for video and Frequency Modulation (FM) for audio.

Amplitude Modulation gives a double side-band signal each containing identical information, and theoretically only one side-band is required for demodulation. To conserve the spectrum analog TV that uses VSB modulation, VSB truncates the high frequency part of the lower side-band. The lower side-band is not completely cut to avoid the extremely abrupt filtering.

In Figure 2.2.16, the PAL-B spectrum is plotted. As can be seen, the video information is divided into Vision Carrier (Luminance) and the Chroma (Colour) with two different carriers. The first one is modulated in VSB format while the second one in a single AM format. Out of the video band, there is the sound information modulated in FM. There are two sound carriers also to allow the stereo reception.

Each analogical channel has a 6 MHz bandwidth and can be multiplexed together with other channels at different carriers. There are some RF bands defined where to put all the TV channels:

- VHF Band I 45–70 MHz
- VHF Band III 174–230 MHz
- UHF Band IV 520–582 MHz
- UHF Band V 582–820 MHz

Digital television. A new era about digital television is coming and some standardization process has begun depending on the country. The Digital Video Broadcasting (DVB) is the one concerning Europe. The organization responsible is the European Telecommunications Standards Institute (ETSI). The standard extension concerning the cable transmission medium is named DVB-C, and in it, the modulation formats for digital television can be found.

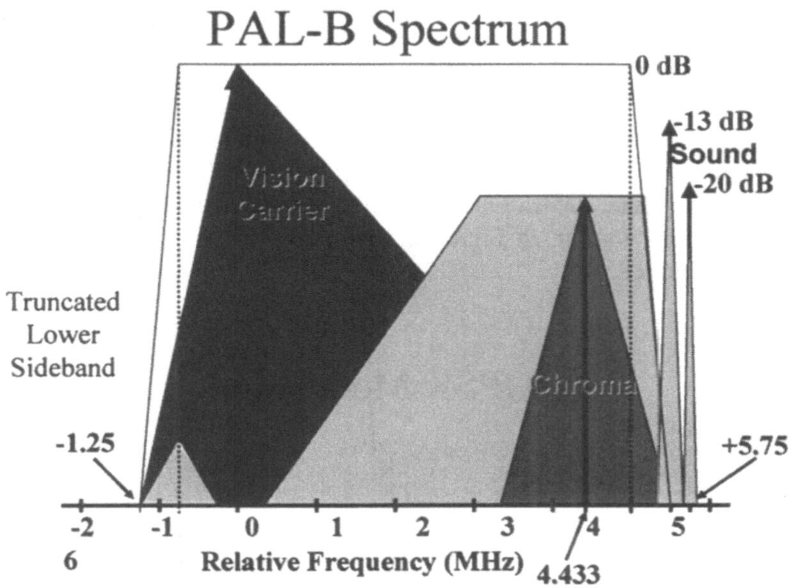

Figure 2.2.16. PAL-B spectrum.

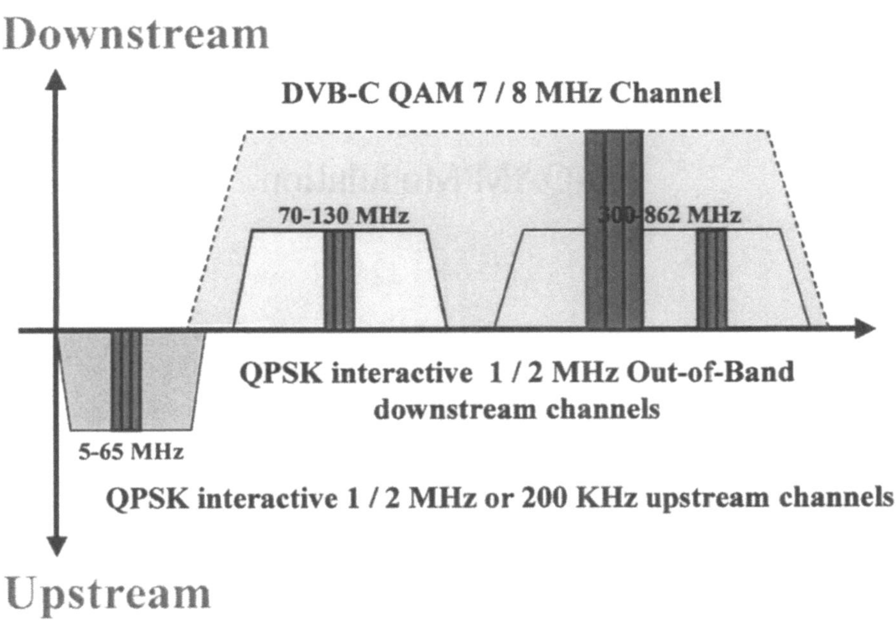

Figure 2.2.17. DVB-C spectrum planning.

In Figure 2.2.17, the spectrum allocation is plotted. As can be seen, the modulation formats are different for the Upstream and Downstream channels. For the Downstream, the modulation format used is Quadrature Amplitude Modulation (QAM) at different levels, from 16 to 256, in an 8 MHz channelization. For the Upstream case, the modulation scheme used is based on the Quadrature Phase Shift Keying (QPSK) in a 2 MHz channelization. Furthermore, some 2 MHz Out-of-Band upstream channels are defined modulated in QPSK format. In Figure 2.2.18, both 16-QAM and QPSK constellations are plotted.

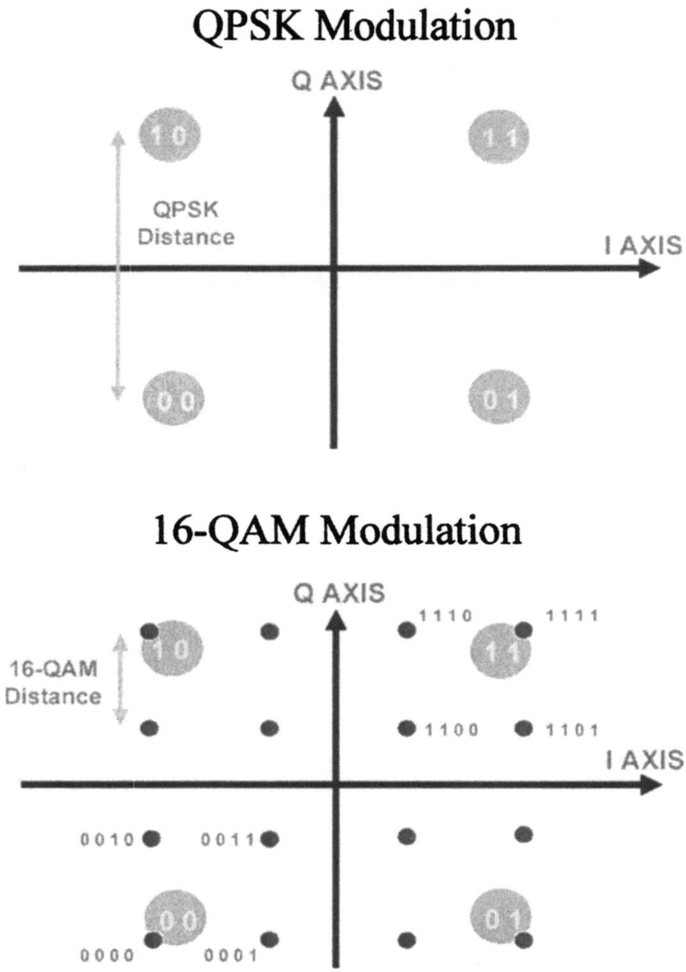

Figure 2.2.18. PAL-B spectrum.

Asymmetric Digital Subscriber line (ADSL)
Another technology used together with the fiber is the old cooper telephony plant. This techniques are mostly extended in places where there is no coaxial infrastructure. The first step is the Asymmetric Digital Subscriber Line (ADSL). The future trend is to standardize the Very-High-Data-Rate Digital Subscriber Line (VDSL). In this section our interest is focused on which kind of electrical modulation format is used in ADSL technology, as a widely extended solution.

The modulation technique used in the ADSL system is based on the multi-carrier techniques called Discrete Multi-Tone (DMT) together with single channel QAM. The QAM technique has been reviewed in the last section; but this new technique DMT is just to increase the spectrum efficiency by dividing the information into several sub-carriers called multi tones. Each one is modulated individually in n-QAM format as can be seen in Figure 2.2.19.

The asymmetric nature of the ADSL system is pointed out in Figure 2.2.18. There are 25 Upstream channels in front of 250 Downstream channels. The data rate can be calculated as follows: Data Rate = Number of Ch × Bits per Ch × Modulation Rate (1.5 Mb/s Upstream – 150 Mb/s Downstream).

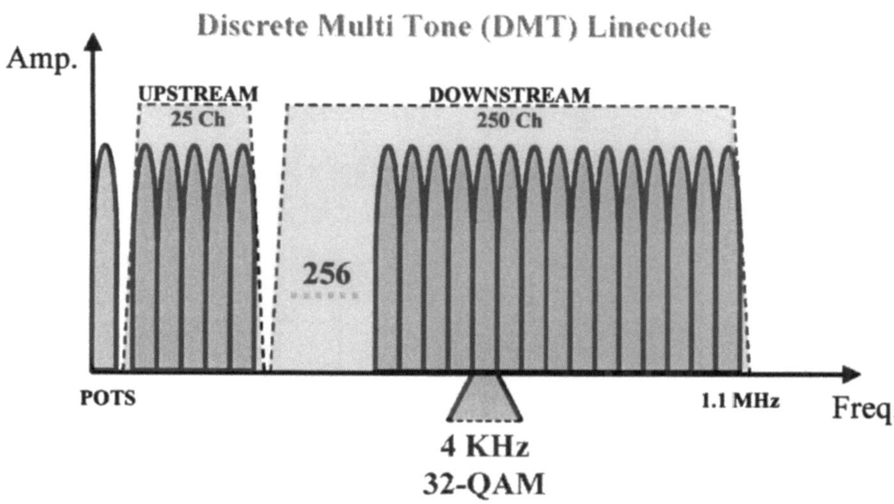

Figure 2.2.19. ADSL spectrum plan.

2.3. MULTIPLEXATION TECHNIQUES

Introduction

In this point the matters under discussion are those concerning the different multiplexation techniques available. In the same manner as previous section, the relationship between the metropolitan network and the access network increases the relevance of the multiplexation mechanism used.

Different techniques can be summarized as follows: Multiplexation concerning the time (OTDM, ETDM), concerning the frequency (FDM, SCM, WDM) or concerning the code (CDM, CM).

A system can be not only based on one of these multiplexation techniques but in some of them. In Figure 2.3.1 **[Pfeiffer 00]**, this idea is showed in a complete system where three different mechanisms, WDM, CDM & TDM, are used together.

This investigation does not pretend to analyze all the possible multiplexation techniques, but review the most attractive in a FTTH environment recently published.

The main subject of interest is about OCDM because of its attractive asynchronous operation mode. A lot of novel approaches are reported and studied in order to have a broad review on this promising technology for the near future solutions. Nevertheless, the most used techniques are those based on TDM and WDM.

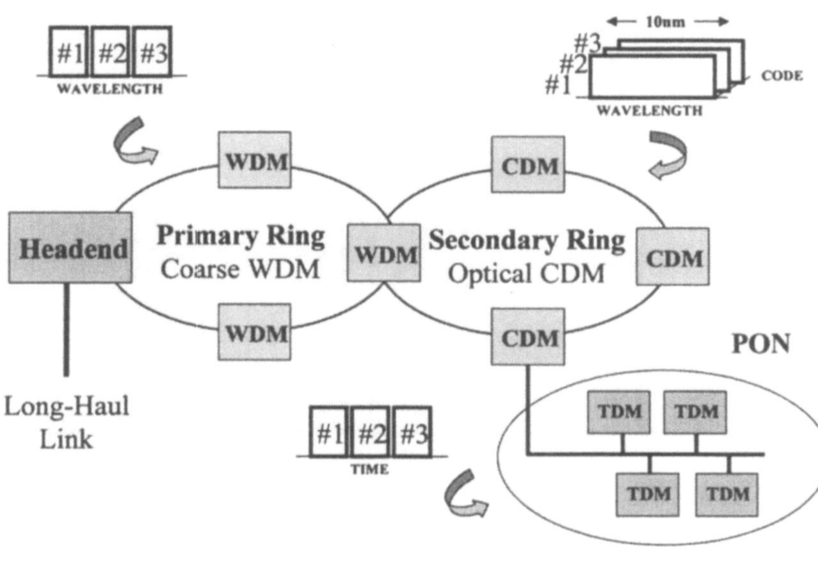

Figure 2.3.1. Multiple multiplexation techniques used in a complete system **[Pfeiffer 00]**.

2.3.1. Time division multiplexing

This kind of multiplexation technique makes use of the time domain in order to separate the optical channels. There exist two ways of dividing the time into temporal slots: in the electrical domain (ETDM) and the optical domain (OTDM). The first one is the classical one used in conventional communications directly translated to optical domain. The second one performs the multiplexation in the optical domain and takes advantage of optical properties.

ETDM is well known, so we will deal only with OTDM due to its originality. The most important optical property used is the pulse compression. It is relatively easy to compress the optical pulses by means of some kind of special fiber. The pulse-width that can be reached is on the order of few pico-seconds, extremely difficult to reach in electrical domain.

2.3.1.1. *Optical Time Division Multiplexing (OTDM)*

In the last few years, there has been a lot of articles concerning the Optical Time Division Multiplexing technique (OTDM) as a promising alternative to WDM. This multiplexing technique can also be seen as a solution for access networks.

PRINCIPLE OF WORK

- Objective: All optical multiplexation / demultiplexation
- Multiplexer based on pulse compressor (fiber) and delay lines
- Demultiplexer based on:
  - High speed electronics
  - Non-linear effects
  - Dispersion
- Advantage: Pulse compression in the optical domain
- Disadvantage: Synchronism and complexity

The electrical pulses are translated to optical domain, and then are compressed by means of concatenation of special fibers. Once compressed they can be time multiplexed by means of delay lines. The number of channels depends on the compression ratio and can be as high as 10 or 100. This procedure can be seen as a non time-sharing multiplexation since the philosophy behind it is different from that in ETDM. The time is not divided into time slots sharing the capacity among the users, but the information of all users are put together thanks to optical pulse compression property. This idea is plotted in Figure 2.3.2.

The main effort is focused in the demultiplexer design. The first step was to implement an electrical demultiplexer working at high speed but the electronic solution is intended to be avoided for simplicity and economic reasons.

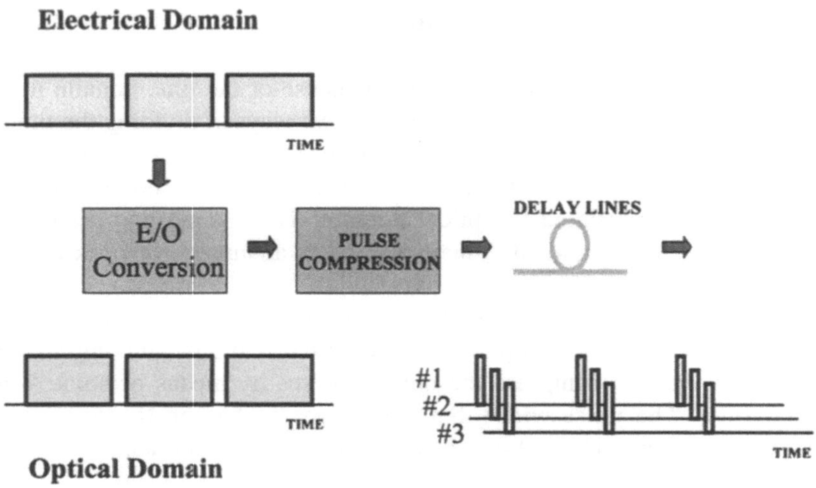

Figure 2.3.2. OTDMA multiplexation mechanism.

So there are several proposals to implement an all-optical demultiplexer based on different optical properties as non-linear effects or dispersion. In this study, we will deal with some solutions recently published, but the definitive technique is far to be defined.

In Figure 2.3.3, a typical OTDMA scheme is plotted **[Yu 01]**. In this case, the demultiplexing technique used is based on the Self-Phase Modulation (SPM) and the Cross-Phase Modulation (XPM). Qualitatively, the idea of the method is as follows. When the control pulses and the OTDM signals propagate simultaneously in a DSF, the optical spectrum of the OTDM signals will be broadened because of SPM and XPM. Assuming that the input OTDM signals have the spectral bandwidth and that dispersion can be ignored, the optical spectrum of the desired channel to be demultiplexed from the OTDM signals broadens due to SPM and XPM. The other channels in the aggregated TDM signal only suffer SPM.

Because of the XPM, the optical spectrum of the desired channel to be demultiplexed will be separated in frequency from that of other channels. After the DSF, the signals pass through an optical filter whose center frequency is shifted with respect to the input OTDM signal frequency. If the spectral broadening of a pulse is small enough, then the pulse is rejected by the filter. This happens when the pulse intensity is too small (noise in 0 bits). If the control pulse intensity is high enough (1 bit) so then a part of the XPM and SPM-broadened spectrum passes through the filter. The bandwidth of the filtered pulse is determined by the filter bandwidth. In this way, simultaneous demultiplexing and regeneration is realized.

Another example using the dispersion instead of non-linearities to imple-

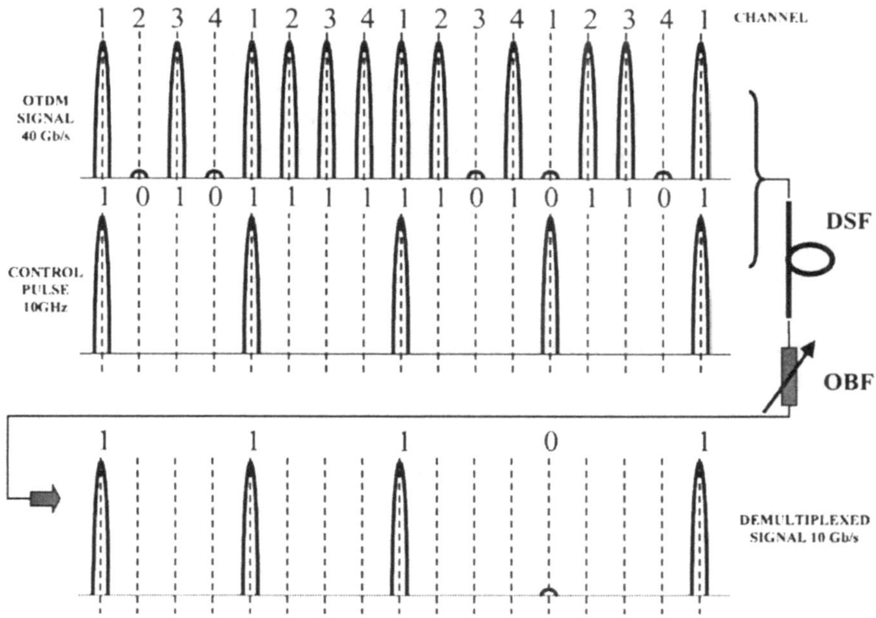

Figure 2.3.3. OTDMA technique based on SPM and XPM effects **[Yu 01]**.

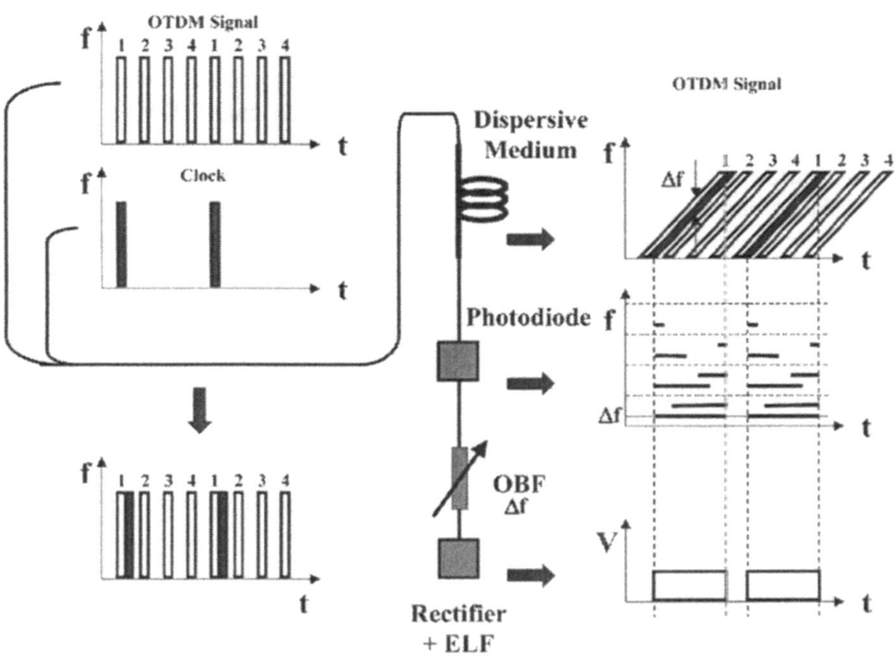

Figure 2.3.4. OTDMA technique based on GVD effects **[Tomoyuki 01]**.

ment the demultiplexing process is presented in **[Tomoyuki 01]**. Figure 2.3.4 shows the operation principle of the proposed demultiplexer. A four-channel OTDM signal and an extracted clock signal are combined with a slight time difference between the clock pulse and the signal pulse of a channel (e.g., channel 4) to demultiplex. The broad spectra of the short pulses are depicted as bars spread along the y-axis (defined as the optical frequency f). The dispersive medium stretches each pulse in the time domain due to chirping which is expressed by the tilt of the bars. At each time, there exists more than one frequency component which generates beat signals in the radio-frequency (RF) signal after photodetection. The beat component between the clock and channel 4 has a center frequency equal to the difference in the optical frequencies of the channel-4 pulse and the clock pulse (Δf) as depicted. This beat component appears only when channel 4 is a high level which corresponds to a demultiplexed signal of channel 4. This signal is converted into the standard voltage signal by using a band-pass filter (BPF) having a pass-band of around (Δf) and an envelope detector achieved by a rectifier and a low-pass filter (LPF).

In **[Ulmer 00]** a new all optical solution is shown. A ultra-fast single-wavelength optically time-division multiplexed (OTDM) link is described. The link exploits a unique integrated, all-optical serial-to-parallel (S/P) converter based on second-harmonic generation that demultiplexes multiple high-speed optical channels with a single operation. The link is composed of five major components: 1) a high-repetition-rate picosecond-pulse source; 2) a planar wave-guide multiplexer that incorporates electro-absorption modulators with integral spot-size converters (SSCs); 3) a dispersion-managed (DM) short-pulse fiber channel; 4) a quasi-phase-matched, resonant-cavity-enhanced AlGaAs wave-guide designed for surface-emitted second-harmonic generation (SESHG); and 5) a 775-nm receiver optimized for return-to-zero (RZ) operation. We describe our recent advances with resonant cavity enhancement of the all-optical demultiplexer and the first bit error rate (BER) measurements for this demultiplexing scheme. Figure 2.3.5 summarizes this technique.

As can be seen above, this kind of modulation technique is very interesting due to the pulse compression possibility in optical domain. In electrical domain, it is difficult to achieve those compression rates. If a compression rate of 10 is given, 10 channels can be multiplexed independently.

The main drawback is just the demultiplexation scheme that is difficult to implement in the optical domain and requires a precise synchronization.

2.3.2. Frequency/wavelength division multiplexing

Another multiplexation domain is the frequency/wavelength. In fact, they are two different domains; frequency stands for electrical frequency and wave-

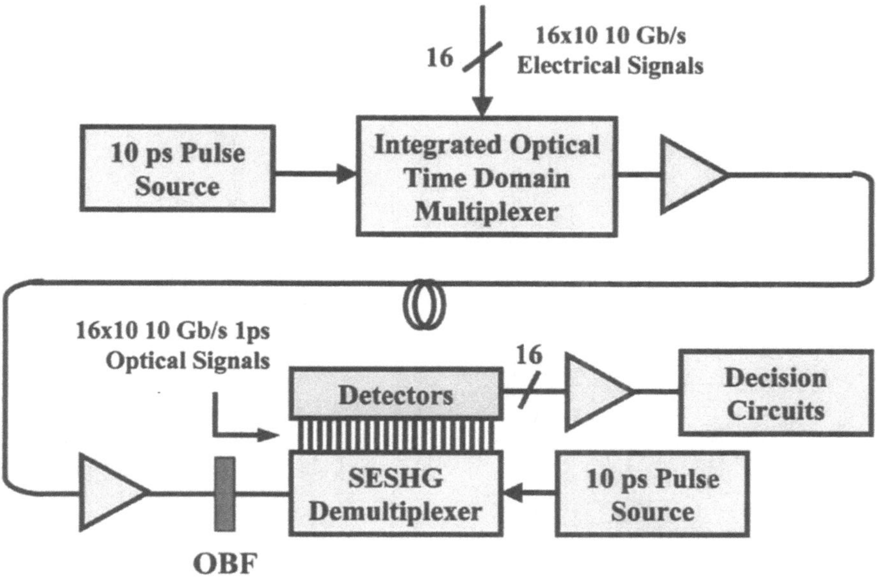

Figure 2.3.5. OTDM technology based on SESHG **[Ulmer 00]**.

length stands for optical wavelength. One example of each one is presented here.

2.3.2.1. *Sub-carrier Multiplexing (SCM)*

The sub-carrier multiplexing technique makes use of the multiplexation in the electrical frequency domain. Then this is translated to optical domain and then recovered at the receiver. Figure 2.3.6 shows a Passive Optical Network (PON), see section 2.5, based on SCM. Each user (Optical Network Termination, ONT) has assigned an electrical frequency in order to multiplex all channels.

PON provides upstream SCMA for a number of remote optical network units (ONUs), with standard lasers transmitting on their wavelengths. In each ONU, the optical output mean power is controlled via the laser bias current by the circuit C which also controls the laser modulation current via the variable amplifier A. The sub-carrier frequencies are modulated with the input signals in the modulators (M).

In the optical line terminal (OLT) of the head-end, the photo-diode converts the sum of optical signals to an electrical signal which is amplified in and separated by a set of filters or heterodyne receivers on the sub-carrier frequencies. The output signals are recovered from the sub-carriers by the demodulators.

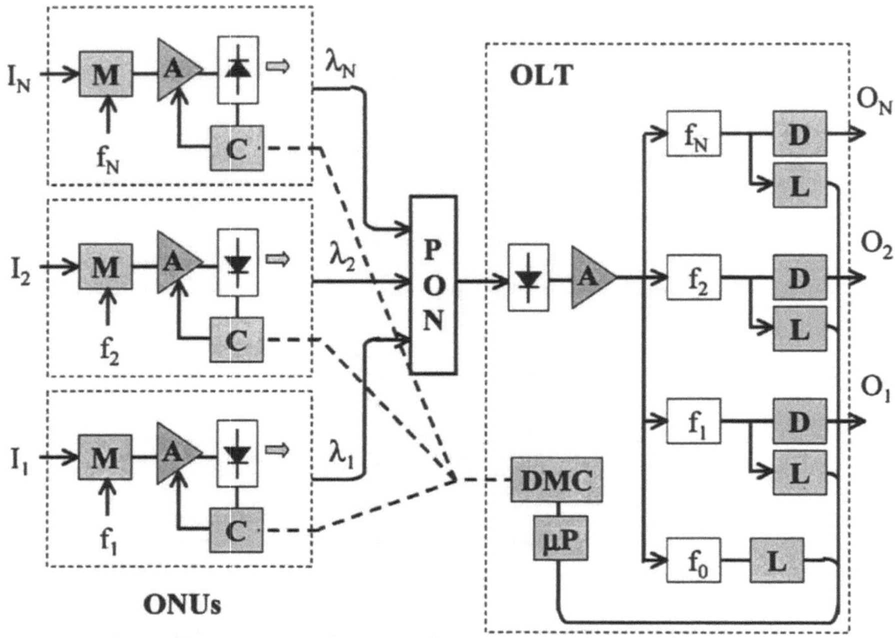

Figure 2.3.6. PON model using SCM technology **[Sorensen 01]**.

Sub-carrier power equalization is done at system start-up by means of level detectors on the filter outputs, providing feedback to the power and modulation controls of the ONUs, via the microprocessor ((P) and the Downstream Management Channels (DMCs). Subsequently, the receiver noise floor level is continuously measured in some free bandwidth not used for sub-carriers and separated from these by a filter. The filtered noise is applied to another level detector the output of which is used to control the laser mean powers of the ONUs. This is also done via the µP and the DMC in accordance with a noise suppression algorithm.

The main limiting factor of SCM multiplexation technique is the so-called Optical Beat Interference (OBI) noise. OBI arises due to the quadratic conversion of optical to electrical power by mixing of optical fields when two or more light sources illuminate a receiver photodiode at the same time (see Chapter 6 for more information about OBI factors).

The signal-to-interference ratio (SIR) is proportional to the square of the modulation index and inversely proportional to the receiver bandwidth. Having chosen the number and type of optical sources and the receiver bandwidth for a given data bit rate and modulation format, the remaining methods of minimizing OBN (Optical Beat Noise) may be divided into two main categories.

1) Spectral Broadening:
   - Light-emitting diodes (LEDs) as optical sources. The low output power limits the split ratio and transmission distance in the PON.
   - Direct intensity laser modulation index. Limits the usable sub-carrier frequency range due to harmonics.
   - Very high sub-carrier modulation frequencies. Increases the cost of the transmitter and receiver.
   - Very high frequency laser chopping. Increases the cost of the transmitter and receiver.
2) Wavelength Separation:
   - Laser selection by wavelength. Expensive, requires laser temperature control for multi-channel field systems.
   - Laser temperature control. Expensive, requires a record of wavelengths/temperatures for replacement or addition.
   - Automatic wavelength control. Requires feedback from the receiver to each of the transmitters.

The method described in Figure 2.3.6 **[Sorensen 01]** belongs to the second category. It is aimed particularly at low cost bi-directional PONs with up-stream SCMA, requiring low sub-carrier harmonic levels over a large optical dynamic range, using direct current/intensity modulated standard multimode Fabry-Perot lasers without temperature control or special wavelength selection. However, it may also be used in systems where the modulation index is larger than 1 and in systems with temperature stabilization of the lasers.

The dynamic range of a SCMA-PON using Fabry-Perot lasers becomes very small in the presence of OBN when the laser modulation index is less than 1. Furthermore the attenuation may be different for each branch in a practical PON. The method therefore includes initial power equalization of the received sub-carrier levels at system start-up. This is followed by automatic fine-tuning of the laser wavelengths, via downstream feedback control of their optical mean powers, performed on the basis of noise measurements in a frequency band not used for sub-carriers.

This procedure minimizes the receiver noise floor level continuously during operation. It will improve the SIR and reduce the number of transmission errors caused by OBN in systems where the following assumptions hold true. The small relative changes in the optical mean powers will increase the distance between interfering modes sufficiently without substantially changing the sub-carrier power level. Additionally, the laser line-widths are large enough to cause a noise reduction in the sub-carrier frequency band as well as the measurement frequency band.

This technology is very interesting in CATV environments where the HFC protocol is used. The analog transmission is translated to optical domain and then recovered at the receiver having a transparent optical layer.

2.3.2.2. *Wavelength Division Multiplexating (WDM)*

Once reviewed the electrical domain, the next step is to study the multiplexation in optical domain where the parameter used is the wavelength not the frequency. The advent of WDM technique allowed increasing heavily the capacity of optical systems.

In High-Capacity environments the extension of WDM is the so-called Dense Wavelength Division Multiplexing (DWDM). The channel separation can be as small as 0,8 or 0,4 nm at 10 Gb/s. This technology is very expensive so its application to access networks is difficult.

Coarse wavelength-division multiplexing (CWDM) is emerging as a robust and economical solution. The advantage of CWDM technology, as a whole, is its ability to leverage conventional technology to make the jump to the next-generation data rates. In metro and access markets, CWDM will offer solutions for 850-, 1300-, and 1500-nm applications at 10 and 40 Gbit/s. Based on a parallel optical subassembly architecture, it will support the eventual 100-Gbit/s Ethernet standards **[LaHa 01]**.

CWDM differs from dense wavelength-division multiplexing (DWDM) in that the optical channel spacing between the light sources that are multiplexed into a single fiber is much wider. Additionally, CWDM transceivers use optical multiplexing for achieving serial-equivalent data rates, whereas DWDM multiplexes many serial data streams to achieve bandwidths on the order of hundreds of gigabits per second. Generally speaking, this is achieved by using temperature to control the DWDM channel spacing to 100 GHz. This exact control over the spacing allows up to 80 separate channels to be multiplexed.

Typical CWDM spacing is on the order of many nanometers (several Terahertz) and does not require temperature control. Consequently, CWDM transceivers with no temperature control, having directly modulated lasers, and incorporating lower-speed components to achieve higher data rates are typically lower in cost, size, and power consumption than their conventional serial counterparts.

Initial implementations of CWDM in optical transceivers focus on the LAN and SONET very-short-reach (VSR) market. Pushing this effort is the availability of low-cost optical sources – namely 850-nm multimode vertical-cavity surface-emitting lasers (VCSELs) – and the significant amount of multimode fiber already installed in the LAN and enterprise environments. Rather than install new single-mode fiber to handle links of less than 2 km, system vendors and site managers seek low-cost, high-data-rate solutions that take advantage of the installed multimode fiber. CWDM fits this bill in its ability to transmit a higher aggregate data rate farther, with lower-speed multimode components.

Gigabit ethernet via eight wavelengths: An application example of this new technology can be found in **[LaHa 01]**. A Gigabit Ethernet system is described based on CWDM for a LAN environment. This idea can be extended to access networks.

On the transmitter side, eight wavelengths from directly modulated VCSELs are multiplexed optically into a single multimode fiber. The eight wavelengths are sent through the multimode fiber to the receiver with the demultiplexer that separates the channels out to individual photo-detectors. Each of the eight channels transmits data at 156 Mbit/s. Although the aggregate signal rate is 1.25 Gbit/s, accounting for 8-bit/10-bit encoding, the lower data rate on the individual channels allows data transmission over 2 km on 62.5-μm multimode fiber. As for 850-nm serial transmission of Gigabit Ethernet data, the IEEE specification in the 802.3z standard is only 220 m over 62.5-μm diameter fiber (see Figure 2.3.6).

The wavelengths of the individual lasers are nominally 778, 789, 800, 812, 825, 837, 850, and 864 nm at room temperature. As the temperature changes in the module, the eight wavelengths will not overlap since the optical spacing between each channel is wide. This spacing prevents the overlap of data between adjacent channels. Multiplexing and demultiplexing the eight channels into and from a single fiber is performed via a zig-zag geometry. The eight optical elements, VCSELs at the transmit end, and photodiodes at the receive end, are mounted onto a substrate. Next, the optical multiplexer/demultiplexer (mux/demux) block is placed on top of the substrate. The mux/demux block has eight band-pass filters on the bottom that line up over the active components and a broadband coating on the top to reflect all of the wavelengths. As the laser light zig-zags through the filter block, channels are injected into the data stream in the mux and ejected from the data stream in the demux. The width of the band-pass filters ensures that as the laser wavelength drifts over the transceiver's operating temperature, the entire signal is captured (see Figure 2.3.7).

Four-channel architecture: Coarse wavelength-division multiplexing is also well-suited for transceivers with data rates at 10 Gbit/s. Because of the abundance of low-cost 850-nm VCSELs at 2.5 Gbit/s (OC-48), it makes sense to build a CWDM device based on a four-channel architecture (4 × 2.5 Gbit/s). Using a similar zig-zag mux/demux as described for the eight-channel gigabit per second device, four 850-nm variety VCSELs and photodiodes may be integrated to yield 10-Gbit/s aggregate data transmission. However, in this case, the optical spacing is set at 20 nm between channels. This spacing, in conjunction with the four corresponding band-pass filters, ensures enough "elbow room" for the lasers to move optically with temperature as the mux heats or cools.

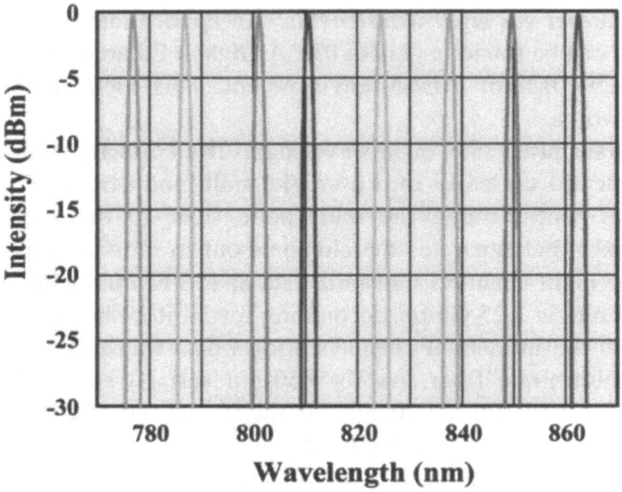

Figure 2.3.7. Eight-channel CWDM transceiver is standardized at 778, 789, 800, 812, 825, 837, 850, and 864 nm at room temperature **[LaHa 01]**.

Coarse wavelength-division multiplexing transceivers provide access to higher bandwidths with conventional active optical components which are lower-power than serial solutions, and typically result in shorter development times. The 850-nm solutions available today range in performance from 1 to 10 Gbit/s and are offered in various industry-standard footprints and electrical interfaces.

Three variables will define the new CWDM data links coming to the market in the future: longer distance, faster speeds, and higher density.

Short reach is really only the beginning for CWDM. Long-wavelength modules at 1300 and 1500 nm will carry low-cost CWDM solutions into the metro environment where 10 km will be the minimum link length. In fact, one of the 10 Gigabit Ethernet standards in the IEEE 802.3ae is a four-channel, 1300-nm CWDM solution. While the focus of Ethernet has historically been on the LAN environment, the adoption of single-mode solutions in conjunction with work being done on the wide-area network physical layer will push Ethernet into areas where SONET has traditionally dominated.

While a large number of applications can utilize such as a budget, a more adequate light source for OC-768 VSR will be 1300- or 1500-nm sources. The 1300-nm sources, whether distributed-feedback lasers or the newly emerging long-wavelength VCSELs, will ensure transmission throughout the VSR link specification. In addition, 1500-nm lasers, robust from years of use in the DWDM market, will propel 40-Gbit/s CWDM out into the metro area with transmission distances in the range of 40 km.

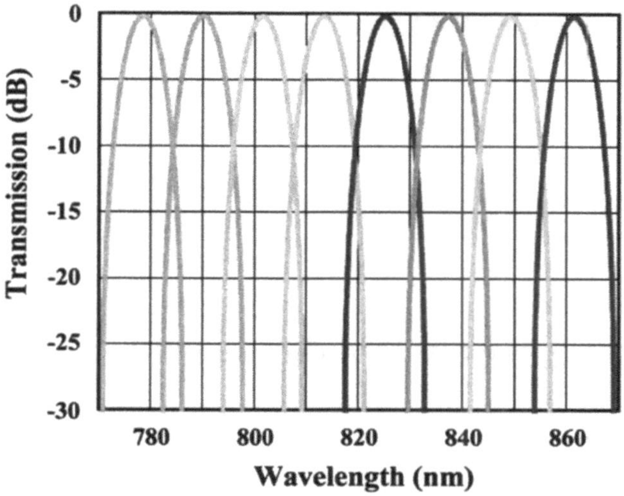

Figure 2.3.8. The output of band-pass filter spacing on an eight-channel CWDM **[LaHa 01]**.

IEEE CWDM standardisation: In Access Networks, there exist an IEEE Study Group called IEEE Ethernet First Mile (EFM). Their work is oriented to standardize the use of Ethernet Protocol in Access Networks (see Chapter 3). There are several matters under standardization process, for instance the Upstream MAC mechanism, but a special interesting one is about Coarse Wavelength Division Multiplexing (CWDM).

The most important IEEE standard concerning the Ethernet technology is the **IEEE 802.3**. Ethernet is the most used technology for Local Area Networks (LAN) mainly because of its simplicity and low cost. There is a tendency to extend the use of the Ethernet protocol toward Access and Metropolitan networks with the Gigabit Ethernet (**IEEE 802.3z**) and 10 Gigabit Ethernet (IEEE 802.3ae) solutions. Those protocols are described in section 2.5. There is a new standard approach about CWDM that defines 20nm bands: 10GBASE-LX4 (IEEE802.3).

In Figure 2.3.9, the bandwidth allocation proposed by the EFM Study Group in a de facto standard can be appreciated . They define fifteen 20 nm channels in a very wide frequency margin (from 1300 nm to 1600 nm) in order to allow different Ethernet connections.

The difference with the ITU-983.3 is pointed out. Nowadays ATM-PON (see Chapter 3) systems use one wavelength for the upstream channel and another one for the downstream channel. In the future, the standard allows implementing WDM solutions and maybe CWDM solution will be adopted by ITU-T organization.

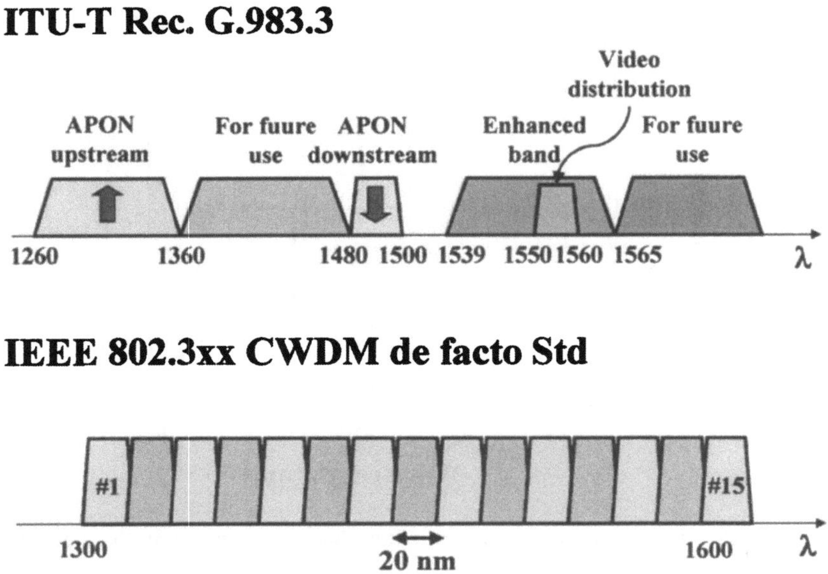

Figure 2.3.9. CWDM de facto standard **[EFM Sep01]**.

Conclusion. CWDM technology seems to be a promising solution to increase the capacity of optical access networks. The main drawback of WDM devices is the high cost, but it can be strongly reduced by using coarse specifications. New WDM components had appeared recently searching the cost-effectiveness, like Vertical Cavity Surface Lasers (VCSEL) or integrated optical devices as modulators or multiplexers.

2.3.3. Code division/coherence multiplexing

As in radio communications a new multiplexing domain is defined. Spread spectrum techniques are well known in order to define new orthogonal channels. In optical communications, there exist some different approaches to this philosophy using delay lines in temporal domain or phase amplitudes masks in spectral domain. This family is known as Optical Code Division Multiplexing (OCDM).

There are several OCDM techniques that can be divided into four categories: OCDM in temporal domain, where the code is applied to the pulse in time, OCDM in spectral domain, where the code is applied to the spectrum of the pulse, 2D OCDM uses two different techniques and 3D OCDM where the number of techniques used is three.

Intensity modulation and direct detection (IM/DD) has been established as the most suitable signal modulation and detection scheme in optical communications systems. Although coherent systems are also possible, it is very costly to have a local oscillator at the optical frequency. In contrast, most radio communications systems makes use of coherent schemes which detect the electrical field. While the optical signal intensity is a positive-only quantity, the electrical field is a bipolar quantity which can be both positive and negative. In order to preserve the simplicity of IM-DD, optical CDMA systems are designed very differently from their radio versions.

Another family put into this category is the Coherence Multiplexing (CM). This technique uses high incoherent sources with a very short coherence time to create several virtual coherence paths.

2.3.3.1. OCDM in temporal domain

With delay lines:
This kind of OCDM technique makes use of some delay lines to build a code into the bit period. In Figure 2.3.10, the working principle is shown. The modulator consists of a short pulse optical source and a module with several delay lines. After traveling across this encoder there are several pulses in a bit period each one with its characteristic delay.

The decoder is an intensity correlation process complementary to the encoder just to equalize the delays and reach a constructive interference

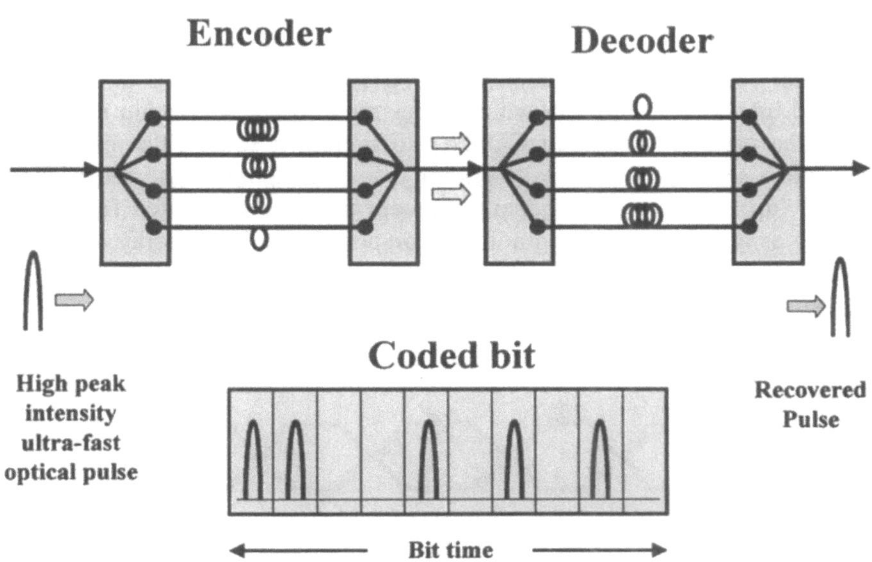

Figure 2.3.10. Temporal Domain OCDM with delay lines [**Fung 99**].

between all the pulses. The pulses of the rest of the users are not equalized and remain with random delays at a low power level. Here the drawback of unipolar modulations can be seen. The pulses cannot compensate between them in order to reduce the interference.

The research in these non-coherent optical CDMA schemes leds to the invention of a few major code groups such as optical orthogonal codes (OOC) **[Salehi 89-I]**, **[Salehi 89-II]**, **[Chung 89]**, prime sequence **[Kwong 91]**, **[Shaar 83]** codes and so son.

The pulses must be very short in order to reach a great number of orthogonal codes. A sub-picosecond pulsed optical source having a pulse width much smaller than the bit duration is required. Even with carefully designed codes, co-channel user interference due to non-orthogonality gives a severe performance penalty. The number of allowable co-active users is very limited.

Another important issue of these OCDM techniques is the power inefficiency. As can be seen in Figure 2.3.11, the power splitters and combiners introduce high losses (10 logN , where N is the number of ports). Just to improve the power efficiency, some alternatives can be used, for instance, the scheme shown in Figure 2.3.11 where the loss is ideally reduced to 6 dB.

Phase-amplitude mask:
Here a robust, passive, and all-fiber approach to Light-wave CDMA in which the devices are designed to operate in a direct-sequence version of optical CDMA is studied **[Grumnet 99]**. The Light-wave CDMA encoders and decoders used here provide flexible phase- and amplitude-encoding in the temporal domain. Sparse Light-wave CDMA, i.e., channel multiplexing without temporal code overlap is demonstrated **[Grumnet 99]**. This is in contrast to dense Light-wave CDMA in which codes are allowed to overlap in time.

In the present version of CDMA, diffracting a short pulse of duration Tchip of a segmented composite grating creates temporal phase codes. Figure 2.3.12 (up) illustrates the general concept using a segmented fiber Bragg grating as an example. The grating is composed of a linear array of uniform sub-gratings all having the same grating spacing, constant, but different

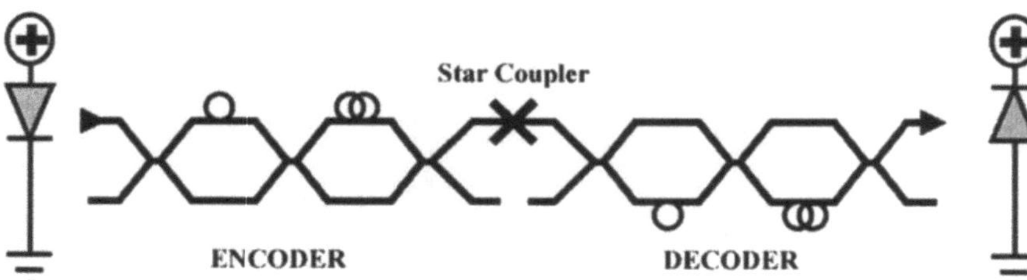

Figure 2.3.11. Time domain encoding using cascaded ladder encoder **[Fung 99]**.

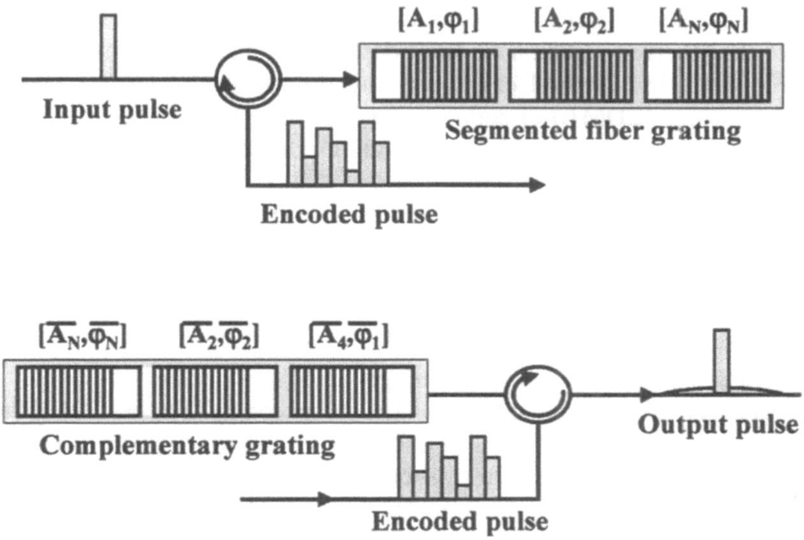

Figure 2.3.12. (up) Encoder based on a segmented fiber grating design. (down) Schematic diagram of the decoder **[Grumnet 99]**.

sub-grating index amplitudes and spatial phases relative to a fixed coordinate system.

When a short pulse is incident on this grating, the light diffracted into the counter propagating direction will consist of a train of time-delayed pulses of width T_{chip}. The relative amplitudes, temporal phases, and bandwidth of the pulses are determined by the specific amplitudes, phases, and length of the individual sub-gratings comprising of the grating. The time-spread encoded waveform is separated from the incident pulse by an optical circulator and can now be transmitted along a transport fiber to the receiving station. Here, a matched grating decodes (de-spreads) the pulse and regenerates the original input pulse of duration, provided the transfer function of the decoder is the inverse of that of the encoder. Complete de-spreading of the encoded pulse will not occur if the encoder and decoder gratings are not matched. This operation is illustrated in Figure 2.3.12 (down).

A proof-of-principle technology demonstration has shown the implementation of compact, polarization-insensitive, cost-effective encoders and decoders in a two-channel sparse Light-wave CDMA transmission link. Sets of dense Light-wave CDMA encoder/decoder devices with four to eight channels within a 30-GHz spread spectrum, by combining with WDM and using the full 3–6-THz EDFA bandwidth, hybrid systems are anticipated that can support hundreds of independent optical communication channels.

There is a limit, of course, to the amount of dispersion that is tolerable in a system before dispersion compensation becomes necessary for proper detection of the received signal. This limit depends on many factors including the specific detection method employed and the presence (dense CDMA) or absence (sparse (CDMA) of code overlap. There is no evidence of non-linear impairments in the system; this observation is consistent with the low peak power in the encoded waveforms.

Another drawback is the need of short optical pulses but, this thick characteristic is alleviated with respect to the delay lines technique.

2.3.3.2. *OCDM in spectral domain*

Coherent spectral phase mask encoding:
Coherent spectral phase encoding was first proposed by **[Weiner 90]**, **[Weiner 98]**. An ultra-short optical pulse that contains a broad spectrum is spectrally encoded. The spectrum of the ultra-short pulse is first spatially dispersed using a diffraction grating [Figure 2.3.13]. A spectral phase encoding grid is employed to apply different phase shifts at various spatially resolved spectral components. As a result of the phase encoding, the original sharp ultra-short pulse [Figure 2.3.14(c)] in the time domain is transformed into a low intensity signal with longer duration. The broadened pulse duration is approximately inversely proportional to the spectral resolution of the phase encoder.

To decode the signal, a conjugate phase shifter matched to the one used at the transmitter is employed. For matched transmitter output, the phase encoding will be removed at the receiver and the high-peak intensity sharp

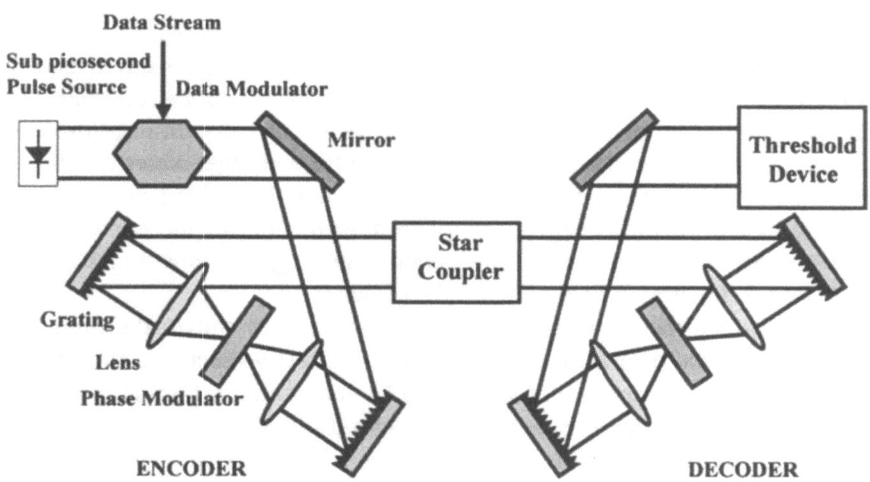

Figure 2.3.13. Coherent spectral phase encoded optical CDMA system using ultra-short pulses **[Weiner 90]**.

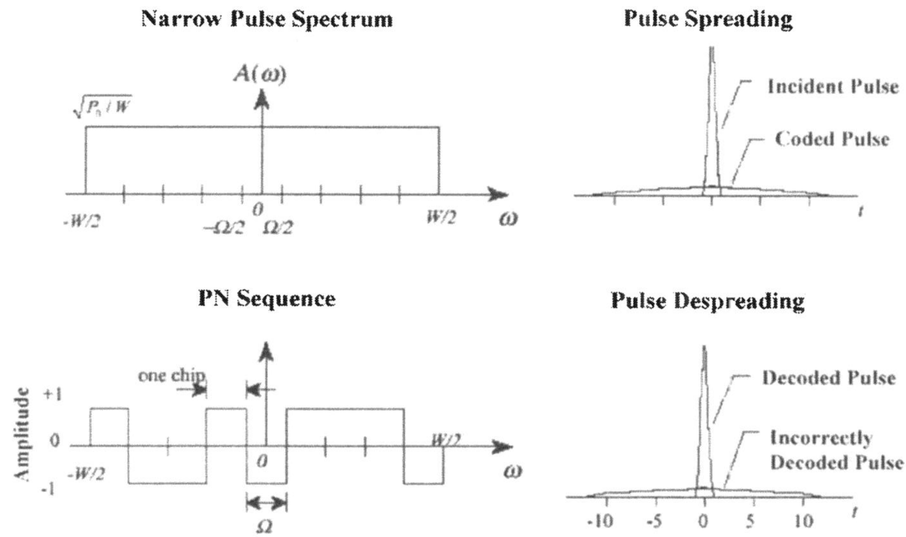

Figure 2.3.14. Frequency and time domain waveforms of coherent spectral phase encoded system: (a) narrow pulse spectrum, (b) pseudo-random (PN) sequence in frequency domain, assuming 0 and p p phase encoding which corresponds to +1 to –1 multiplication in amplitude here, (c) pulse spreading due to phase encoding and (d) pulse de-spreading for matched and unwanted channels.

pulse will be reconstructed. The signals from unmatched transmitters will be further scrambled [Figure 2.3.14 (d)] at the receiver and they form background interference. The reconstructed sharp pulse is distinguished from the background signal by an optical threshold device, or by non-linear optics.

Salehi and Weiner **[Weiner 90]** investigated codes consisting of arbitrary 0 and pi phase shifts which can be considered as +1 and –1 multiplication in amplitude in the frequency domain [Figures 2.3.14 (a) and (b)]. The encoding and decoding concepts have also been demonstrated experimentally **[Weiner 98]**.

Similar to the time domain encoded system described previously, the background interference generated by co-channel users degrades the network performance. Moreover, the phase encoding operation is very sensitive to dispersion in optical fibers. To obtain good results, dispersion compensation or dispersion shifted fiber needs to be used. The non-linear optical threshold detection scheme is also very cumbersome to implement.

The need of ultra-short pulses is also present in this technique.

Non-coherent spectral amplitude mask encoding:

Zacarrin and Kavehrad first described this approach **[Zaccarin 93]**, **[Zaccarin 95]**. It is similar to the coherent phase encoded system in the sense that the frequency components from a broadband optical source are resolved first. Each code channel then uses a spectral amplitude encoder to selectively block or transmit certain frequency components [Figure 2.3.15].

A balanced receiver with two photo-detectors is used as part of the receiver. The receiver filters the incoming signal with the same spectral amplitude filter (called the direct filter) used at the transmitter as well as its complementary filter. The outputs from the complementary filters are detected by the two photo-detectors connected in a balanced fashion. For an unmatched transmitter, half of the transmitted spectral components will match the direct filter and the other half will match the complementary filter. Since the output of the balanced receiver represents the difference between the two photo-detector outputs, unmatched channels will be cancelled, while the matched channel is demodulated.

Since there is a subtraction between two photo-detectors, it is possible to design codes so that full orthogonality can be achieved with the non-coherent spectral intensity encoding approach. In principle, orthogonality eliminates the cross-talk from other users. Bipolar signaling can also be obtained by sending complementary spectrally encoded signals. There is a 3-dB power advantage for bipolar signaled systems. However, the performance of this type of systems is spoiled by the intensity fluctuations arising from the beating between optical waves at the same wavelength, but coming from different users, which is called speckle noise.

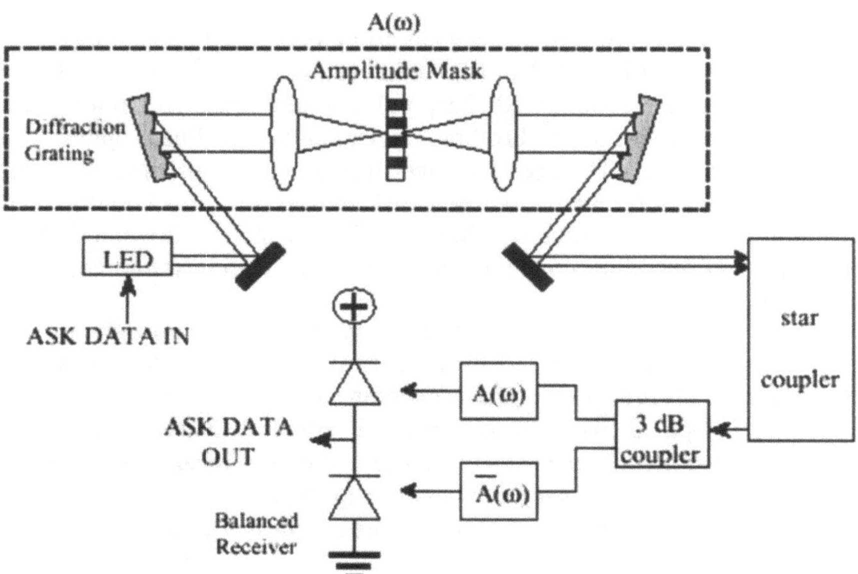

Figure 2.3.15. Non-coherent spectral intensity encoded optical CDMA system **[Zaccarin 93].**

All optical CDMA networks are generally broadcast and select systems. In a broadcast and select network, the receiver receives the signals from all the transmitters. Ideally, all the unmatched channel signals are cancelled due to orthogonal encoding. Nevertheless, a receiver detects the optical energy from the unmatched transmitters. In spite of signal orthogonality, shot noise does not subtract but is always additive increasing with the total detected signal intensity. Therefore, detecting the signals from all users gives rise to cumulative shot noise which amounts to cross-talk and interference.

Another important issue is the speckle noise. In a CDMA system, all the users simultaneously occupy the same frequency-band. They all transmit using the same wavelengths. When light waves of the same wavelengths from different users fall onto the photo-detectors simultaneously. They will interfere with each other and give rise to large fluctuations in signal intensity. This is called speckle noise and can be even more severe than cumulative shot noise, as we will see.

Code orthogonality, shot noise and speckle noise, are the three important fundamental issues in the design of optical CDMA systems and they form the foci of this dissertation.

Non-coherent spectral intensity bipolar encoding:
The block diagram of a non-coherent spectral intensity encoded bipolar CDMA system is shown in Figure 2.3.16 **[Fung 99]**. As mentioned before, one of

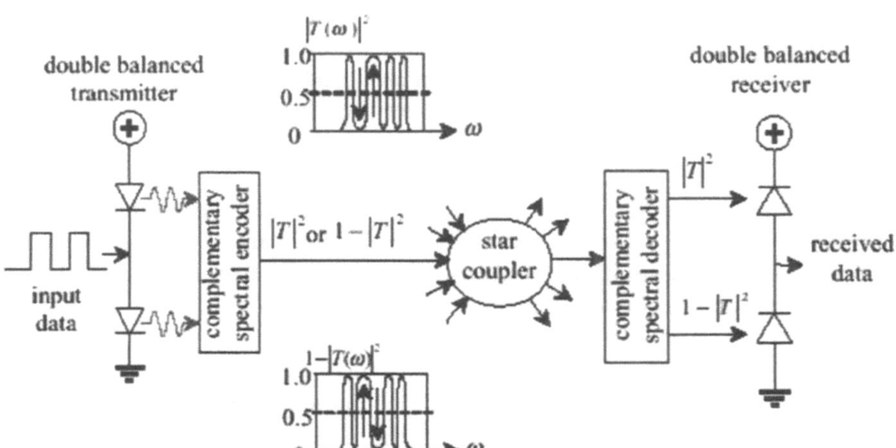

Figure 2.3.16. System structure of bipolar spectrally encoded optical CDMA. A balanced broadband source is spectrally encoded and received by a balanced receiver. The complementary spectra produced by the spectral filter, $|T|^2$ and $1 - |T|^2$. represent the complementary intensity transfer functions of the spectral filter outputs **[Fung 99]**.

the important goals here is to obtain full orthogonality. The optical signal intensity is a positive quantity. However, to achieve full orthogonality, both positive and negative quantities are required. The way to obtain negative quantities with the positive optical intensity signal is using an analogical subtraction operation. A balanced receiver does the analogical subtraction.

The output from the balanced receiver is the difference of the signals detected by the two photo-detectors which is sometimes positive and sometimes negative. The 0 and 1 binary digits are represented by two complementary encoded spectra. So if a 0 bit is transmitted, a directly encoded spectrum is transmitted and if a 1 bit is transmitted, its complementary spectrum is transmitted.

The filter in front of the balanced receiver sends all the frequency components in the direct spectrum (of the matched transmitter) to the upper photo-detector and those in the complementary spectrum to the lower photo-detector. Therefore, the complementary encoded spectra are interpreted as positive and negative (bipolar) outputs. Unmatched channels are encoded such that half of its spectral components are matched to either of the two complementary spectra of the transmitter filter, no matter whether a 0 or 1 is transmitted. Therefore, the balanced receiver cancels the unmatched channel signals, and full orthogonality is achieved.

A novel electrically driven balanced transmitter is used to generate complementary spectra. It consists of a pair of broadband light sources connected in a balanced fashion. The input data differentially modulate the intensity of the two balanced light sources. Examples of the broadband sources are super luminescent light emitting diodes (SLDs), erbium doped fiber amplifiers (EDFA) biased in super-luminescent mode, mode-locked lasers, and multi-wavelength laser diode arrays. The spectra of the broadband sources pass through a spectral encoder which selectively transmits or blocks certain spectral components.

This complex architecture increases dramatically the orthogonality of the codes so the performance is expected to be very much better. The main drawback of such solutions is always the speckle noise. Balanced receiver reduces its impact but still remains as a limiting factor.

Spectral domain phase-frequency:
Although bipolar signaling achieves full orthogonality and has the simplicity of being non-coherent, the capacity of the non-coherent spectral intensity encoded system is severely limited by speckle noise. In order to avoid speckle noise, the multi-wavelength spectral phase encoded optical CDMA system is proposed **[Fung 99]**. A broadcast and select star network is assumed as the architecture of this multi-wavelength optical CDMA system. A mode-locked laser is used as the broadband optical source for encoding. The mode-locked pulses from each user are synchronized using a time synchronization signal

generated by the receiver so that they all arrive at the star coupler at the same time. Absolute optical phase synchronization is not needed.

Transmitter: The transmitter is shown in Figure 2.3.17. A mode-locked laser output is used as the broadband optical source. The output of the mode-locked laser consists of frequency components equally separated in the spectral domain. Because of the phase locking relationship between the spectral components in the time domain, a mode-locked laser output consists of periodic sharp pulses. If we encode a mode-locked laser output by adding different phase shifts to the various frequency components, the output will come to resemble noise-like periodic bursts. More precisely, the encoded output will look like a multi-mode laser output. This phase encoded mode-locked laser output is our "secret" CDMA carrier.

The mode-locked laser output at the transmitter is separated into two fibers using a 3-dB beam splitter. For the kth user, the first fiber passes through a spectral filter which phase encodes the carrier components. The other fiber's spectral components (for the kth user) are encoded by a different code C2 in a similar way. A modulator is placed after the encoded carrier C1 to impose data modulation. Either phase or intensity modulation could be applied. For Amplitude Shift Keying (ASK) signal, it takes the values 0 and +1 while for Phase Shift Keying (PSK), it is either +1 or −1. Modulation generates information-carrying side bands around the pure carrier tones. The side bands and the encoded carrier C2 are combined using a carrier add/drop filter, and then transmitted to the receiver.

The idea of artificially combining a carrier with side bands is emulating amplitude modulation (AM). AM has a neat feature that the carrier and the side bands are transmitted together to the receiver. The receiver, usually an envelope detector, demodulates the signal by mixing the side bands with the carrier. The transmitted carrier functions like a transmitter supplied local oscillator (LO) removing the need to have an LO at the receiver. Also, since the carrier and the side bands travel through the same path, they experience the same phase shifts and are always phase synchronized. For the same reason, the carrier is transmitted to the receiver as an externally supplied LO. However, the carrier and side bands are separately encoded so that the receiver will not be able to demodulate the signal by simple envelope detection. First, this enhances the security of the system. Secondly, encoding the carriers suppresses the interference of carriers from different transmitters by imposing an orthogonal code onto each carrier.

Receiver: The side bands and the encoded carrier are broadcast to the receivers through a star coupler. Figure 2.3.18 shows the receiver structure. As shown in Figure 2.3.18, the same carrier add-drop filter is used to separate the side bands and the encoded carrier tones. The carrier tones go through another

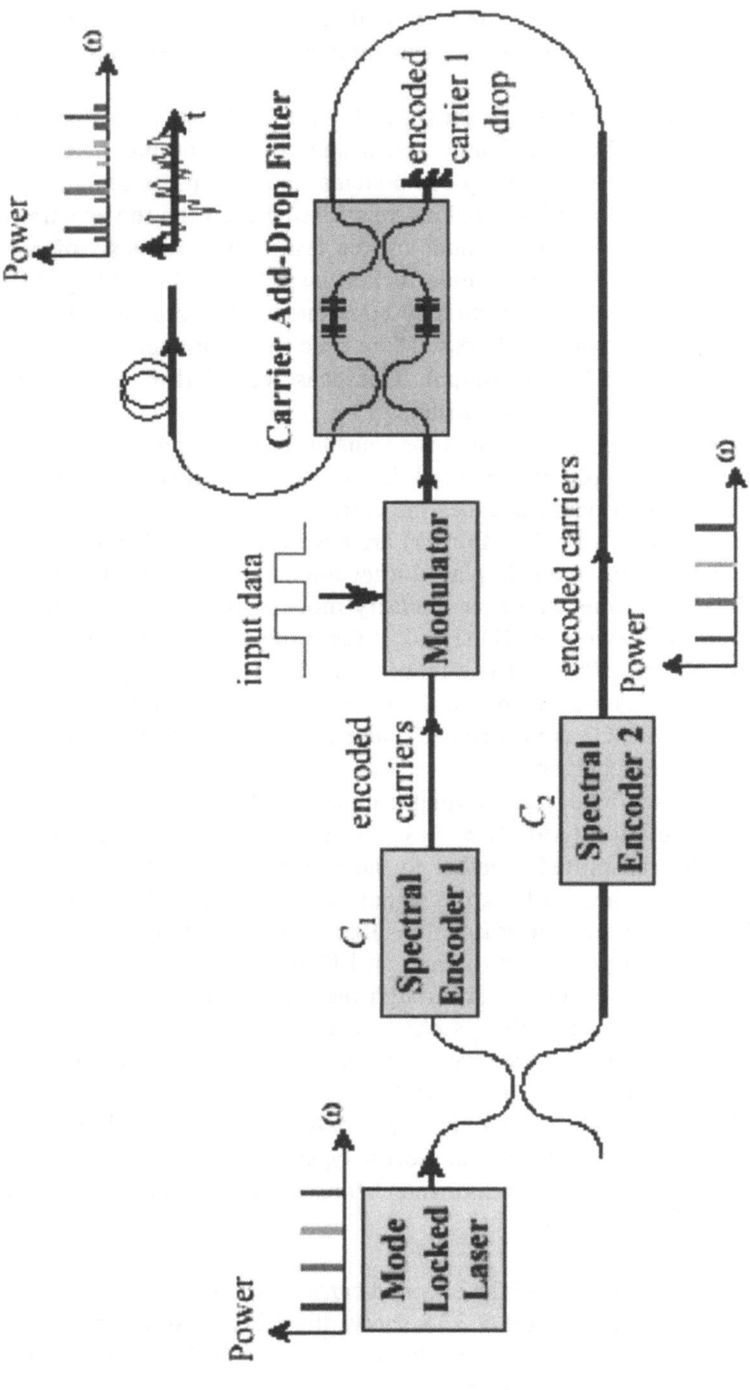

Figure 2.3.17. Spectral Domain Phase-Frequency Transmitter **[Fung 99]**.

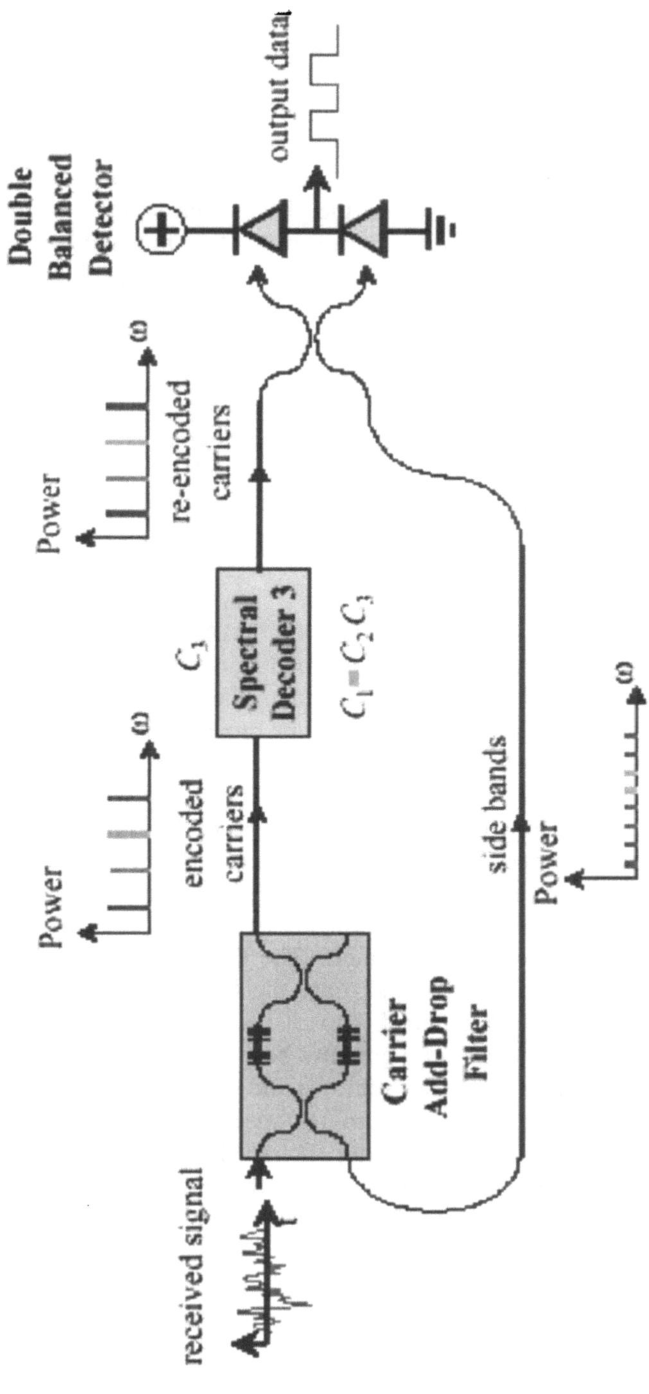

Figure 2.3.18. Spectral Domain Phase-Frequency Receiver [**Fung 99**].

spectral encoder C3 at the receiver such that C1 = C3C2, and C1 is orthogonal to C3C2. Thus the receiver reconstructs the original encoded carrier C1 which is modulated by the transmitted signal. The side bands and the reconstructed carriers mix at a second 3-dB splitter which feeds a double balanced detector.

By maintaining time synchronization and code orthogonality, speckle noise does not appear in this system. Since the phase information is used, a reference "local oscillator" is required. However, it is managed to avoid the complexity of having an optical local oscillator by supplying it from the transmitter, and sending it over the same optical path with the data. This method allows for large numbers of users (100 users). While the signals channel interference can be cancelled by using proper orthogonal codes at the balanced receiver, the proposed system is still limited by shot noise which is associated with the aggregate light being detected. This new proposed optical CDMA system resembles radio CDMA systems most closely.

This is an example of a method with a good performance but difficult to implement because of its complexity and cost.

Spectral domain frequency filter:
Another approach to this large variety of OCDMA methods is using a periodical frequency filter together with a broadband source **[Pfeiffer 00]**. The optical output power of a thermal light source, like a LED, is intensity modulated by the electrical data. After emission the broadband spectrum is optically shaped using passive filters with periodic transmission functions like Mach–Zehnder (MZ) or Fabry–Perot (FP) filters Figure 2.3.19. The periodicity of the transmission function for these type of filters is given in terms of the free spectral range (FSR). In the network, different FSR are allocated to different optical transmitters and define the codes in the system. At the receiver, an optical filter with periodic transmission function is tuned by matching its FSR to the desired channel. The filter types at transmitter and receiver do not have to be identical, only the FSR or the round trip times and , respectively, must be matched.

To achieve nearly perfect orthogonality between channel codes, a differential receiver set-up is used that removes the offset of the received signal power in Figure 2.3.19. The optical source spectra are assumed to be much broader than the FSR of the transmitter filters. Then the relative positioning of the source spectrum center frequency and the filter transmission function are irrelevant which leads to the first benefit of this approach: the selection of the source spectra and the filter transmission are not critical as long as $T_C << T_{Tx}$ and the difference of the round trip times for every pair of transmitters, and is in the order of the source coherence time or larger. [In practice, the gain spectra of the optical amplifiers define the coherence time, since usually they are narrower (30 nm) than the source spectra (50 nm)] .

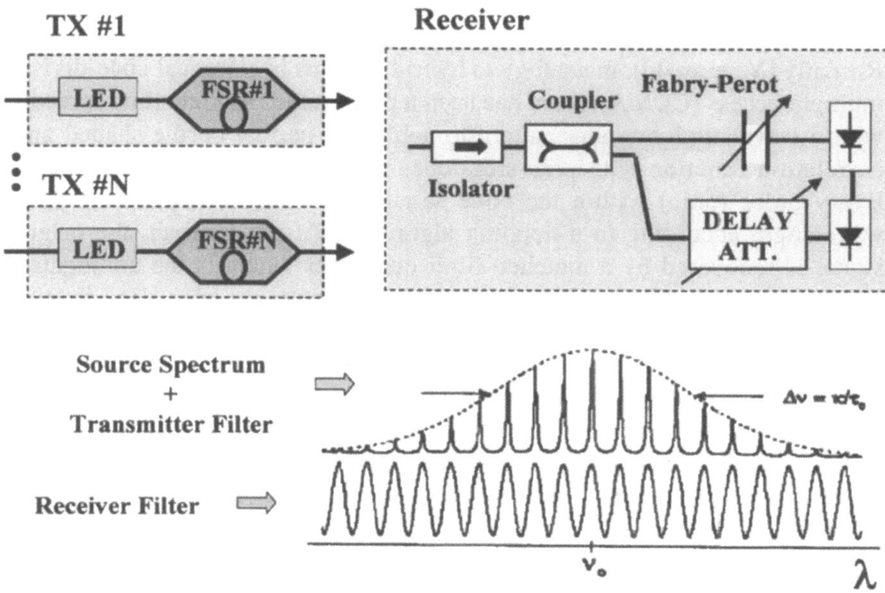

Figure 2.3.19. Spectral Domain Frequency Filter **[Pfeiffer 00]**.

The second benefit of periodic encoding is that only the difference in filter roundtrip times for any two transmitters needs to be larger than some picoseconds. The absolute value of the T_{Tx} is not important as long as they are much larger than T_C. From that, it follows that even temperature changes of more than 100 C do not affect the system performance as proven by experiment **[Pfeiffer 00]**. The reason for this insensitivity to temperature changes is the low temperature coefficient of the refractive index of glass (about 10^{-5}/C for fused silica).

The robustness of the system with respect to component specifications and drift is unique to the approach applying periodic spectral encoding with broad source spectra and small filter FSR. Other code multiplex systems like m-sequence encoded systems are not expected to show similar robustness due to the requirement of the exact matching of the a-periodic spectra at transmitter and receiver.

The main contributions to BER degradation originate from (thermal) receiver noise, shot noise, optical intensity noise, and cross-talk. The variance of the latter two disturbances increases linearly with electrical signal power and are responsible for the potential occurrence of a BER floor, hence they decide on the feasibility of a system configuration: value for MZ or for FP with finesse.

2.3.3.3. *2D OCDMA time-frequency*

Recently **[Yegnan 00]**, in analogy to hybrid systems in electrical code-division multiple-access (CDMA), there has been a proposal to integrate time spreading with a wavelength-hopping pattern to achieve a perfect needle-shaped auto-correlation function with zero side-lobes and a very low cross correlation. Every pulse (chip) within the code sequence is transmitted at a different wavelength according to a hopping algorithm. At the receiver, the original signal is recovered by a matched-filter correlator which is the conjugate of the transmission filter. Since every wavelength occurs only once in every hopping pattern, for a prime-hop sequence of order p, the autocorrelation is a maximum of p at zero time-shift and side-lobes are completely absent, while cross-correlation gives individual coincidences of at most one. For the worst case cross correlation (when chips are synchronous), the average variance can be shown to be very low. Consequently, the number of stations in the network and the number of simultaneous users is expected to be greatly increased.

The experiment set-up is shown in Figure 2.3.20. An actively mode-locked fiber laser, generating 3 ps optical pulses at a 1-GHz repetition rate, is used as the light source. To achieve a spectral width of 2 nm (for three chips in spectrum spaced by 0.8 nm) the pulses are amplified and super-continuum

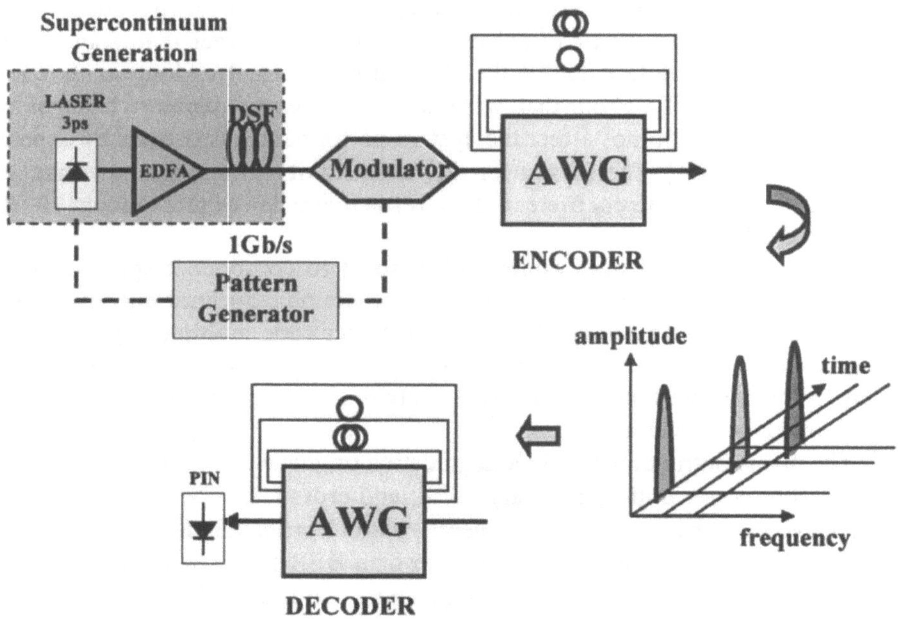

Figure 2.3.20. 2D OCDMA Time-Frequency **[Yegnan 00]**.

generation is used in 4 km of DS fiber followed by a tunable band-pass filter to select a 2-nm band. A 10-nm spectral bandwidth after super-continuum broadening are obtained. While this is sufficient for this experiment, it has been shown that more than 200-nm super-continuum broadening with relatively flat spectral response can be obtained using a fiber with optimized dispersion profile.

Pattern generator data at 1 Gb/s is used to modulate the 2-nm 3-ps pulses. The encoder consists of a wave-guide grating (AWG) to de-multiplex the input spectrum followed by fiber loops of different lengths. The delayed and amplitude weighted spectral components are then recombined using the same wave-guide grating in a feedback configuration. The delays and the amplitude of the spectral components are used to define wavelength-hopped and time-spread optical codes. The use of switched delay lengths would enable to program the encoder to a different prime-hop code. Figure 2.3.20 shows a three-chip time-spread and spectral-coded output of the encoder that was obtained from a 3-ps 2-nm spectral width input optical pulse. The transmission link consists of 15-km single-mode DS fiber with a total loss of 4 dB. Because the encoded signal is optically broadband, DS fiber is needed to minimize dispersion-induced penalty. The received signal is passed through the decoder which is identical to the encoder with conjugate delay loops. Here, the time-spread and spectral-coded data are collapsed and the original data is recovered.

Conventional time-spread incoherent OCDM systems are limited chiefly by the nonzero multiple-user cross correlation. However, in this case, the time-spreading combined with the wavelength-hopping pattern achieves a perfect needle-shaped autocorrelation function with zero side-lobes and a very low cross-correlation, resulting in a very low average variance and hence improved capacity compared to traditional time-spread OCDM system. Nevertheless, since all the code-words share the same optical spectrum, beat noise that occurs between: 1) the signal and the multiple user interference and 2) between the different multiple user interference signals, needs to be considered. Assuming that the phase of the individual spectral components are random variables uniformly distributed over [−180, 180], it has been recently shown that the beat noise due to multiple-user interference in such two-dimensional (2-D) OCDM systems appears as a dominant source of noise and degrades the system performance. The proposed technique avoids the need for a fast wavelength-tunable optical source.

2.3.3.4. *3D OCDMA frequency-time-frequency*

With simple extension of the 2-D optical code, 3-D optical code in which optical pulses are spread in wavelength^2 and time domains can be easily constructed, and the performance is further improved **[Kim 00]**.

This scheme can be easily implemented with the help of periodic characteristics of AWG. When a number of wavelengths are de-multiplexed in an AWG, a group of wavelengths, which are separated by free spectral range of the AWG, are coupled into the same output port. The scheme can be implemented by the structure shown in Figure 2.3.21. Each output port of the AWG in the encoder/decoder corresponds to a spatial channel, and the center wavelengths of gratings in each fiber belong to a group of wavelengths that can be coupled to the corresponding output port of the AWG. Figure 2.3.21 depicts the structure for wavelength^2/time scheme for the 3-D code. We can implement wavelength^2/time scheme for the 3-D code by using only one grating in each output of the AWG in encoder/decoder. In either structure, lengths of delay lines between gratings and the order of gratings are determined according to the 3-D codes.

Note that coherent light sources such as laser diodes (LD's) are not needed. Instead of a bunch of LD's, one incoherent broadband light source such as light emitting diode (LED) is used for each user which is advantageous over wavelength division multiple access system. Within the bandwidth of the light source, we can increase the number of wavelengths by simply increasing the number of the AWG ports or employing more kinds of fiber gratings.

The advantage of the wavelength/time scheme is that an AWG with a smaller number of ports is used in the implementation of the codes. Moreover, if the wavelength^2/time scheme is employed, we will have system extendibility and can accommodate more simultaneous users depending on system requirements.

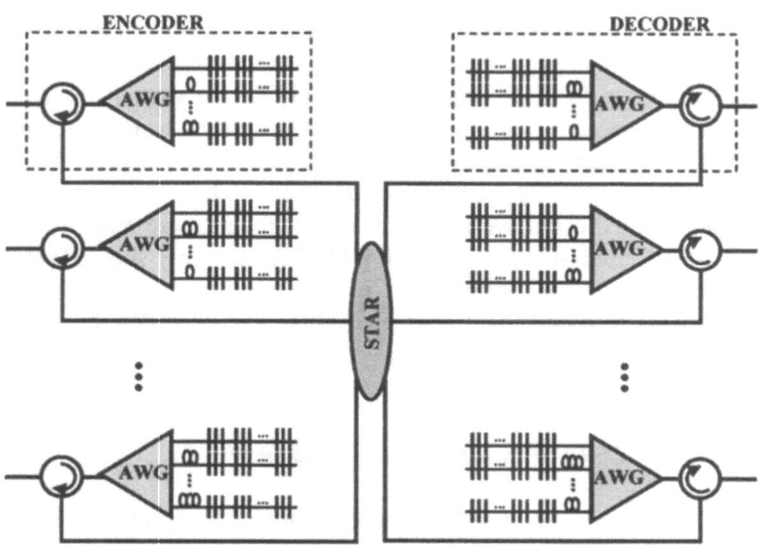

Figure 2.3.21. 3D OCDMA Time-Space-Frequency **[Kim 00]**.

Another advantage of the wavelength²/time scheme is that if we use tunable filters and reflectors instead of fiber gratings, code-words can be selectively transmitted or received with one encoder/decoder by tuning the filters.

2.3.3.5. *Coherence multiplexing*

Coherence Multiplexing Techniques:
Figure 2.3.22 shows a unidirectional transmission system based on coherence multiplexing (CM) techniques using on–off shift keying **[Kani 01]**. Each transmitter consists of a direct modulated low-coherence light source (LS), two 3-dB optical couplers (1:1), and an optical delay line l_k. Each transmitter has a delay line with different length. The difference between any two delay-lines is designed to be much larger than the coherence length of the light sources used. The receiver consists of two 3-dB couplers; an optical delay line, and a differential detector consisting of two balanced photodiodes (PDs). The receiver can decode the signal from a transmitter whose delay-line length equals to that of the receiver.

If delay-line length of the receiver equals to the one of the transmitter, the signal is recovered by the interference between the two light paths. One is the path that transits the arm with delay line in the encoder and the arm without delay line in the decoder, and the other is the converse (the path that transits the arm without delay line in the encoder and the arm with delay line in the decoder). Effective interference does not occur between any other two light paths, because the differences of the light path lengths are much longer than the coherence length of the light sources. Unwanted optical power from such light paths is equally divided into two PDs and is extinguished in the differential reception.

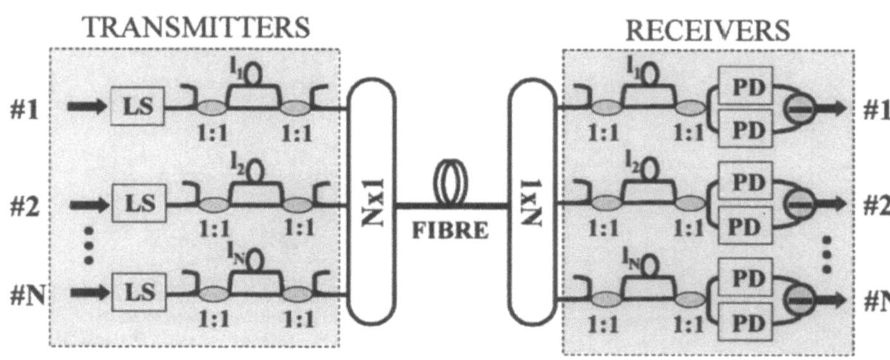

Figure 2.3.22. Coherence Multiplexing schemes **[Kani 01]**.

Proposed CM Access Network

A simple bi-directional PON, see section 2.5, is proposed based on CM techniques using the half-duplex communication procedure and a novel configuration: hybrid transceiver units. Figure 2.3.23 shows the configuration of the proposed PON. Each ONU consists of a direct modulated light source (LS), one optical tap, two PDs, two 3-dB couplers, and an appropriate optical delay line. This configuration reduces the number of optical components used, and thus ONU cost. The OLT simply consists of almost the same components as the ONU. A bi-directional optical amplifier (BOA) may be

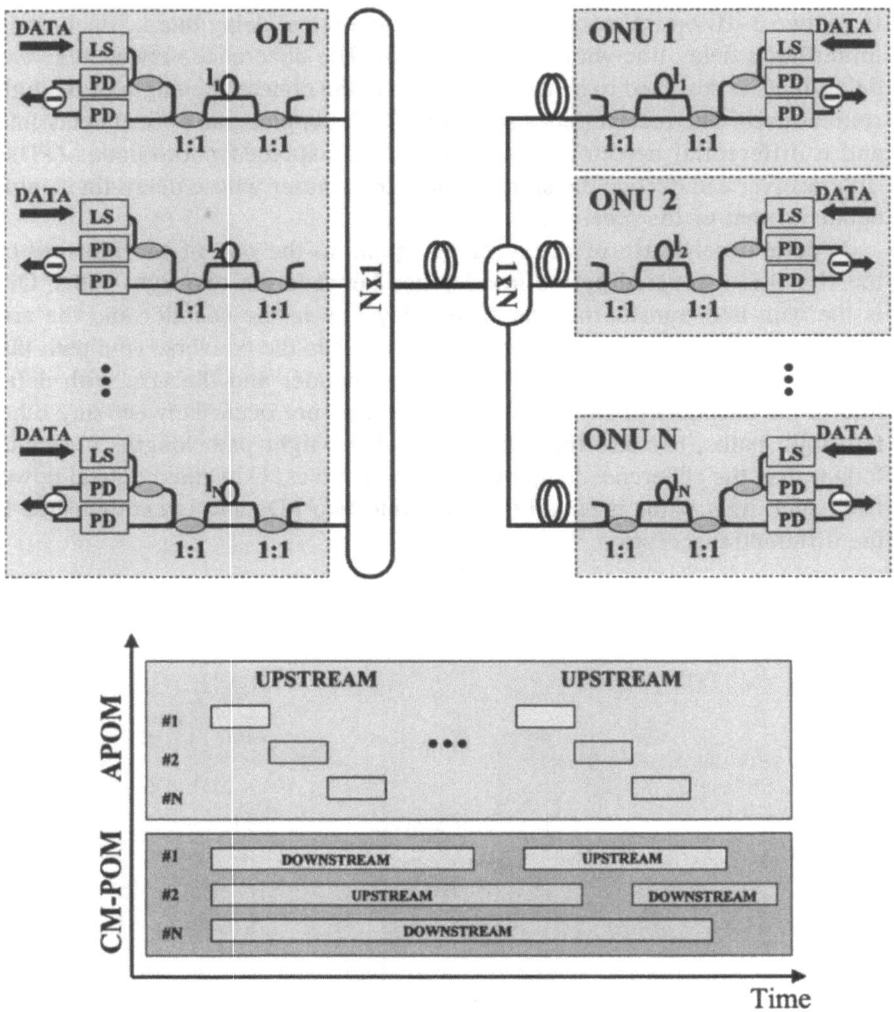

Figure 2.3.23. Coherence Multiplexing Passive Optical Network **[Kani 01]**.

employed in the OLT. Figure 2.3.23 shows time charts of the proposed system and the TDMA scheme in the conventional ATM-PON. As shown in the figure, users can independently send or receive signals using the half-duplex communication procedure. The ratio of up-to downstream bandwidth can be changed as demanded by each user. The proposed concept also suits a CM system using phase-shift keying **[Kani 01]**. The bi-directional CM access network features fewer taps and photodiodes in the transceiver units but requires external phase modulators (PMs). System performance is expected to be worse because unbalanced receivers are used.

Scalability of the Proposed Network:
The degradation due to fiber chromatic dispersion increases with the line-width of the light sources while the inter-path beat noise decreases as the spectrum line-width of light sources increases, so there is an optimum spectrum width of light sources. The numerical results by the waveform simulation matched the theoretical results, 35-nm line-width light sources are found to allow a 10-km G. 652 access line to carry 16 coherence-multiplexed 100-Mb/s signals, while a 7-km G. 652 access line can carry 12 coherence-multiplexed 155-Mb/s signals.

Acceptable access-line loss in the CM network that uses 35-nm spectrum width light sources was estimated with the multiplexing number-limiting noise factors and the loss-limiting noise factors taken into account. Calculation conditions were as follows: the output power of the light sources was 1 mW to simulate the use of low-cost super luminescence diodes (SLDs). The balanced receiver was assumed to consist of two p-i-n PDs with 50-Ohm electrical resistance. The analysis considered the principle losses of the 3-dB taps, couplers (1/2), and 1:N splitters.

A novel and simply configured broadband optical access network that uses coherence multiplexing, half-duplex bi-directional transmission, and appropriate hybrid transceiver units are proposed. It was found that we can design a network in which each of 16 users can flexibly use the bandwidth of 100 Mb/s on demand using a standard access line that is under 10 km length with loss no longer than 9 dB. With a 7-km 9-dB standard access line, 12 × 155 Mb/s users were also found to be acceptable. Experimental results confirmed that eight-channel 155-Mb/s half-duplex optical packet signals can be multiplexed and transmitted through a 7-km standard access line with bit error rates better than 10^{-9}. This study showed that the proposed access network could significantly enhance the capacity of existing fiber-to-the-home access lines that employ power splitters.

2.4. MULTIPLE ACCESS TECHNIQUES

Introduction

Probably one of the most critical issues when designing a communications system is the definition of the Medium Access Control (MAC). Optical Access Networks (OAN) in general, but Passive Optical Networks (PON), in particular, are in standardization process for MAC solution. In fact Asynchronous Transfer Mode Passive Optical Networks (APON), standardized by International Telecommunications Union (ITU), see chapter 3, has adopted the ATM MAC mechanism.

IIn this section, there is a summarized review of several medium access techniques. Section 2.1 deals with the point-to-point solution in order to understand the cost efficiency of every solution. In section 2.2, the channel sharing technologies are summarized. The shared channel can be the frequency (or the wavelength) and the code (or the coherence).

The main goal of sharing a channel between the users is to optimize the utilization and to reduce the cost. There are several protocols to decide how to distribute the channel among the users but basically they can be put into the following categories: deterministic, random and request-grant (Figure 2.4.1).

2.4.1. Point-to-point

The easier solution for an optical access connection is to use as many fibres as necessary in a so-called point-to-point link avoiding, in this way, to share

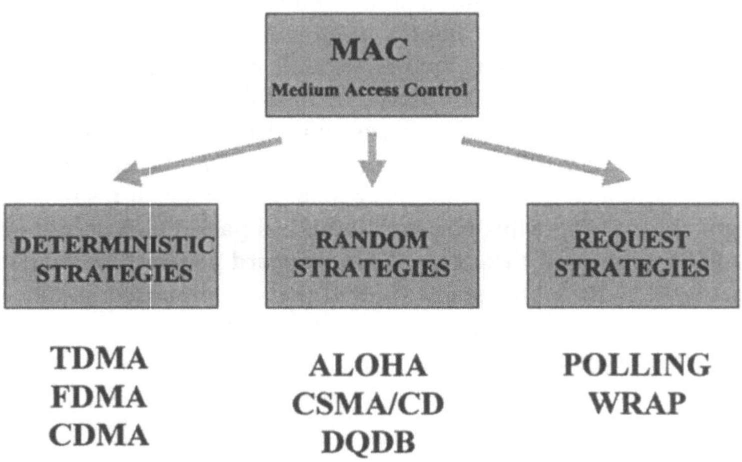

Figure 2.4.1. MAC strategies.

the channel. The main drawback is however the cost of this kind of systems. As can be seen in Figure 2.4.2, the amount of fiber and the number of transceivers (2N) is excessive.

The shared medium solution, Figure 2.4.3, gives an important reduction in costs but also an increase of the complexity. In this case, there is one fiber to be shared between all users, the amount of fiber is extremely reduced as the number of transceivers (N + 1). An economical analysis is performed in chapter 7.

More than the physical channels (fibers), logical channels can be considered. At the same fiber several logical channels can be defined concerning each multipexation format defined (see section 2.3) frequency or wavelength domain, code or coherence domain, or both of them. In Figure 2.4.4, this

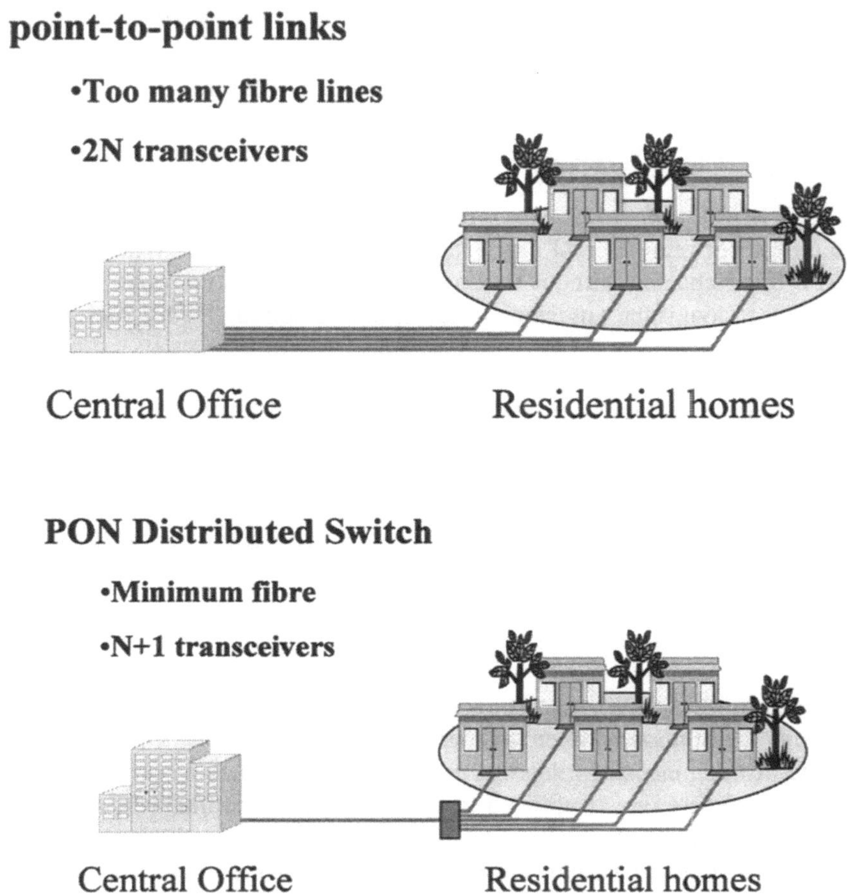

Figure 2.4.3. Shared medium advantages.

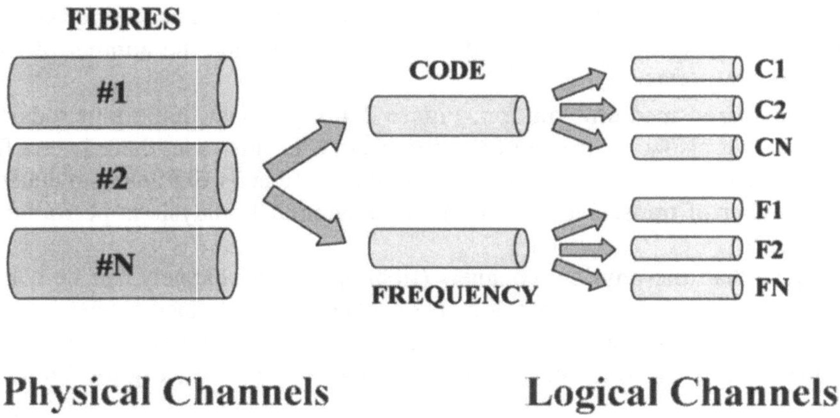

Figure 2.4.4. Physical channels and Logical channels.

idea is shown considering only two multiplexation schemes, code and frequency. There exist other possibilities and more than one can be used, for instance frequency and code, in a cascaded scheme giving more logical channels.

The MAC protocol can be seen as a strategy to share one of the logical channels between several users to avoid time collisions. When every user has a fixed channel without sharing, the collisions are not possible. Another strategy to avoid the collisions is to fix one temporal slot for each user.

2.4.2. Shared medium: time frequency and code

Some examples of each strategy are given. The idea is to figure out the philosophy behind the method. The examples given have been chosen from recent publications allowing in this way to discuss about useful technologies.

2.4.2.1. *Deterministic strategies*

The deterministic case, also known as fixed case, consists of a fixed medium capacity distribution between the users. The typical example is the fixed TDMA applied to one channel (physical or logical) where each user has a slot time assigned and no one is allowed to transmit but himself/herself. Clearly this method has a poor efficiency when the traffic model is not uniform. The advantage is however that there are no collisions and the latency delay is very short. Furthermore the simplicity is very high.

One solution to increase the efficiency is to extract some traffic statistics and divide the time not in equal time slots but different according to the needs of each user. The performance of this kind of methods is well known **[Tanembaum 97]**.

The example chosen for this case is the Coherent Multiple Access (CMA). This technology has drawn a lot of attention recently because of their simplicity **[Kani 01]**. In Figure 2.4.5, the architecture is shown. A more detailed explanation can be found in section 2.3.3. According the above discussion, this method can be classified as a deterministic strategy where the physical channels are divided into logical channels in the coherence domain. Then each one is applied to only one user so there is no time-sharing and hence no collisions.

The principle of work of this method is very simple: each transmitter consists of a directly modulated low-coherence light source (LS), two 3-dB optical couplers (1:1), and an optical delay line. Each transmitter has an optical delay line with different length (l_1, l_2, . . .) in a Mach-Zehnder configuration. The difference between any couple of delay lines is designed to be much larger than the coherence length of the light sources used. The receiver consists of two 3-dB couplers, an optical delay line, and a differential detector consisting of two balanced photodiodes (PDs). Each receiver decodes the signal from a transmitter whose delay-lines length equals to that of the receiver.

Every user has a virtual channel (coherence channel) deterministically assigned that allows an asynchronous communication. So this kind of methods (another example is a WDMA-based system) can be seen as a virtual point-to-point configuration. The number of users can be as high as 16 or 32. The shared medium is the fiber but also the virtual channels can be taken advantage of. Another way to assign each channel could be randomly or by request, in these cases, the transmitter and/or the receiver should be tunable.

A comparison between a system using TDMA over a single physical channel and another one using CMA over the same physical channel but creating a logical channel for each user is plotted in Figure 2.4.6. As can be seen, the

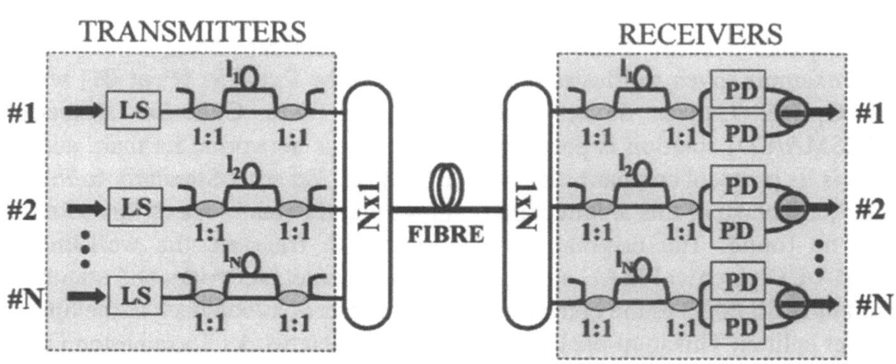

Figure 2.4.5. Coherence Multiple access scheme **[Kani 01]**.

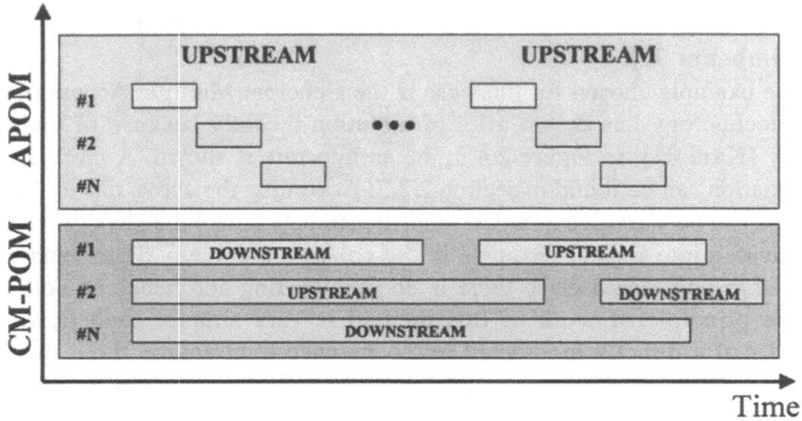

Figure 2.4.6. CMA vs TDMA **[Kani 01]**.

fiber is used all the time by all users allowing to implement both upstream and downstream channel over the same logical channel.

The performance parameters like throughput and delay for this kind of methods are well-known and can be found in **[Tanembaum 97]**.

2.4.2.2. *Random strategies*

The second family of solutions for the upstream MAC is that where the random access techniques are used. The Aloha family is the most known example. This kind of solution allows two different packets to collide in a contention period. The performance is very poor, but the system is very simple. The main interest is when using the random access in combination with the request-grant mechanisms.

Optical carrier sense multiple access with collision detection

The example given to illustrate this family can be found in **[Prat 01]** where an Optical Carrier Sense Multiple Access with Collision Detection (OCSMA/CD) solution is proposed for the access networks. Its main advantage is its protocol compatibility with IEEE 802.3 Ethernet Standard. In Figure 2.4.7, a scheme of this architecture is shown and the main idea of this solution can be found. The original contribution is to translate the well-known CSMA/CD electrical MAC strategy used in Ethernet Networks and standardized in IEEE 802.3 to the optical domain by means of a loop back at the optical power splitter. This loop-back allows the Optical Networks Termination ONTs to receive a feedback from the network and the Carrier Sense/Collision Detection to take place.

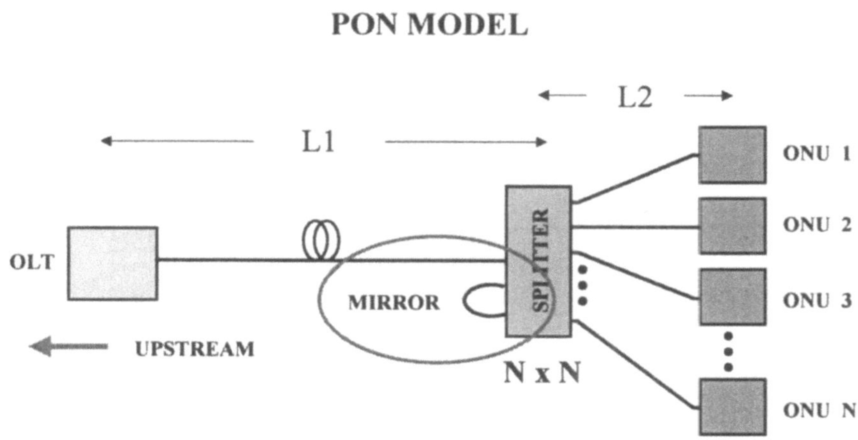

Figure 2.4.7. OCSMA/CD solution for the upstream MAC problem **[Prat 01]**.

A more detailed scheme of the possible implementation of the OCSMA/CD mechanism can be found in Figure 2.4.8. There are two possible solutions in order to avoid in line amplification: the first one (a) uses one fiber per user, but allows one ONT less and the second one (b) uses two fibers between user and an extra optical power splitter.

The performance of this kind of solution is difficult to predict analytically, but using the classical and pessimistic Poisson models, the minimum throughput for a 16 users and L_2 = 200 m 1 Gb Ethernet scenario can be found to be as high as 10 Mb/s/user **[Prat 01]**, capacity that allows to include more than one video-on-demand channel. Note that L_1 is not a limiting factor.

The main drawbacks are:

- The non-guaranteed quality of service (delay)
- The limited distance

But this technique can be seen as a first step toward an improved full-duplex Ethernet solution using the same configuration or with little changes. Another application of this technology can be a request-grant solution with a random strategy for the request mechanism.

Other advantages are the intrinsic transparency and the scalability that will allow to use different multiplexation or modulation formats and to grow in an easy manner.

WDM multi-channel CSMA/CD

Another possible application of the CSMA/CD protocol is the one presented in **[Rodellar 98]**. The study presents a comparison between single channel and multi-channel MAC protocols for broadband optical LAN. The multi-channel protocol exploits the WDM techniques, each channel corresponds to a different

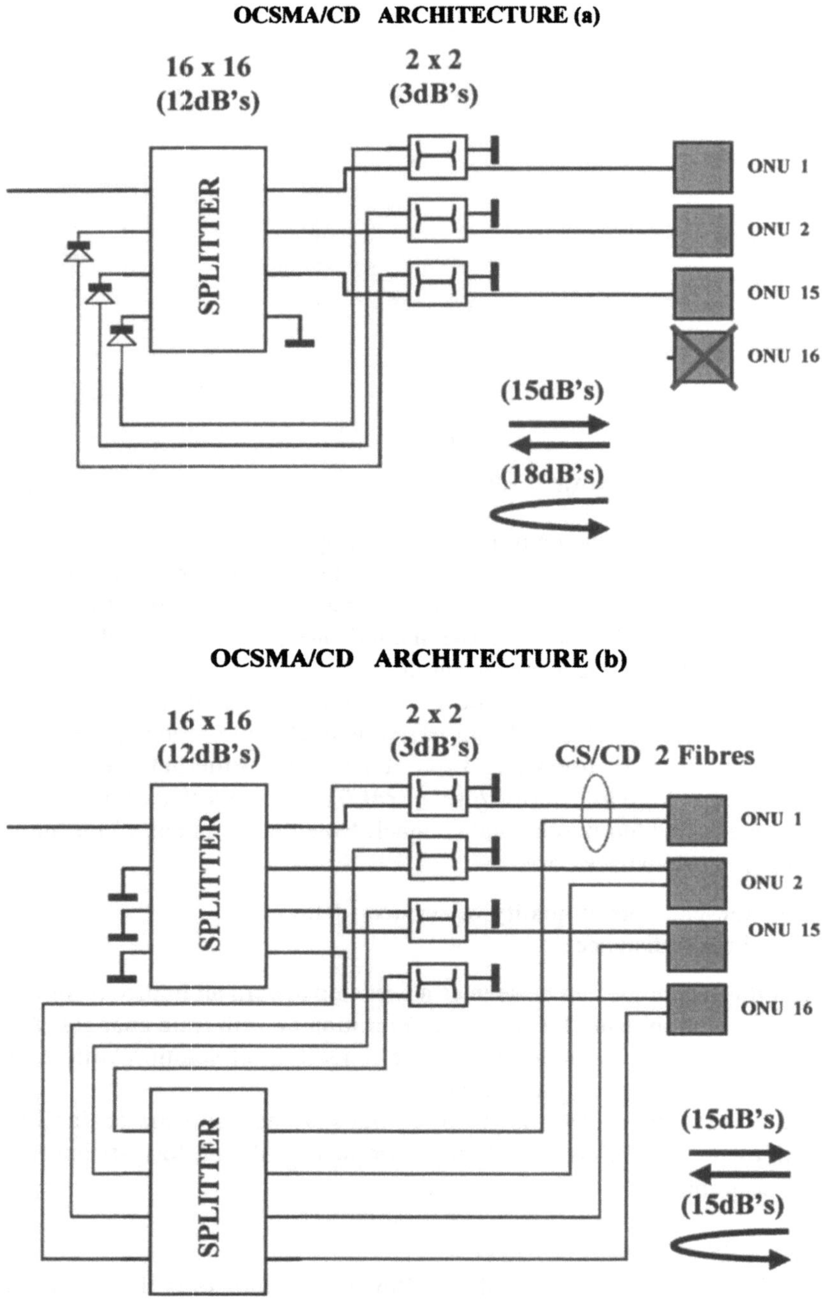

Figure 2.4.8. OCSMA/CD architecture implementations: (a) one fiber and one optical splitter, (b) two fibers and two optical splitters **[NOC 01 Prat]**.

optical wavelength allocated dynamically, and it is based on the Ethernet standard, using CSMA/CD scheme.

The proposed network architecture is based on a star coupler configuration (Figure 2.4.9). Each transmitter has a tunable laser module and an external modulator allowing each user to choose one of the possible wavelength channels. The receiver is based on a coherent detection and has multiple branches, one per channel. After the photo-detection, each optical channel corresponds to each electrical band-pass filer and envelope detectors. Note that the use of a local oscillator is needed.

The MAC protocol is based on the commercial Ethernet protocol standard. The multi-channel protocol behaves according to more specific criteria:

- The bus is no more a single channel but a group of channels. Full connection from the stations to all the channels is assumed, both for transmission and for reception.
- There is a hopping time (which is the laser tuning time) to jump from one channel to another.
- A new parameter is introduced for the number of consecutive collisions encountered by the packet before jumping to another wavelength. This parameter can be adjusted such that slow tuning lasers will stay longer on the same wavelength. But for the fast tuning lasers, it is no longer necessary to wait to change the emitting wavelength with virtually no influence on the delay measure.

To optimize the protocol parameters the entire network activity has been performed by real Ethernet traces on the simulations. This traffic has a long-range dependent statistic also called self-similar traffic.

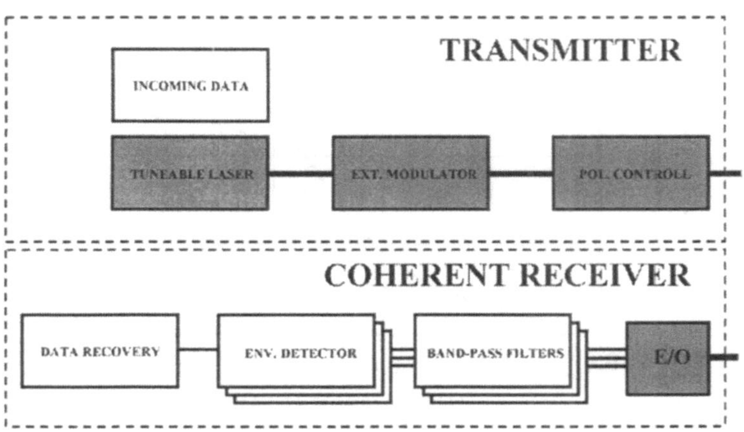

Figure 2.4.9. Multi-channel OCSMA/CD transmitter and receiver [**Rodellar 98**].

The performance of the multi-channel method is compared with the single channel for the same aggregated throughput 100 Mb/s (Figure 2.4.10). One channel of 100 Mb/s for the single case and 10 channels of 10 Mb/s for the multi-channel case. When increasing the capacity, the major goals are to obtain an adequate delay, and throughput results in order to support real-time services and to satisfy the high demand of the large numbers of connected users. The difference between two protocols at congestion (which correspond to maximum throughput) is about 0.2 times the total capacity, which is 100 Mb/s in this case. The mean bit rate per station is about 2 Mb/s, and the difference corresponds to about 10 more stations connected per multi-channel protocol.

Single channel congestion for 100 Mb/s occurs for normalized throughput of 0.65, while for multi-channel this congestion is reached for 0.85. Delay comparison only makes sense before the single-channel congestion, and below a throughput of 60 Mb/s, multi-channel delay is 0.1 ms below Fast Ethernet delay.

This WDM configuration is very attractive for the high bandwidth given, but the main drawback is the cost of such components like tunable lasers. In principle the solution using one single wavelength is more cost-effective, but when new and cheaper WDN components appear, they could be of great interest.

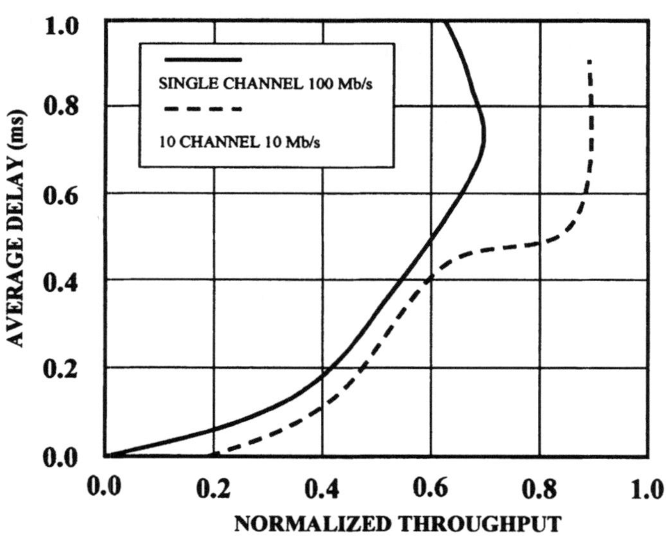

Figure 2.4.10. Multi-channel OCSMA/CD performance in front of single-channel **[Rodellar 98].**

2.4.2.3. *Request/Grant strategies*

This kind of solutions makes use of the Self-Similar traffic characteristics of current data traffic, for example video streams. The main characteristic of this kind of traffic is that its shape looks the same when increasing the time scale (Figure 2.4.11). The Poisson processes do not have this behavior; the shape trend is towards a uniform distribution. In this scenario some studies **[Kramer 01]** conclude that the best mechanism for the medium access is the request-grant one.

A central element is mandatory in order to process the medium grants to different users taking into account some aspects like delay, priority and so on.

Two recent examples have been chosen to illustrate the philosophy of this family of MAC techniques. The first one is known as Interleaved Polling Access Technique (IPACT) and this technique is a proposal given by the IEEE Ethernet First Mile (EFM) Study Group in the meeting of March **[Kramer 01]**. The second one is a recent technology found in the Journal of Lightwave Technology, December 2001 **[Spencer 00]**: WDM Request/ Allocation Protocol (WRAP).

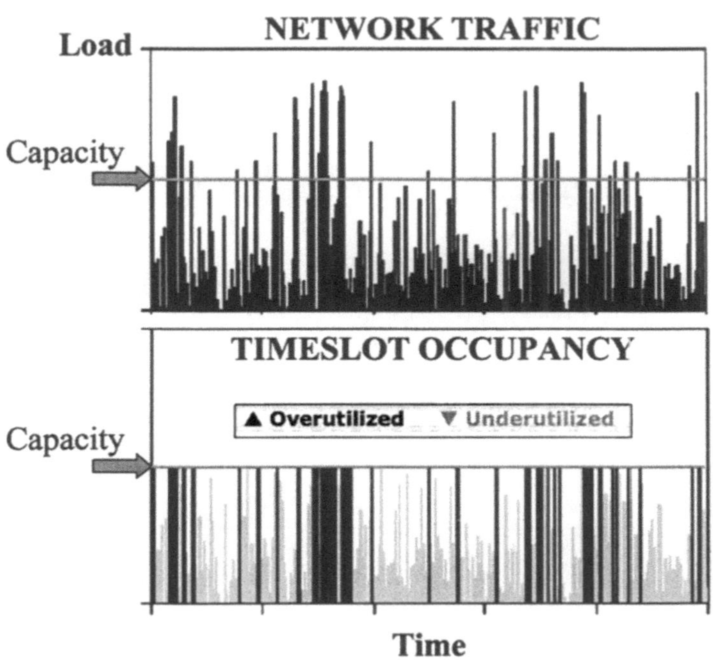

Figure 2.4.11. Example of measured network traffic **[Kramer 01]**.

Interleaved Polling Access Technique (IPACT)

In Figure 2.4.12, the IPACT Request-Grant scheme is shown. As can be seen the Optical Line Termination (OLT) element sends the grants messages to the users, also known as Optical Network Termination (ONT), and then each one sends their data and their request messages in a synchronous way.

The authors stand in favor of for this technique in future access networks. Data/Video traffic is Self-Similar, i.e., bursty at many timescales. Flow aggregation does not help, i.e., no optimal fixed timeslot exist. Burst-Size distribution is Long-Range Dependant (heavy tailed): most bursts are small, but most bits arrive in large bursts.

The proposed solution consists of interleaving polling schemes in time. This kind of technology is well-known and the performance well-studied **[Tanembaum 97]**, in Figure 2.4.13 the delay versus offered load is plotted for both Time Division Multiple Access (TDMA) and Interleaved Polling Access Technique. The TDMA technology works quite good under heavy load but not in the opposite case, i.e., bursty traffic. The IPACT method performance is nearly the same as in the first case and very much better in the bursty case.

In conclusion, the IPACT technology and the Request-Grant technologies in general can be seen as a promising solution to solve the Upstream problem in future optical access networks. In fact the APON (ITU G.983) solution based on ATM technology uses it.

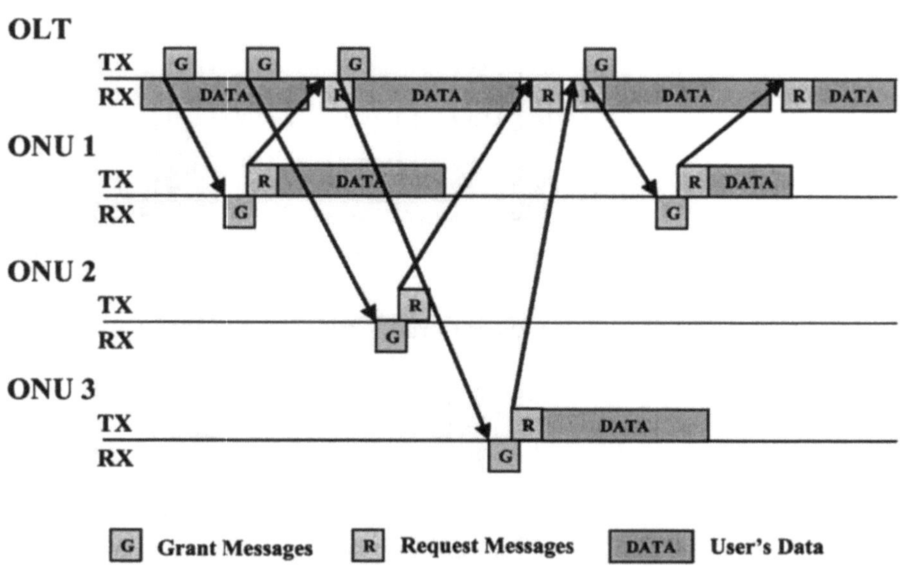

Figure 2.4.12. Interleaving Polling Access Technique (IPACT) **[Kramer 01]**.

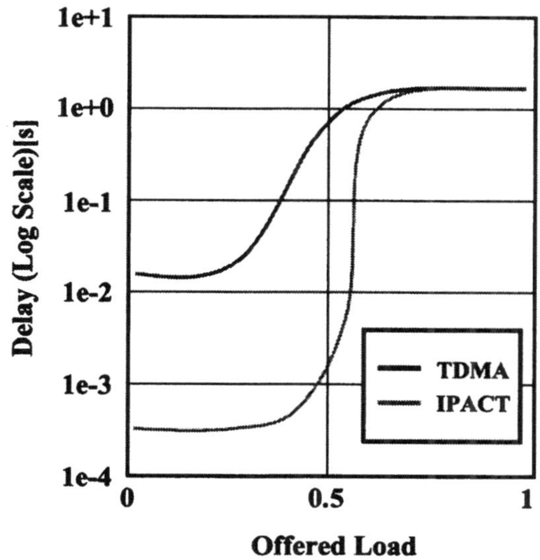

Figure 2.4.13. Simulated packet Delay comparison between TDMA and IPACT techniques [**Kramer 01**].

WDM Request/Allocation Protocol (WRAP)
The second example for this kind of MAC solution is named by the authors as WDM Request/Allocation Protocol (WRAP) [**Spencer 00**]. The scenario in this work is different because a WDM Passive Optical Network MAC is the subject of study. In Figure 2.4.14, the scheme of this architecture is shown and a Fixed Receiver and Tunable Transmitter user configuration can be seen.

Each user has an assigned wavelength to receive all traffic and the users attempting to send data to it must use a MAC mechanism. The WRAP protocol stream is shown in Figure 2.4.15. The main idea is the use of two kind of minislots: Request/Allocation minislots & Data minislots.

If the number of nodes is N there is N – 1 R/A minislots, one per user except the receiver one. If the number of Data minislots is M, the number of bits of a R/A minislot is Log2(M) + M, for request, and allocation respectively. In other words, each user has a reserved R/A minislot where he/she can request until M Data minislots . The Receiving user can allocate the available Data minislots among the asking users. An example for N = 3 nodes is shown in Figure 2.4.16.

The performance comparison with respect to other applicable methods for WRAP technique can be found in [Spencer 00]. In Figure 2.4.17, a delay comparison with TDMA protocol is presented in order to have an idea about the advantages of the mechanism that can be summarized as:

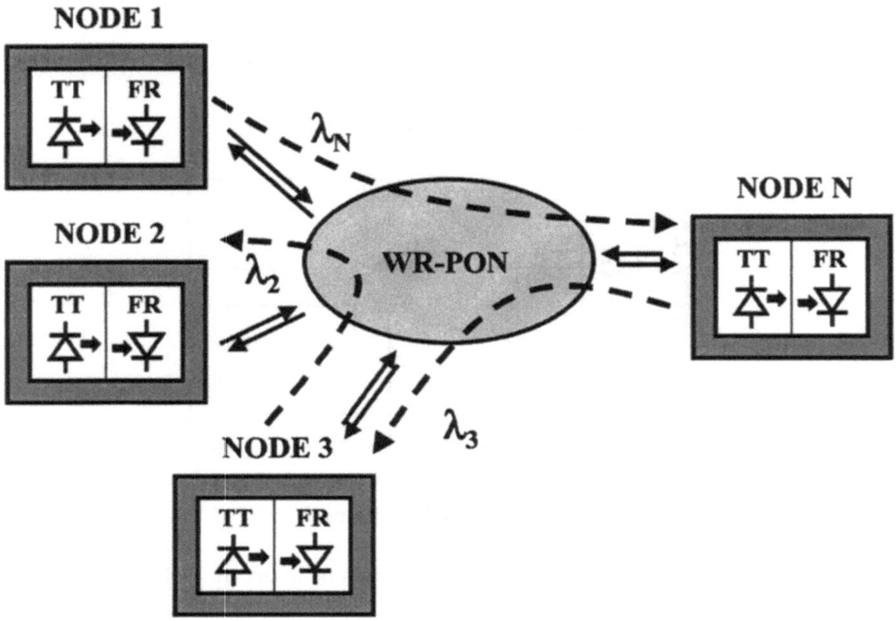

Figure 2.4.14. WRAP MAC mechanism for WR-PON networks **[Spencer 00]**.

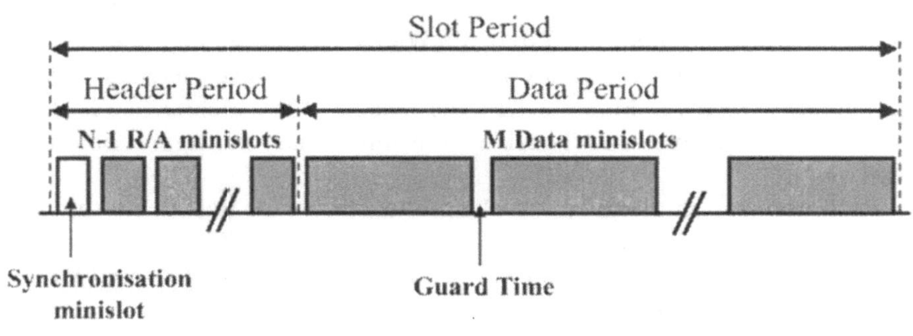

Figure 2.4.15. WRAP stream **[Spencer 00]**.

The main advantages of the WRAP solution are the following:

- It does not require reservation requests from all source nodes to be distributed globally within the network, nor does it require a centralized scheduler to handle requests. This means that it does not require a shared control channel, or a broadcast facility.
- It is inherently fair, under any well-defined fairness criterion, since access to any specified destination can be regulated completely by the allocation algorithm used by the destination node.

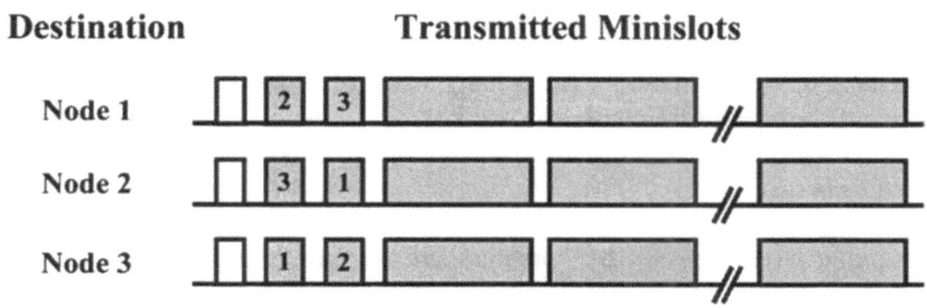

Figure 2.4.16. WRAP routing **[Spencer 00]**.

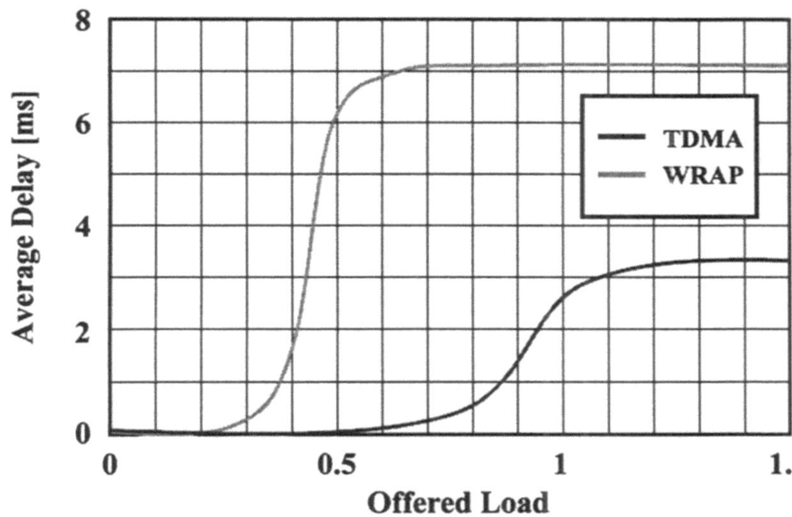

Figure 2.4.17. Simulated performance comparison of WRAP protocol **[Spencer 00]**.

With a suitable choice of allocation algorithm, performance can be made equivalent to that of Interleaved TDMA under conditions of uniform heavy loading.

2.4.2.4. *Mixed technologies*

Finally a mix between any of the three technologies discussed above can be taken into account. One of the most used methods in the last publications is that which makes use of the request-grant technology together with the random access methods. The example given for this family of solutions can be found in **[Sala 01]**.

Contention-based reservation mechanism
In this solution contention slots are shared poll opportunities for all the users. The head-end assigns slots for transmission of requests and any CPE can transmit. A contention resolution mechanism is needed to resolve collisions between different users attempting to send their requests messages at the same time.

In Figure 2.4.18 the performance of this technology is simulated and compared with static TDMA mechanism. As can be appreciated, the contention-based mechanism works quite well.

The main issues pointed out are as follows:

- Less need to predict individual needs: statistical multiplexing of request opportunities;
- Adds the need for a contention resolution algorithm;
- Decide allocation of contention slots.

In Figure 2.4.19, a summary of features of different multiple access technologies is given in a table in order to compare it.

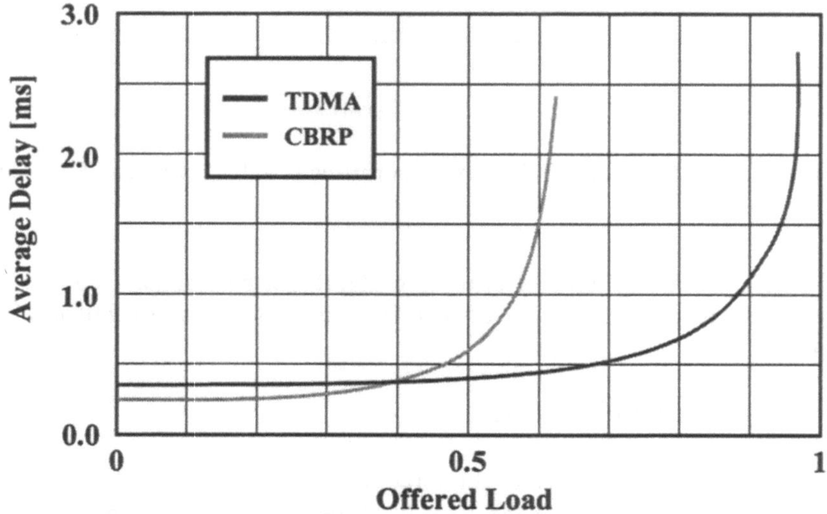

Figure 2.4.18. Contention-based mechanism performance for video traffic **[Sala 01].**

	Grant Msg	Ranging	Request Msg	Poll Msg	Conten-tion	Service BW
CSMA/CD					*	On demand
Unslotted TDMA	*					Fixed
Adaptive TDMA	*	*				Fixed
Adaptive TDMA	*	*	*			Variable
Polling	*	*	*	*		On demand
Contention based	*	*	*		*	On demand

Figure 2.4.19. Performance comparison **[Sala 01]**.

SUMMARY

PON (Passive Optical Study) seems to be the best way to give bandwidth services to the customers. Different parameter networks as scalability, upgradability, power consumption and others have been considered in the study. Several optical access networks have been studied from different point of view (initial cost, life, evolution, security, reliability). The topology that has theorically achieved the best performance in the study is physical star-logical star (the *physical topology* of a network segment refers to the physical attributes of that segment, the *logical topology* refers to how the information is accessed). In the near future OLTs will have three main parts: OADM (Otpical Add Drop Multiplexer) that select a frequency band pass, wavelength converters and IP router. With this features, the OLT (and all the optical network) is looking for a fast way to deliver the information to the customers.

A lot of modulation formats have been reviewed. They are classified as Basic Modulation Formats, advanced Modulation Formats and Electrical Modulation Formats. Concerning the Direct Detection aspects, the first modulation formats considered are the well-known basic ones, i.e., the Non Return to Zero (NRZ) and the Return to zero (RZ) formats. A comparison between both techniques shows that RZ is more robust to non-linear effects whereas the NRZ is more robust to dispersion. So depending on the data rates and link distances one or another is chosen. Another reason can be the architecture used.

Concerning the most sophisticated Direct Detection formats Multilevel Signaling has been studied. The most promising technique is the AM-PSK Duobinary Modulation due to the high dispersion tolerance and spectral efficiency. The main drawback is the poor non-linear effects robustness that limits the application to metropolitan networks.

Coherent Detection technology has been also reviewed as an advanced solution. This technology gives the best sensitivity at the receiver and the

	Stream multiplex	User multiplex	Capacity	Number of users	Characteristics
ATM-PON	WDM	TDM/TDMA	< 155 Mb/s total (up) < 622 Mb/s total (down)	< 32	Bandwidth shared by users
Dynamic WDM-PON	WDM, etc.	WDM-based TDMA	< 155 Mb/s total (up) < 2.5 Gb/s total (down)	16–32	Bandwidth shared by users Low-loss splitter (WR) → long distances Requires wavelength-tunable lasers
WDM-PON	WDM	WDMA	> Gb/s per user	16–32	Ultra-wide bandwidth Low-loss splitter (WR) On demand access to the network Expensive
CDMA	WDM, CDM, etc.	CDMA	> 100 Mb/s per user > 1 Gb/s total	< > 16	On demand access to the network Low interference due to the reflection
Coherence multiplexing					On demand access to the network User existing star couplers Low costs Low interference due to the reflection

Figure 2.4.20. Multiplexation Techniques comparison **[Kani 01]**.

best spectral efficiency, but there are some severe drawbacks, which are the frequency stabilization, polarization dependence and the need of high speed RF electronics, that increases the cost dramatically. So it can be seen as a future solution if the cost is reduced.

Finally, several electrical modulations have been studied as real application cases. One is the analogical TV over fiber, PAL system. The second is the European standard for digital television over cable DVB-C and the last is the ADSL format. These techniques will coexist the next years together with the all-optical networks in order to have a gradual transition.

Figure 2.4.20 categorizes the currently standardized, previously reported, and proposed passive optical networks (PONs) **[Kani 01]**. The widely standardized asynchronous transfer mode (ATM)-based PON provides a total bandwidth of 155 Mb/s upstream and 622 Mb/s down-stream. The maximum number of users is up to 32. In the ATM-PON, signals from users are multiplexed by time-division multiple access (TDMA) techniques, and up- and downstream signals are multiplexed by wavelength-division multiplexing (WDM) techniques. Dynamic WDM-PON, in which each user uses a different wavelength and different time-slot by the use of TDMA techniques, increases the number of users by replacing existing star couplers with wavelength routers (WRs). The typical number of available wavelengths is 16 to 32. However, PONs that use TDMA techniques place strict limits on the total bandwidth and communication flexibility because the bandwidth is shared by all users under synchronous time-slot control.

One candidate for the next-generation access infrastructure is WDM-PON using wavelength-division multiple access (WDMA) techniques in which each user accesses the network independently from the other users, thus leading to on-demand use of broad bandwidth. The critical issue is to minimize system cost: it requires laser sources with accurate wavelength control, replacement of existing star couplers by wavelength routers, and so on. Optical code-division multiple access (CDMA)] and coherence multiplexing (CM) access] also provide broad-band and on-demand access to all users. With these techniques, there is no need to replace the installed star couplers with wavelength routers. In particular, the CM system uses simple components. Note that the CM systems can be categorized in optical CDMA systems using incoherent light sources and simple periodic spectrum encoding]. In addition to unidirectional transmission reports of CM systems], a bi-directional CM access system that used WDM for up- and multiplexing has been reported.

A lot of technologies have been reviewed about Medium Access Control mechanism. The actual trend seems to be toward a Request-Grant solution like used in APON (ITU G.983), for instance the IEEE 802.3 study group attempts to standardise the Ethernet protocol for PONs with a multiple access method based on those techniques.

But the random strategies should not be forgotten because when they are used together with the polling schemes, they give a good performance reducing the complexity.

REFERENCES

[Agrawal 97] Govinf P. Agrawal, 'Fiber-Optic Communication Systems'. John Wiley & Sons, Inc.

[Antona 00] J.C. Antona et al., 'Enhanced Phase-Shaped Binary Transmission modulation format for Dispersion-Managed WDM systems' – ECOC '00.

[Bachus 96] E.J.Bachus, et al., 'Coherent optical systems implemented for business traffic routing and access: the RACE COBRA project', Journal of Lightwave Technology, vol. 14, no. 6, pp. 1309–1319, June 1996.

[Bhatt 01] Vipul Bhatt (Finisar), David Cunningham (Agilent); EFM Study Group Meeting, Portland, July 2001.

[Bissessur 00] H. Bissessur et al., 'Experimental demonstration of Enhanced Phase-Shaped Binary Transmission' – ECOC'00 Wed, 6 8.3.2.

[Chang 98] C.C. Chang, H.P. Sardesai and A.M. Weiner, 'Code-Division Multiple-Access Encoding and Decoding of Femtosecond Optical Pulses over a 2.5-km Fiber Link', IEEE Photonics Tech. Let., Vol. 10, No. 1, pp. 171–173, Jan 1998.

[Chung 89] Chung et al., 'Optical orthogonal codes: design, analysis and applications', IEEE Transactions on Information Theory, vol. 35, no. 3, p. 595 – May 1989.

[Fung 99] C. Fung, 'Multiwavelength Optical Code Division Multiple Access Communications Systems', PhD Dissertation – University of California 1999.

[Grunnet 99] A. Grunnet-Jepsen, et al., 'Demonstration of All-Fiber Sparse Lightwave CDMA Based on Temporal Phase Encoding', IEEE Photonics Technology Letters, Vol. 11, No. 10, p. 1283, October 1999.

[Hayee 99] M. I. Hayee, et al., 'NRZ Versus RZ in 10-40-Gb/s Dispersion-Managed WDM Transmission Systems', IEEE Photonics Technology Letters, vol. 11, no. 8, p. 991, August 1999.

[Hinkov 95] I. Hinkov, V. Hinkov, K. Iversen and O. Ziemann, 'Feasibility of Optical CDMA Using Spectral Encoding by Acoustically Tunable Optical Filters', Electronics Letters, Vol. 31, No. 5, pp. 384–386, March 1995.

[Iversen 95] K. Iversen and O Ziemann, 'An All-Optical CDMA Communication Network By Spectral Encoding of LED Using Acoustically Tunable Optical Filters', Proc. 1995 Int. Symposium on Signals, Systems and Electronics (ISSSE '95), San Francisco, pp. 529–532, 1995.

[Kaiser 00] G. Kaiser et al., 'SPM Limit of Duobinary Transmission', ECOC '00 Wed, 6 7-2-2.

[Kani 01] Jun-ichi Kani et al., 'A Simple Broad-Band Coherence Multiplexed Optical Access Network and Its Scalability', Journal Of Lightwave Technology, Vol. 19, No. 4, p. 456, April 2001.

[Kavehad 95] M. Kavehrad and D. Zaccarin, 'Optical Code-Division-Multiplexed Systems Based on Spectral Encoding of Noncoherent Sources', Journal of Lightwave Technology, Vol. 13, No. 3, pp. 534–545, March 1995.

[Kim 00] Sangin Kim et al., 'A New Family of Space/Wavelength/Time Spread Three-Dimensional Optical Code for OCDMA Networks', Journal Of Lightwave Technology, Vol. 18, No. 4, p. 504, April 2000.

[Kramer 01] G. Kramer, 'Multiple Access Techniques for ePON', Alloptic IEEE EFM March 2001.

[Kuznetsov00] Mark Kuznetsov, Nan M. Froberg, Scott R. Henion and Hemonth G. Rao (MIT Lincoln Laboratory); Jeff Korn (Axsun Technologies); Kristin A. Rauschenbach (Photonex Corp.); Eytan H. Modiano and Vincent W. S. Chan (Massachussetts Institute of Technology; 'A Next-Generation Optical Regional Access Network'; IEEE Communications Magazine, January 2000.

[Kwong 91] W.C. Kwong et al., 'Performance comparison of asynchronous and synchronous Code Division Multiple Access Techniques for fiber optic local area networks', IEEE Transactions on Communications, vol. 39, no. 11, p. 1625 – November 1991.

[LaHa 01] Michael LaHa, 'Coarse WDM opens the road beyond very-short-reach markets', WDM Solutions October, 2001.

[Lin 01] Wen-Piao Lin et al., 'The Modified Star-Ring Architecture for High-Capacity Subcarrier Multiplexed Passive Optical Networks" Journal Of Lightwave Technology, Vol. 19, No. 1, January 2001

[Lumetta 01] Steven Lumetta, University of Illinois Urbana-Champaign; 'Architectural issues for robust optical access'; Muriel MÈdard (Massachusetts Institute of Technology), IEEE Communications Magazine, July 2001.

[Mochida94] Yukoa Mochida, 'Technologies for Local-Access Fibering'; IEEE Communications Magazine; February 1994.

[Nguyen 95] L. Nguyen, B. Aazhang and J.F. Young, 'All-Optical CDMA with Bipolar Codes', Electronics Letters, Vol. 31, No. 6, pp. 469–470, 16 March 1995.

[Nguyen 97] L. Nguyen, T. Dennis, B. Aazhang and J.F. Young, 'Experimental Demonstration of Bipolar Codes for Optical Spectral Amplitude CDMA Communication', Journal of Lighwave Tech., Vol. 15, No. 9, Sept. 1997, pp. 1647–1653.

[Ono 98] T. Ono, 'Characteristics of Optical Duobinary Signals in Terabit/s Capacity, High- Spectral Efficiency WDM Systems', Journal of Lightwave Technology, vol. 16, no. 5, pp. 788–1998.

[Pennincks97] D. Penninckx, 'Enhanced-Phase-Shaped Binary Transmission', Electronics Letters, vol. 36, no. 5, pp. 478–1997.

[Penninckx97] D. Penninckx et al., 'The Phase-Shaped Binary Transmission (PSBT): a new technique to transmit far beyond the Chromatic Dispersion limit', IEEE Photonics Technology Letters, vol. 9, no. 2, pp. 259–1997.

[Penninckx98] D. Penninckx et al., 'Experimental verification of the Phase-Shaped Binary Transmission (PSBT)', IEEE Photonics Technology Letters, vol. 10, no. 4, pp. 612–1998.

[Penninckx98] D. Penninckx, 'Effect of Electrical Filtering of Duobinary Signals on the Chromatic Dispersion Transmission Limitations', ECOC '98 vol. 1, p. 537.

[Pfeiffer 00] Thomas Pfeiffer et al., 'Coarse WDM/CDM/TDM Concept for Optical Packet Transmission in Metropolitan and Access Networks Supporting 400 Channels at 2.5 Gb/s Peak Rate', Journal Of Lightwave Technology, Vol. 18, No. 12, p. 1928, December 2000.

[Prat 01] J. Prat et al., 'Optical CSMA/CD Strategy for an Ethernet Access Network', NOC Thursday Morning 9:00 Room 2, 2001.

[Prat 95] J. Prat, J. Comellas, G. Junyent, 'Experimental demonstration of an all-fiber endless polarization controller based on Faraday rotation', IEEE Photonics Technology Letters, vol. 7, no. 12, December 1995.

[Prat 98] J. Prat et al., 'Minimum channel spacing in an OFDM CPFSK Optical Coherent System', Optical and Quantum Electronics, Vol. 30 No. 3, 1998.

[Rodellar 98] D. Rodellar, 'A comparison between single and multi-channel CSMA/CD protocols of equivalent capacity', NOC '98.

[Royset 98] Royset et al., 'Symmetry requirements for 10 Gb/s Optical Duobinary Transmitters', IEEE Photonics Technology Letters, vol. 10, no. 2, p. 273, 1998.

[Sala 01] D. Sala, 'PON Functional Requirements: Services and Performance', Broadcom IEEE EFM July 2001.

[Salehi 89-I] J.A. Salehi et al., 'Code Division Multiple Accessin optical fiber networks – Part I: Fundamental principles", IEEE Transactions on Communications, vol. 37, no. 8, p. 824, August 1989.

[Salehi 89-II] J.A. Salehi et al., 'Code Division Multiple Accessin optical fiber networks – Part II: System performance analysis', IEEE Transactions on Communications, vol. 37, no. 8, p. 834, August 198.9

[Salehi 90] J.A. Salehi, A.M. Weiner and J.P. Heritage, 'Coherent Ultrashort Light

Pulse Code-Division Multiple Access Communication Systems', J. Lightwave Tech., Vol. 8, no. 3, pp. 478–491, Mar. 1990.

[Shaar 83] A.A. Shaar et al. 'Prime sequences: Quasi-Optimal sequences for OR channel Code Division Multiplexing', Electronics Letters, vol. 19, no. 21, p. 888, October 1983

[Shtaif 99] M. Shtaif et al., 'The Relation Between Optical Duobinary Modulation and Spectral Efficiency in DWDM Systems', IEEE Photonics Technology Letters, vol. 11, no. 6, p. 712, 1999.

[Sorensen 01] S. Soerensen et al., 'Optical Beat Noise Suppression and Power Equalization in Subcarrier Multiple Access Passive Optical Networks by Downstream Feedback', Journal Of Lightwave Technology, Vol. 18, No. 10, p. 1337, October 2000.

[Spencer 00] M.J. Spencer et al., 'WRAP: A Medium Acces Control Protocol for Wavelength-Routed Passive Optical Networks', Journal of Lightwave Technology, vol. 18, no. 12, p. 1657, 2000.

[Tanembaum 97] A. S. Tanembaum 'Computer Networks', 3rd. Edition, Prentice Hall 1997.

[Tomesen 94] M.T. Tomesen, 'Novel heterodyne CPFSK receiver allowing dispersion equalization in a narrow IF bandwidth starting from nearly DC', ECOC '94, p. 73, Florence, 1994.

[Tomoyuki01] Tomoyuki Akiyama and Osamu Wada, 'Beat-Detect OTDM Demultiplexer', Journal Of Lightwave Technology, Vol. 19, No. 9, p. 1326, September 2001

[Tseng 01] Emy Tseng, Massachusetts Institute of Technology, September 2001.

[Ulmer 00] Todd G. Ulmer et al., '160-gb/s Optically Time-division Multiplexed Link With All-optical Demultiplexing', Journal Of Lightwave Technology, Vol. 18, No. 12, p. 1964, December 2000.

[Walker 90] N.G.Walker and G.R.Walker, 'Polarization Control for Coherent Communications', Journal of Lightwave Technology, Vol. 8, No. 3, 1990.

[Walklin 97] S. Walklin et al., 'On the Relationship Between Chromatic Dispersion and transmitter Filter Response in Duobinary Optical Communications Systems', IEEE Photonics Technology Letters, vol. 9, no. 7, p. 1005, 1997.

[Walklin 99] S. Walklin et al., 'Multilevel Signalling for Increasing the Reach of 10 Gb/s Lightwave Systems', Journal of Lightwave Technology, vol. 17, no. 11 p. 2235, 1999.

[Wedding 96] B. Wedding et al., 'Multilevel Dispersion Suported Transmission at 20 Gb/s over 46 km Installed Single Mode Fiber', ECOC '96 MoB.4.4.

[Weiner 90] J.A. Salehi, A.M. Weiner and J.P. Heritage, 'Coherent Ultrashort Light

Pulse Code-Division Multiple Access Communication Systems', J. Lightwave Tech., Vol. 8, no. 3, pp. 478–491, March 1990.

[Weiner 98] C.C. Chang, H.P. Sardesai and A.M. Weiner, 'Code-Division Multiple-Access Encoding and Decoding of Femtosecond Optical Pulses over a 2.5-km Fiber Link', IEEE Photonics Tech. Let., Vol. 10, No. 1, pp. 171–173, Jan 1998.

[Woodward97] S. L. Woodward, X. Lu and A. H. Gnauck, 'Bidirectional, Subcarrier-Multiplexed Transmission Using 1.3-(m Fabry-Perot Lasers', IEEE Photonics Technology Letters, Vol. 9, No. 10, October 1997.

[Woodward98] S. L. Woodward, J. W. Stayt, D. M. Romero, J. M. Freund and G. J. Przybylek, 'A Study of Otpical Beat Interference Between Fabry-Perot Lasers', IEEE Photonics Technology Letters, Vol. 10, No.. 5, May 1998.

[Yegnan 00] S. Yegnanarayanan et al., 'Fast Wavelength-Hopping Time-Spreading Encoding/Decoding for Optical CDMA', IEEE Phototonics Technology Letters, Vol. 12, No. 5, p. 573, May 2000.

[Yu 01] Jianjun Yu and Palle Jeppesen, 'Simultaneous All-Optical Demultiplexing and Regeneration Based on Self-Phase and Cross-Phase Modulation in a Dispersion Shifted Fiber', Journal of Lightwave Technology, Vol. 19, No. 7, p. 941, July 2001.

[Zaccarin 93] Zaccarin and M. Kavehrad, 'An Optical CDMA System Based on Spectral Encoding of LED', IEEE Photonics Technology Letters, Vol. 4, No. 4, pp. 479–482, April 1993.

[Zaccarin 95] M. Kavehrad and D. Zaccarin, 'Optical Code-Division-Multiplexed Systems Based on Spectral Encoding of Noncoherent Sources', Journal of Lightwave Technology, Vol. 13, No. 3, pp. 534–545, March 1995.

[Zheng 98] X. Zheng et al., 'Receiver optimization for 40 Gb/s Optical Duobinary Signal', IEEE Photonics Technology Letters, vol. 13, no. 7, p. 744, 2001-09-25.

Chapter 3

PROTOCOLS AND STANDARDISATION

INTRODUCTION

In this chapter several protocols have been studied from different point of view. Nowadays, several field trials have been deployed with the ATM technology and others are going to use Ethernet technology (Gigabit and 10 Gigabit Ethernet). A comparison between both technologies is presented.

3.1. APON: ATM PASSIVE OPTICAL NETWORK

APON answers to the acronym of ATM PON (passive optical network based on ATM technology). The ATM technology for multiplexing and the PON architecture for the optical media network. Nowadays, there is an ITU's standard: G.983 ("High-speed Systems Based on PON Techniques"); this standard has several sections:

- G.983.1 focuses on the physical layer and transmission convergence layer.
- G.983.2 includes the control and management interface. It specifies managed entities of a protocol-independent Management Information Base (MIB) that models and controls the exchange of information between the OLT and ONU, and the protocol and messages.
- G.983.3 introduces a new optical channel or wavelength in the downstream direction that can be used for downstream video broadcast or bi-directional transport using DWDM.
- G.983.4 comprises of a dynamic bandwidth allocation.
- G.983.5 enumerates several protection schemes for PONs.

J. Prat et al., Fiber-to-the-Home Technologies, 117–147.

ITU-T Recommendations (G.983.1, G.983.2, and G.983.3) have been completed. FSAN (Full-Service Access Network) group is actively working to control and modernize this standard. The FSAN group is composed of the world's leading telecommunications operators that meet to develop and discuss common system specifications to introduce a quick commercial deployment of optical access systems. In the last section, the APON system configuration is shown according to the ITU-T Recommendation. The APON configuration functional aspect is being explained here.

In the downstream channel, the TDM technology is used whereas the TDMA technology is used in the upstream channel. From the physical point of view, the APON is constituted by an OLT, a signal distributor and several ONUs. The signal distributor can split the input signal up to 32 ONUs. The use of the passive distributors gives necessary considerations to privacity and security of the information in downstream channel since downstream frames are multicast to all the ONUs from the OLT. Only these cells addressed to an ONU are extracted by that ONU and the others are discarded. One fiber can be used for each transmission direction, but it is common to only use one fiber for both directions using WDM technology.

The specific information required to describe the passive optical components is given in Recommendation G.671 and for optical fibers and cable are explained in Recommendation G.652.

ATM is a Layer 2 technology that establishes dedicated, connection-oriented, and Virtual Circuits (VCs) across the network. VCs can be manually configured using Permanent Virtual Circuits (PVCs) or setup or torn down dynamically, as needed, using Switched Virtual Circuits (SVCs). Once a connection is established, data packets are segmented into 53-byte cells (= ATM cell). These cells are transmitted across the VC to the egress ATM switch, then reassembled into the original packet, and delivered to the intented destination. This process is called segmentation and reassembly (SAR).

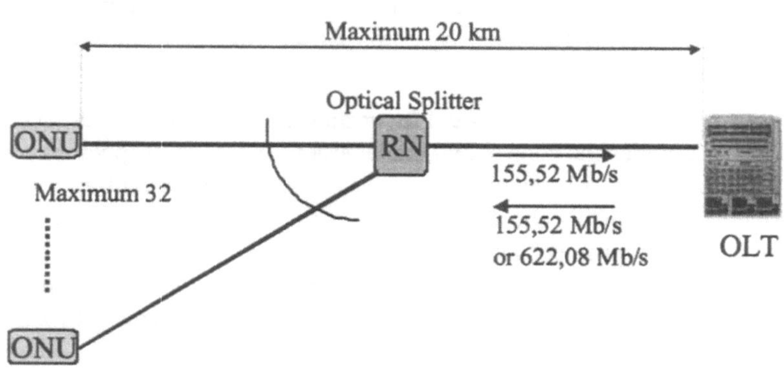

Figure 3.1.1. APON System configuration.

Since ATM is a protocol at Layer 2, it is unbiased as to what can be sent on it. Data, voice, video, or vitually anything else can be sent across an ATM network. All data packets are converted into ATM cells for transport, then reassembled for end-node delivery. ATM is a key technology to offer QoS over an important range of traffic from several sources. However, due to the limitations of the efficiency of video over ATM, the G.983.3 standard is created to allow video to be delivered in analogical format or ATM format on a wavelength separated from voice and data.

FSAN limits the distance between OLT and ONU to 20 km. Higher distances can cause a bad operation of the ranging protocol which is used by the OLT to determine the distance between ONUs, and to be able to assign an optimum time slot to each ONU to transmit (without interference) to the other ONUs. Therefore, ONUs are synchronized to emit in the upstream channel. The OLT is in charge of sending grant messages of TDMA time slot to every ONU by means of maintenance, administration, and operation cells from the physical layer. Then, the ONU sends the response ranging cell to the OLT. Finally the OLT tells the ONU to insert the appropiate delay so that all equivalent OLT-ONU distances are 20 km. To do this, the OLT sets a ranging window in the upstream frame whose size is equivalent to 40 km (maximum distance × 2) of round-trip delay (see Figure 3.1.2).

PLOAM cells, in addition to delivering the information about bandwidth allocation, are responsible for the synchronization, error control, security, ranging, and maintenance system.

Currently there are two APON options: a symmetric 155,52 Mb/s, and an

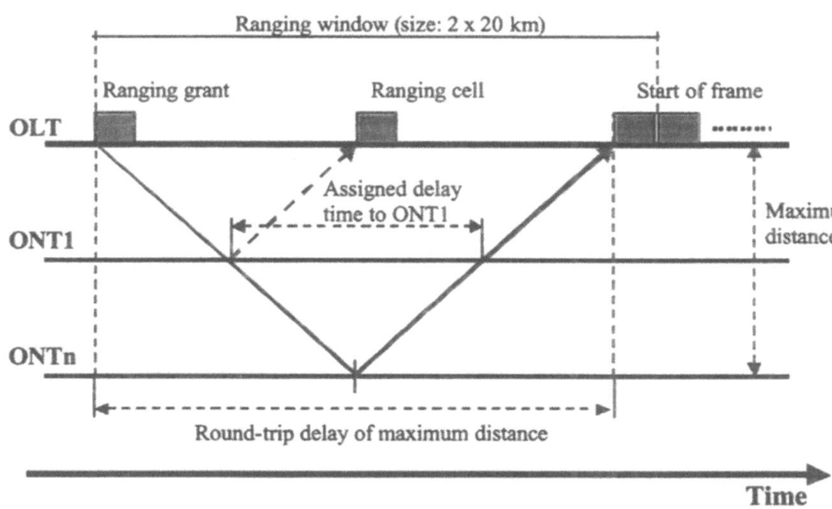

Figure 3.1.2. The ranging method in ITU-T Recommendation G.983.1.

asymmetric 622.08 Mb/s for downstream channel and 155,52 Mb/s for upstream channel. More studies are being done to increase the emission rate in both directions. Following (Figure 3.1.3), the frame formats between OLT and ONU are shown in both directions and symmetrical transmission (155 Mbps). The asymmetrical transmission (622 Mbps/155 Mbps) has similar content.

In the downstream channel, two PLOAM cells are inserted for every 54 ATM cells of load, i.e., two PLOAM cells in every frame. Obviously, the introduction of these cells decrease the emission rate of the information data.

In the upstream channel, there are 3 extra bytes for every ATM cell. These 3 bytes are divided into three fields. The 3 overhead bytes contain a minimum of 4 bits of guard time to provide enough distance in time to prevent collisions with cells from other ONTs. This field length is actually programmable by the OLT. The preamble field extracts the phase of the arriving cell relative to the local timing of the OLT, and it is used to acquire bit synchronization and amplitude recovery. The Delimiter field is used to indicate the start of incoming cell. The guard time length, preamble pattern and delimiter pattern can be controlled by the OLT. The contents of these fields are defined by the Upstream_overhead message in the downstream PLOAM cells.

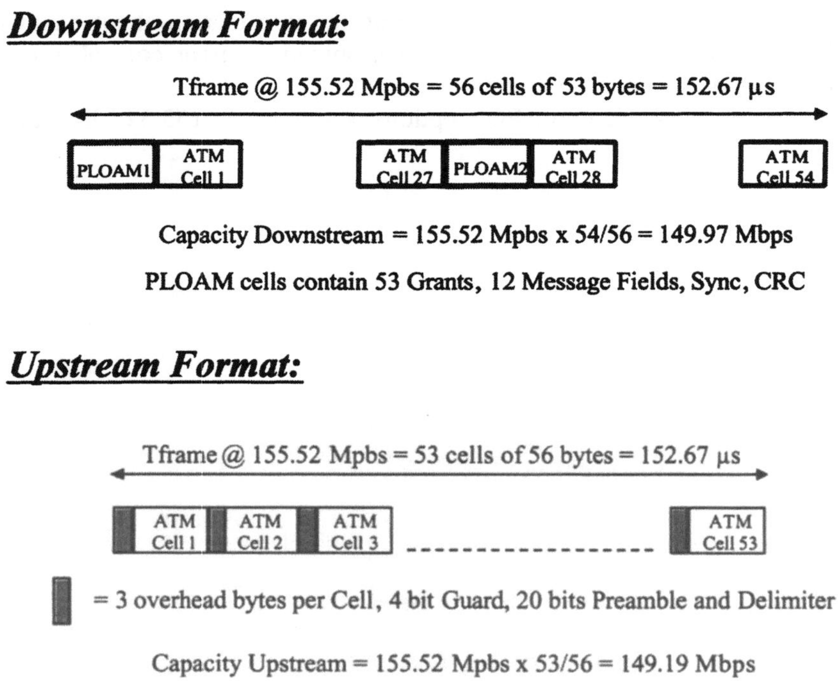

Figure 3.1.3. ATM PON frame formats.

In addition to these cells, PLOAM cells are also sent in upstream direction; the rate of which is defined by the OLT for each ONT or ONU depending on the required functionality. The minimum PLOAM rate in the upstream direction is one PLOAM every 100 ms. This equates to approximately one PLOAM every 655 frames which is negligible. Although the maximum PLOAM rate is undefined, it is also expected to be negligible.

3.1.1. G983.3 standard

G.983.3 standard has been approved in a short period of time. The G.983.3 aim is to increase the system capacity .These recommendations describe an APON system with increased service capability by wavelength allocation. The main points to be considered for the enhancement of APON were:

- Compatibility with existing recommendation G.983.1 and with the optical power budget.
- Usability of optical devices compliant with G.983.1.
- Broadcast/multicast capability for economical video delivery services, including capability for varied deployment scenarios and modulation schemes, and capability for a wide range of high-speed digital services.

The next scheme depicts the WDM enhancement of this recommendation.

As it is shown in Figure 3.1.4, the downstream wavelength band of G.983.1 has been narrowed. This portion is used for transporting APON signals downstream. An additional wavelength band, referred to as the "enhancement band", is now available. There are two options for the enhancement band spectrum, 1539–1565 nm and 1550–1560 nm. The former is to be used for transmitting DWDM private lambda services, while the latter is to be used for video delivery services. A wide guard band separates the basic band from the enhancement band. This helps to minimize the interference signal in these two bands.

3.1.2. G.983.4

The aim of G.983.4 is to increase efficiency. DBA (Dynamic Bandwidth Assignment) is a system of protocols and algorithms that work to deliver bandwidth to customers that need it in real time. This allows bursty data applications to be run in efficient way maintaining the quality of service for voice-oriented and circuit-oriented services.

At high level, DBA works in three phases. In the first phase, the ONUs transmit to the OLT the current status of their traffic buffers. This can be done through an explicit message (which is called minislot transmission) or

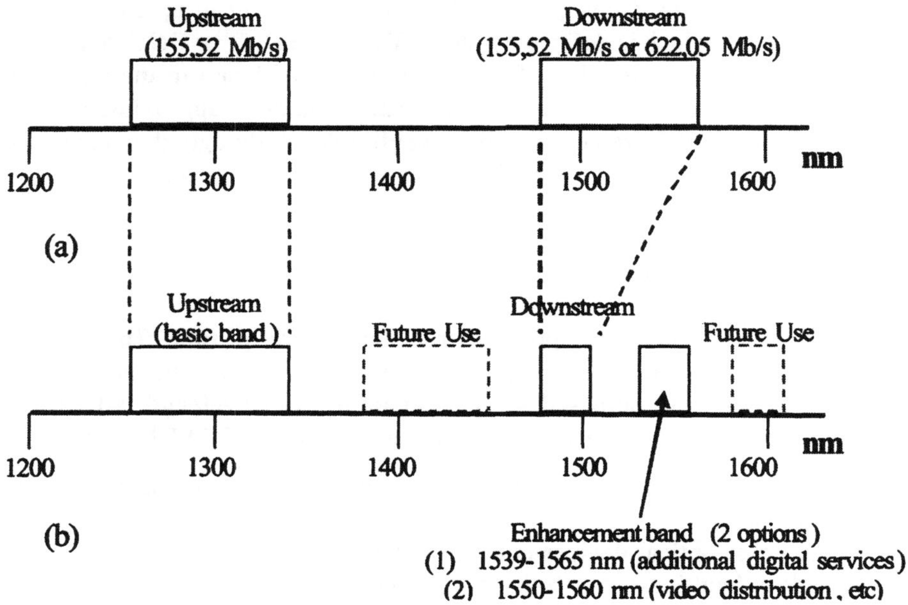

Figure 3.1.4. G.983.3 specification.

implicitly through the volume traffic analysis which each ONU is sending to the OLT. In the second phase, the OLT calculates the better distribution of bandwidth to all the ONU's. In this calculation, both the demands of the ONUs and their traffic contracts are taken into account. Finally, in the third phase, the OLT sends transmission grants to all the ONUs telling them the proportion of assigned bandwidth.

In **[Effenberg 01]**, charts about the increased effective capacity and quality of service applying DBA are shown. Both the loss ratio and the average delays are improved by the use of DBA.

The current work in the FSAN and ITU-T task force is concentrating on defining the formats and procedures used in the DBA process. The main challenge is the definition of the minislot channel that comprises the format of the minislot itself. The semantics of the fields contained within are the control parameters configuration, the system for alarms, and the monitoring of the DBA.

3.1.3. G.983.5

The aim of G.983.5 is to increase the network reliability. Previously, it is interesting to have a look at the ITU-T Recommendation G.9831 appendix

IV to see the types of protection network architectures. There are 4 types of protection nodes defined. They are shown in the following Figures.

Type A

In this type of protection, only the fiber is protected. A fiber switch and the TC (Transmission Convergence) protocols, such as Ranging, are controlled independently. The re-ranging protocol is done after the failure detection and optical switch recovery. Type A has limited protection and uses optomechanical components.

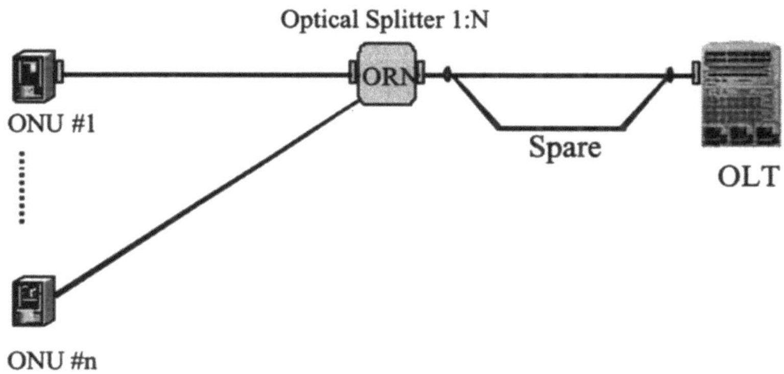

Figure 3.1.5. Protection Network. Type A.

Type B

In Type B protection, the equipment between the OLT and the ORN is duplicated. The primary OLT is actively working while the reserve OLT interface is in cold standby. If the active fiber is cut or the working OLT interface fails, the reserve line is activated. The switching control is only within the OLT, there is no need to adopt a switching protocol.

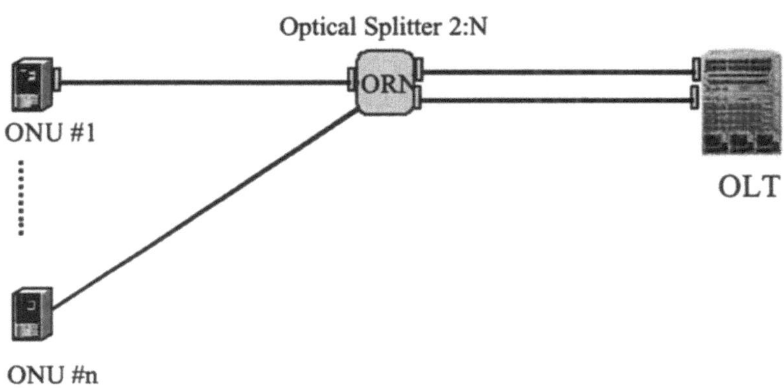

Figure 3.1.6. Protection Network. Type B.

Type C

This type of protection is a full duplication system. Both the primary and backup interfaces are normally working, i.e. hot standby. This allows for fast switching time (50 ms or less). Switching is done at the TC layer for each individual ONU. It can protect failures fibers, branch line failures and ONU failures.

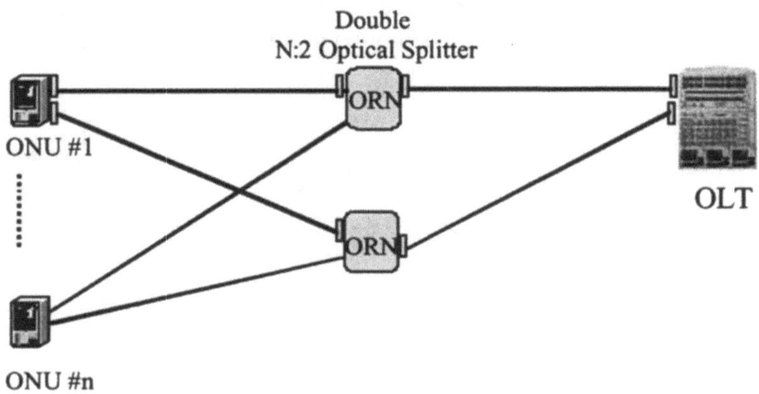

Figure 3.1.7. Protection Network. Type C.

Type D

This protection specifies independent duplication of branch line and common lines. Type D protection cannot offer fast switching. This alternative is not considered attractive.

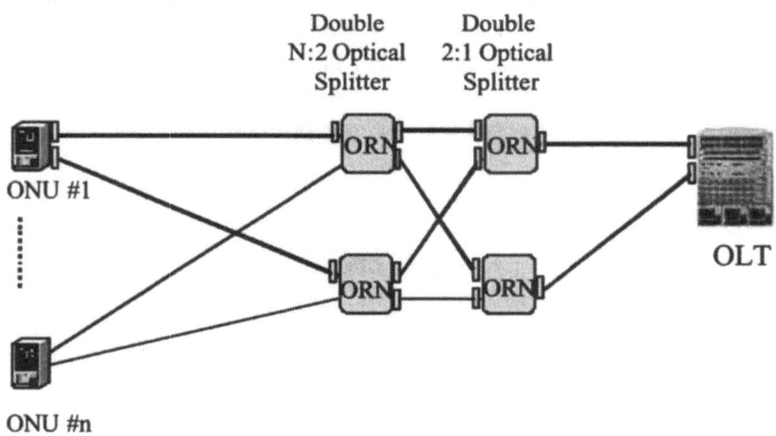

Figure 3.1.8. Protection Network. Type D.

Type B and Type C were selected as the basis of the new studies. The aim of the new approaches is to get fast switching and flexible mixed operation. The TC layer protection group is defined to be the entire VP group running between the OLT and an ONU. The PST (PON section trace) messages carry the switching commands and notifications between OLT and ONU. The PST messages in G.983.1 have been expanded to support this functionality.

The protocols are being analysed by the FSAN consortium. The basic challenge is to reuse a subset of the established protection messages defined in ITU-T G.783 Recommendation.

3.1.4. A brief view in the ITU-T Recommendation G.983.1

In this section, the reference configuration for an ATM based PON is shown according to G.983.1 standard. The ONU functional blocks, the OLT functional blocks and the general physical configuration of the ODN is depicted according to G.983.1 standard. These functions can be compared with the generic functions of each block explained in chapter 2.

ATM PON Architecture
S is the point on the optical fiber just after the OLT [Downstream direction]/ONU [Upstream direction] optical connection point (i.e. optical connector or optical splice). R is the point on the optical fiber just before the ONU [Downstream]/OLT [Upstream] optical connection point. The interface at the

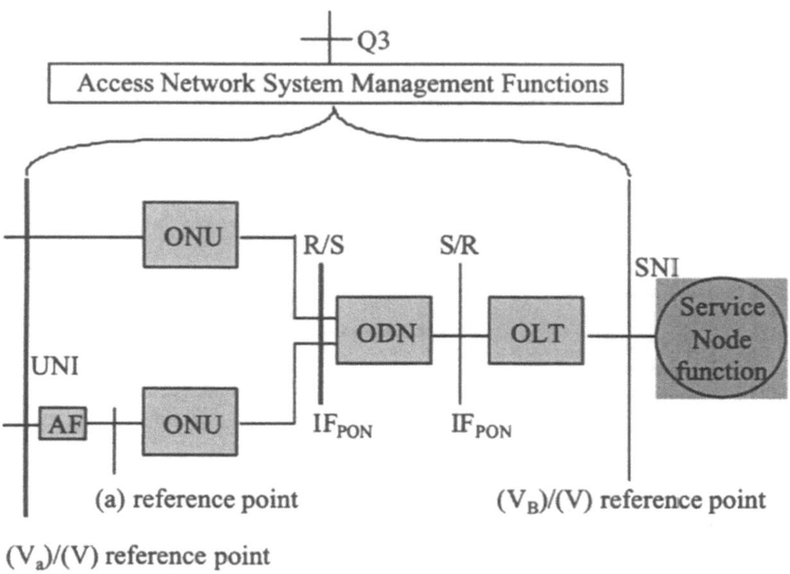

Fgure 3.1.9. Reference configuration for an ATM based PON.

reference points S/R and R/S is defined as IF_{PON}. This is a PON-specific interface that supports all the protocol elements necessary to allow the transmission between OLT and ONUs. In this architecture, the ATM transport protocols at an IF_{PON} are described as consisting of Physical Media Dependent layer, Transmission Convergence Layer and ATM layer. This architecture is only intented to address the transport of ATM. Further detail is contained in Recommendation I.732.

The (a) reference point is added to differentiate the AF (Adaptation Function) from the ONU.

OLT functional Blocks

The OLT is connected to the switched networks via standardized interfaces (VB5.x, V5.x, NNI's). On the distribution side, it presents optical accesses according to the agreed requirements in terms of bit rate, power budget, etc.

The service port function interfaces to service nodes. This function may handle inserting ATM cells into the upstream SDH payload and extracting ATM cells from the downstream payload. The function may be duplicated, then the protection switching function is necessary.

The MUX/DEMUX provides VP connections between the service port function and the ODN interface, and different VPs are assigned to different services at IF_{PON}. Various information such as main contents, signalling, and OAM flows are exchanged by using VCs of the VP.

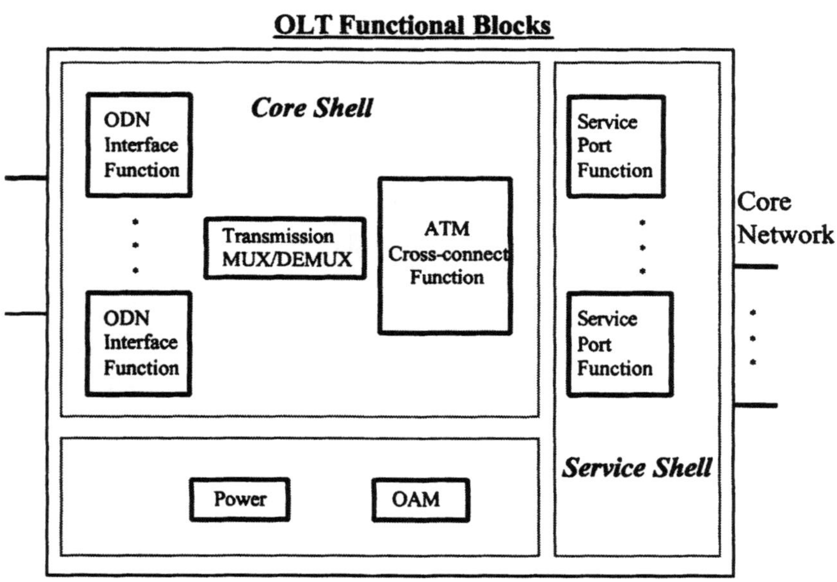

Figure 3.1.10. Example of G.983.1 OLT functional blocks.

Finally, the ODN interface handles inserting ATM cells into downstream PON payload and extracting ATM cells from the upstream PON payload. Also, ODN interface handles the optoelectronic conversion process.

ODN functional Block

The ODN, as it has been mentioned, provides the optical transmission medium for the physical connection from the ONUs to the OLTs.

Transmission in downstream and upstream directions can take place in the same Fiber and components (bi-directional system) or in separate fibers and components (unidirectional system). If additional connectors or other passive devices are needed for ODN rearrangement, they shall be located between S and R and their losses shall be taken into account in any optical power budget calculation.

The ODN offers one or more optical path between one OLT, and one or more ONUs. Each optical path is defined between reference points in a specific wavelength window. The following optical interfaces are defined in Figure 3.1.11:

- O_{ru}, O_{rd} Optical interfaces at the reference point R/S between the ONU and the ODN for the upstream and downstream direction respectively.

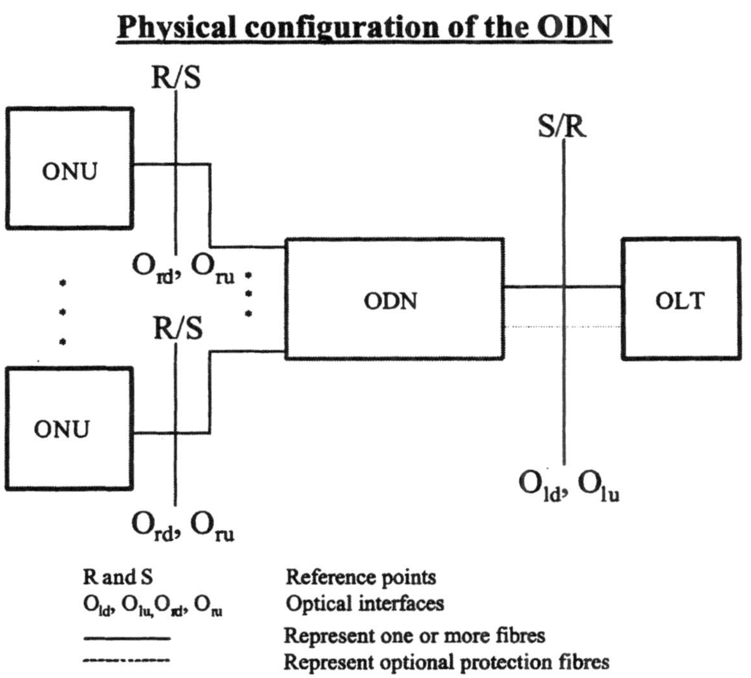

Figure 3.1.11. Generic physical configuration of the Optical Distribution Network.

- O_{lu}, O_{ld} Optical interfaces at the reference point S/R between the OLT and the ODN for the upstream and downstream direction respectively.

At the physical layer, the interfaces may require more than one Fiber, e.g. for separation of transmission directions or different types of signals.

ONU Functional Blocks

The ONU consists of ODN interface, user port, transmission, services, and customers Multiplexing/demultiplexing functions and powering as it's shown in Figure 3.1.12.

The ODN interface handles the optoelectronic conversion process. It extracts ATM cells from the downstream PON payload and inserts ATM cells into upstream PON payload based on the synchronization acquired from the downstream frame timing.

MUX multiplexes service interfaces to ODN interface. Only valid ATM cells can be passed through the MUX. So many VPs can share the assigned upstream bandwidth effectively.

The user port interfaces over UNI to a terminal. The User Port may handle inserting ATM cells into the upstream payload and extracting ATM cells from the downstream payload.

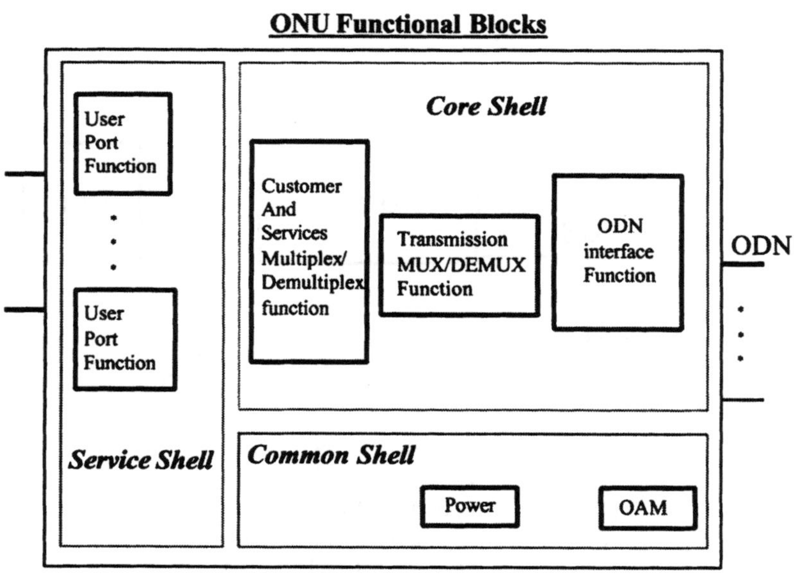

Figure 3.1.12. Example of G.983.1 ONU functional blocks.

Physical medium dependent layer requirements for the ATM-PON

The next tables show the ITU-T G983 specifications for the physical medium dependant layer parameters of ODN, the optical interface parameters of both 155 Mbit/s downstream and upstream direction, and the optical interface parameters of 622 Mbit/s downstream direction.

Physical medium dependant layer parameters of ODN

Item	*Unit*	*Specification*
Fiber Type	-	Recommendation G.652
Attenuation range (Recommendation G.982)	dB	Class B: 10-25 Class C: 15-30
Differential optical path loss	dB	15
Maximum optical path penalty	dB	1
Max differential logical reach	km	20
Maximum fiber distance between S/R and R/S points	km	20
Minimum supported split ratio	-	Restricted by path loss and ONU addressing limits. PON with passive splitters (16 or 32 way split)
Bidirectional transmission	-	1-fiber WDM or 2-fiber
Maintenance wavelength	nm	To be defined

Optical interface parameters of 155 Mbit/s downstream direction

Item	*Unit*	*Single Fiber*		*Dual Fiber*	
		OLT Transmitter (optical interface O_{ld})			
Nominal bit rate	Mbit/s	155.52		155.52	
Operating Wavelength	nm	1480-1580		1260-1360	
Line Code	-	Scrambled NRZ		Scrambled NRZ	
Maximum reflectance of equipment, measured at transmitter wavelength	dB	NA		NA	
Minimum ORL of ODN at O_{ld} and O_{lu} (Notes 1 and 2)	dB	more than 32		more than 32	
ODN Class		Class B	Class C	Class B	Class C
Mean launched power MIN	dBm	-4	-2	-4	-2
Mean launched power MAX	dBm	+2	+4	+1	+3
Launched optical power without input to the transmitter	dBm	NA		NA	
Extinction ratio	dB	more than 10		more than 10	
Tolerance to the transmitter incident light power	dB	more than -15		more than –15	
If MLM Laser- Maximum RMS width	nm	1.8		5.8	
If SLM Laser- Maximum – 20 dB width (Note 3)	nm	1		1	
If SLM Laser- Minimum side mode suppression ratio	dB	30		30	

		ONU receiver (optical interface O_{rd})			
Maximum reflectance of equipment, measured at receiver wavelength	dB	less than -20		less than –20	
Bit Error Ratio	-	less than 10^{-10}		less than 10^{-10}	
ODN Class		Class B	Class C	Class B	Class C
Minimum sensitivity	dBm	-30	-33	-30	-33
Minimum overload	dBm	-8	-11	-9	-12
Consecutive identical digit immunity	Bit	More than 72		More than 72	
Tolerance to the reflected optical power	dB	less than 10		less than 10	

NOTE 1 – The value of "minimum ORL of ODN at point O_{ru} and O_{rd}, and O_{lu} and O_{ld}" should be more than 20 dB in optional cases.
NOTE 2 – The values on ONU transmitter reflectance for the case that the value of "minimum ORL of ODN at point O_{ru} and O_{rd}, and O_{lu} and O_{ld}" is 20 dB.
Note 3- Values of maximum –20 dB width, and minimum side mode suppression ratio are referred to in Recommendation G.957.

Optical interface parameters of 155 Mbit/s downstream direction

Item	*Unit*	*Single Fiber*		*Dual Fiber*	
		OLT Transmitter (optical interface O_{ld})			
Nominal bit rate	Mbit/s	622.08		622.08	
Operating Wavelength	nm	1480-1580		1260-1360	
Line Code	-	Scrambled NRZ		Scrambled NRZ	
Maximum reflectance of equipment, measured at transmitter wavelength	dB	NA		NA	
Minimum ORL of ODN at O_{lu} and O_{ld} (Notes 1 and 2)	dB	more than 32		more than 32	
ODN Class		Class B	Class C	Class B	Class C
Mean launched power MIN	dBm	-2	-2	-2	-2
Mean launched power MAX	dBm	+4	+4	+3	+3
Launched optical power without input to the transmitter	dBm	NA		NA	
Extinction ratio	dB	more than 10		more than 10	
Tolerance to the transmitter incident light power	dB	more than –15		more than –15	
If MLM Laser- Maximum RMS width	nm	NA		1.4	
If SLM Laser- Maximum – 20 dB width (Note 3)	nm	1		1	
If SLM Laser- Minimum side mode suppression ratio	dB	30		30	
		ONU receiver (optical interface O_{rd})			
Maximum reflectance of equipment, measured at receiver wavelength	dB	less than –20		less than –20	
Bit Error Ratio	-	less than 10^{-10}		less than 10^{-10}	
ODN Class		Class B	Class C	Class B	Class C
Minimum sensitivity	dBm	-28	-33	-28	-33
Minimum overload	dBm	-6	-11	-7	-12
Consecutive identical digit immunity	Bit	More than 72		More than 72	
Tolerance to the reflected optical power	dB	less than 10		less than 10	

NOTE 1 – The value of "minimum ORL of ODN at point O_{ru} and O_{rd}, and O_{lu} and O_{ld}" should be more than 20 dB in optional cases.
NOTE 2 – The values on ONU transmitter reflectance for the case that the value of "minimum ORL of ODN at point O_{ru} and O_{rd}, and O_{lu} and O_{ld}" is 20 dB.
Note 3- Values of maximum –20 dB width, and minimum side mode suppression ratio are referred to in Recommendation G.957.

Optical interface parameters of 155 Mbit/s upstream direction

Item	*Unit*	*Single Fiber*		*Dual Fiber*	
		ONU Transmitter (optical interface O_{ru})			
Nominal bit rate	Mbit/s	155.52		155.52	
Operating Wavelength	nm	1260-1360		1260-1360	
Line Code	-	Scrambled NRZ		Scrambled NRZ	
Maximum reflectance of equipment, measured at transmitter wavelength	dB	less than –6 dB		Less than –6 dB	
Minimum ORL of ODN at O_{lu} and O_{ld} (Notes 1 and 2)	dB	more than 32		more than 32	
ODN Class		Class B	Class C	Class B	Class C
Mean launched power MIN	dBm	-4	-2	-4	-2
Mean launched power MAX	dBm	+2	+4	+1	+3
Launched optical power without input to the transmitter	dBm	less than Min sensitivity –10		less than Min sensitivity -10	
Extinction ratio	dB	more than 10		more than 10	
Tolerance to the transmitter incident light power	dB	more than –15		more than –15	
If MLM Laser- Maximum RMS width	nm	5.8		5.8	
If SLM Laser- Maximum – 20 dB width (Note 3)	nm	1		1	
If SLM Laser- Minimum side mode suppression ratio	dB	30		30	
Jitter generation from 0.5 khz to 1.3 Mhz	UI ptp	0.2		0.2	
		ONU receiver (optical interface O_{rd})			
Maximum reflectance of equipment, measured at receiver wavelength	dB	less than –20		Less than –20	
Bit Error Ratio	-	less than 10^{-10}		Less than 10^{-10}	
ODN Class		Class B	Class C	Class B	Class C
Minimum sensitivity	dBm	-30	-33	-30	-33
Minimum overload	dBm	-8	-11	-9	-12

NOTE 1 – The value of "minimum ORL of ODN at point O_{ru} and O_{rd}, and O_{lu} and O_{ld}" should be more than 20 dB in optional cases.
NOTE 2 – The values on ONU transmitter reflectance for the case that the value of "minimum ORL of ODN at point O_{ru} and O_{rd}, and O_{lu} and O_{ld}" is 20 dB.
Note 3- Values of maximum –20 dB width, and minimum side mode suppression ratio are referred to in Recommendation G.957.

3.2. EPON: ETHERNET PASSIVE OPTICAL NETWORK

EPON means Ethernet over PON. In November 2000, the EFM (Ethernet First Mile) study group was created to investigate the viability of developing a standard based on Ethernet to the access market. There are also efforts to work closely with other organizations (IETF, Standards Committee T1) and try to connect with G.983 standards. Their web site is the following: http://grouper.ieee.org/groups/802/3/efm/.

The main key difference between APON and EPON technologies is the way to transmit the information. The former, as it has been explained before, transmits the info in fixed cells (53), while EPON technology uses variable length packets from 72 to 1518 bytes according to the IEEE 802.3 protocol for Ethernet.

With the current trend to transmit all type of information over IP protocols the APON solution has become more difficult and inefficient. IP data packets can carry on until 65.535 bytes. Therefore, IP traffic has to be encapsulated into cells of 48 bytes and add an overhead of 5 bytes. This mechanism is cost-inefficient from both processing time and ONU-OLT designs point of view. Instead, Ethernet picks IP traffic up and minimises the ATM tax or ATM overhead.

Ethernet is the most extended local area technology in the world; more than 85% of LAN's use it. Ethernet corresponds to 802.3 standard of the IEEE. The most popular version supports 10 Mb/s. New versions comprise 100 Mb/s (Fast Ethernet) and 1 Gb/s (Gigabit Ethernet). 10 GB/s is also being studied.

Traffic flow in EPON has the same philosophy as APON. Several medium access techniques is being treated in upstream traffic similar to APON. The CSMA/CD is discarded as access medium protocol. The key difference is the use of variable length packets. The Figure 3.2.1 shows the flows for both upstream and downstream channels.

Downstream traffic is segmented into frames of fixed interval which carry variable length packets (see Figure 3.2.2). Clock information which is needed to synchronize OLT and ONUs is at the beginning of every frame. It is formed by one byte which is transmitted every 2 ms. Every packet is sent to the corresponding ONUs like it is indicated by the numbers 1 to N. Packets, as it is already mentioned, are formatted according to the 802.3 standard.

The upstream traffic is segmented into frames and every frame is divided into specific time slots for every ONU. Frames are sent continuously and their duration is 2 ms. An overhead identifies the beginning of each frame. Every ONU has assigned its time slot in the frame to send packets. In Figure 3.2.3, the frame is divided between all ONUs which compose the PON. The TDM controller of every ONU, in close collaboration with the clock information from the OLT, controls the correct performance of upstream channel.

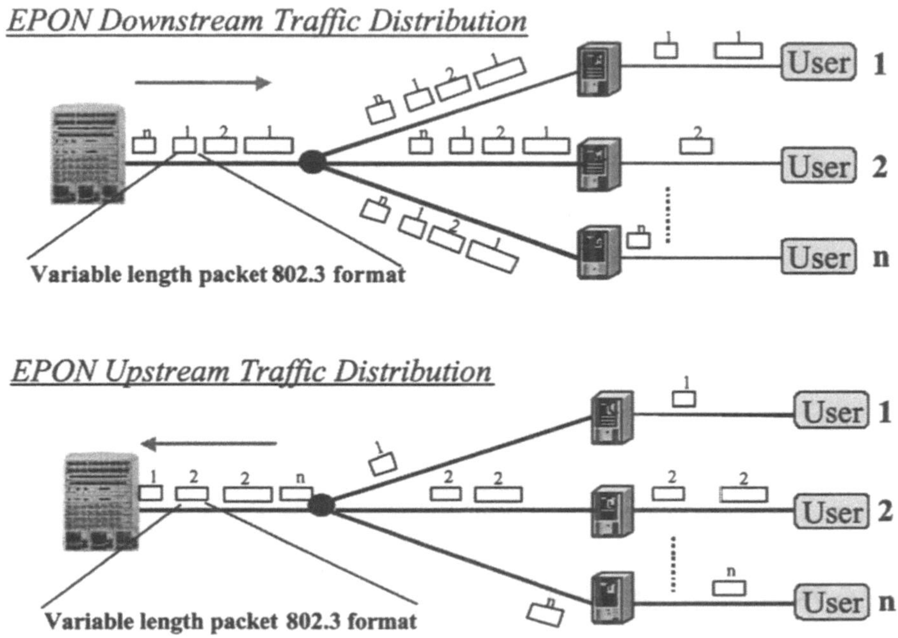

Figure 3.2.1. Traffic flow for both downstream and upstream channels in EPON technology.

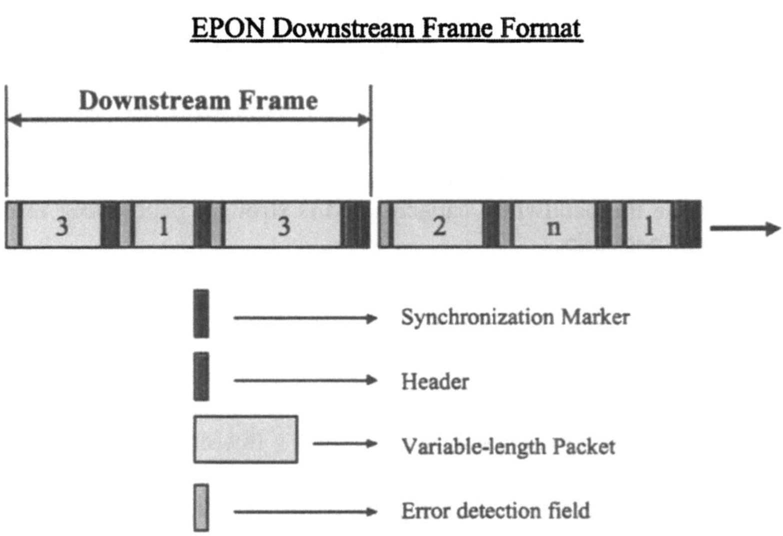

Figure 3.2.2. Frame format for downstream channel in EPON technology.

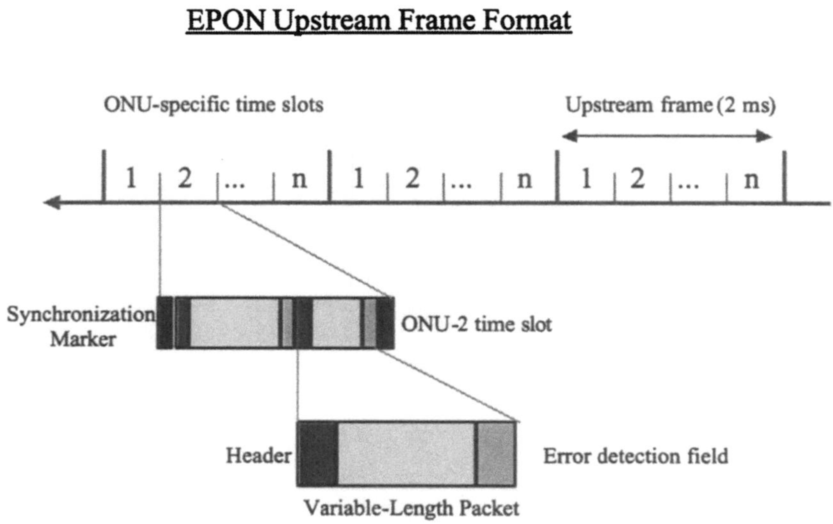

Figure 3.2.3. Frame format for upstream channel in EPON technology.

The overhead of the time slot assigned to each ONU includes a guard time, time indicators, and signal power indicators. If there is no information to transmit from an ONU, its time slot could carry an established signal which could represent "no traffic".

Therefore, the use of Ethernet in FFTH technology includes the frame format and overheads, but the Medium Access Control is not used. CSMA/CD protocol is discarded, due to the distances in FTTH which are higher than the initial ones in CSMA/CD protocol.

Particularities about Gigabit Ethernet and 10-Gigabit Ethernet technologies are analyzed deeper in the next section. The former seems to be currently the best one to implement in the access network. The latter could be the natural way to increase the bandwidth capacity and is strongly penetrating in metropolitan access networks.

3.2.1. Gigabit Ethernet

Gigabit Ethernet is an extension to the highly successful 10 Mbps and 100 Mbps IEEE 802.3 Ethernet standards. Offering a raw data bandwidth of 1000 Mbps, Gigabit Ethernet maintains full compatibility with the huge installed base of Ethernet nodes. The key objectives of the 802.3z Gigabit Ethernet task force were to develop a Gigabit Ethernet standard that does the following:

- Allows half and full-duplex operation at the speed of 1000 Mbps;

- Uses the 802.3 Ethernet frame format;
- Uses the CSMA/CD access method with support for one repeater per collision domain;
- Addresses backward compatibility with 10BASE-T and 100Base-T technologies.

The functional elements of Gigabit Ethernet Technology are shown below.

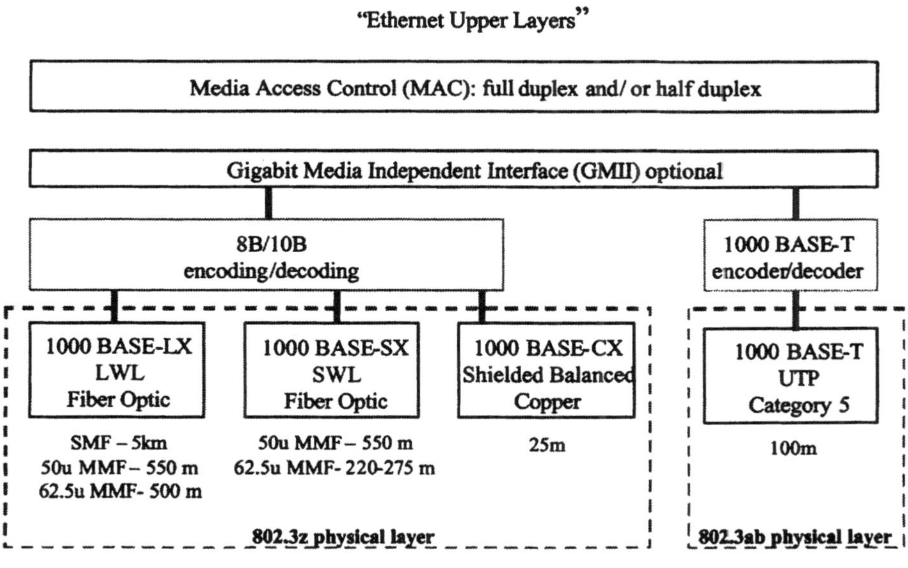

Figure 3.2.4. Functional elements of Gigabit Ethernet Technology **[Gigabit 98]**.

As it is depicted in the Figure 3.2.4, Gigabit Ethernet has two different fiber implementations, 1000BASE-SX and 1000BASE-LX. These technologies are the focus of the next two chapters.

1000BASE-SX

Short wavelength Gigabit Ethernet, or 1000BASE-SX operates at 850 nm over multimode fiber.

1000BASE-SX is designed to support 220 meter link distances using standard 160 MHz-km 62.5 micron fiber, and can support 275 meter links using 200 MHz-km 62.5 fiber. If a user has 500 MHz-km 50 micron fiber, 550 meter link distances can be supported. A number of vendors has certified laser launch characteristics of their fiber cable and can support 300 and even 600 meter link distances over enhanced 62.5 or 50 micron fiber. TIA FO2.2 is completing work on extended distance support for Gigabit Ethernet over 62.5 micron fiber.

Fiber Type	Modal Bandwidth @ 850 nm	Distance
62,5 μm	160 MHz·km	220 m
62,5 μm	200 MHz·km	275 m
50 μm	400 MHz·km	500 m
50 μm	500 MHz·km	550 m

Figure 3.2.5. 1000BASE-SX. Modal bandwidth versus distance.

1000BASE-LX

Long wavelength Gigabit Ethernet, or 1000BASE-LX operates at 1300 nm, and can run over singlemode or multimode fiber.

1000BASE-LX can support 550 meter link distance over 62.5 or 50 micron multimode fiber, or 5.0 kilometer link distance over singlemode fiber.

Fiber Type	Modal Bandwidth @ 1300 nm	Distance
62,5 μm	500 MHz·km	550 m
50 μm	500 MHz·km	500 m
9 μm SM	n/a	5000 m

Figure 3.2.6. 1000BASE-LX. Modal bandwidth versus distance.

Sources for Gigabit Ethernet technology

LEDs are not practical solutions for Gigabit Ethernet for two reasons:

1. Broadband LED sources can suffer from chromatic dispersion problems at 850 nm.
2. Modulating an LED one billion times per second for Gigabit speed is simply not technically possible.

The solution for Gigabit Ethernet is to use lasers as the light source. Fortunately, laser technology is now comparable in priced with LEDs, due to recent development of Vertical Cavity Surface Emitting Laser (VCSEL) and CD laser technology. VCSELs provide a laser quality source at a similar manufacturing cost as an LED. The IEEE standards committee wrote Gigabit Ethernet standard to accommodate these technologies.

Power Budget

The IEEE specifications also detail the maximum loss allowed across a link, or power budget (more information about power budget concept in Chapter 2).

	Maximum Attenuation
Multimode Cable	
• 62,5/125	3,57 dB/km @ 850 nm 1,5 dB/km @ 1300 nm
• 50/125	3,5 dB/km @ 850 nm 1,5 dB/km @ 1310 nm
Single-mode Cable	
• Outside Plant/ Loose Tube	0,5 dB/km @ 1300 nm
Connectors	0,75 per each mated pair
Splicers	
• Fusion or Mechanical	0,3 dB

Figure 3.2.7. Power budget specifications.

3.2.2. 10 Gigabit Ethernet

The overall goals of the IEEE standards effort is to extend the existing IEEE802.3 standards to support 10 Gb/s speeds and to enhance Ethernet to include support for WAN links. Standards for 10 Gb/s will be produced as an extension to the existing IEEE standards with the basic changes being at the physical layer. 10 Gigabit Ethernet uses the IEEE 802.3 Ethernet Media Access Control (MAC) protocol, the IEEE 802.3 Ethernet frame format, and the minimum and maximum IEEE 802.3 frame size. 10 Gigabit Ethernet is controlled by the standard 802.3ae.

As 1000BASE-X and 1000BASE-T (Gigabit Ethernet) remained true to the Ethernet model, 10 Gigabit Ethernet continues the natural evolution of Ethernet in speed and distance. Since it is a full-duplex only and fiber-only technology, it does not need the carrier -sensing multiple-access with collision detection (CSMA/CD) protocol that defines slower, half-duplex Ethernet technologies. In every other respect, 10 Gigabit Ethernet remains true to the original Ethernet model.

An Ethernet PHYsical layer device (PHY), which corresponds to Layer 1 of the OSI model, connects the media (optical or copper) to the MAC layer that corresponds to OSI Layer 2. Ethernet architecture further divides the PHY (Layer 1) into a Physical Media Dependent (PMD) and a Physical Coding Sublayer (PCS). Optical transceivers, for example, are PMDs. The PCS is made up of a coding (e.g. 64/66b) and a serializer or multiplexing functions.

The 802.3ae specification defines two PHY types: the LAN PHY and the WAN PHY. The WAN PHY has an extended feature set added onto the functions of a LAN PHY. These PHYs are solely distinguished by the PCS. There is a number of PMD types.

3.2.2.1. *Physical Media Dependent (PMDs)*

The IEEE 802.3ae Task Force has developed a draft standard that provides a physical layer that supports link distances for optical media as shown in the table below (Figure 3.2.8):

PMD (Optical Transceiver)	Type of Fiber Supported	Target Distance (Meters)
850 nm serial	Multi mode	65
1310 nm WWDM	Multi mode	300
1310 nm WWDM	Single mode	10.000
1310 nm serial	Single mode	10.000
1550 nm serial	Single mode	40.000

Figure 3.2.8. Draft standard about physical layer.

To meet these distance objectives, four PMDs were selected. The task force selected a 1310 nm serial PMD to meet its 2 km and 10 km single-mode fiber (SMF) objectives. It also selected a 1550 nm serial solution to meet (or exceed) its 40 km SMF objective. Support of the 40 km PMD is an acknowledgement that Gigabit Ethernet is already being successfully deployed in metropolitan and private long distance applications. An 850 nm PMD was specified to achieve a 65-meter objective over multimode fiber using serial 850 nm transceivers.

Additionally, the task force selected two versions of the wide wave division multiplexing (WWDM) PMD; a 1310 nm version over single-mode fiber to travel a distance of 10 km and a 1310 nm PMD to meet its 300-meter-over-installed-multimode-fiber objective.

3.2.2.2. *Physical Layer (PHYs)*

The LAN PHY and the WAN PHY will operate over common PMDs and, obviously, will support the same distances. These PHYs are distinguished solely by the Physical Encoding Sublayer (PCS).

Over time, it is expected that the LAN PHY will be used in pure optical switching environments extending over all distances. However, for compatibility with the existing WAN network, the 10 Gigabit Ethernet WAN PHY supports connections to existing and future installations of SONET/SDH circuit-switched telephone access equipment.

<table>
<tr><td></td><td>Serial LAN</td><td>Serial WAN</td><td>4-Lane LAN</td></tr>
<tr><td rowspan="3">PHY (chipsets)</td><td rowspan="2">64/66</td><td>64/66</td><td rowspan="3">XAUI</td></tr>
<tr><td>WIS</td></tr>
<tr><td colspan="2">SerDes</td></tr>
<tr><td>PMD (optics)</td><td colspan="2"><u>Serial</u>
850 nm (mmf; 65m)
1310 nm (smf; 10km)
1550 nm (smf; 40km)</td><td><u>WWDM</u>
1310 nm (mmf; 300m)
1310 nm (smf; 10km)</td></tr>
</table>

Figure 3.2.9. The LAN PHY and the WAN PHY will operate over common PMDs and will support the same distances **[10Gigabit 01]**.

The WAN PHY differs from the LAN PHY by including a simplified SONET/SDH framer in the WAN interface sublayer (WIS). It is because the line rate of SONET OC-192/SDH STM-64 is within a LAN PHY at 10 Gbps or with a WAN PHY payload rate of approximately 9.29 Gbps.

In order to enable low-cost WAN PHY implementations, the task force specifically rejected conformance to SONET/SDH jitter, stratum clock, and certain SONET/SDH specifications. The WAN PHY is basically a cost effective link that uses common Ethernet PMDs to provide access to the SONET/SDH and time division multiplexed infrastructure. This feature enables Ethernet to use SONET/SDH for Layer 1 transport across the WAN transport backbone.

It is important to note that Ethernet remains as asynchronous link protocol. As in every Ethernet network, 10 Gigabit Ethernet's timing and synchronization must be maintained within each character in the bit stream of data, but the receiving hub, switch or router may re-time and re-synchronize the data.

Among the many technical innovations of the 10 Gigabit Ethernet Task Force there is an interface called XAUI. The "AUI" portion is borrowed from the Ethernet Attachment Unit Interface. The "X" represents the Roman numeral for ten and implies ten gigabits per second. The XAUI is designed as an interface extender, and the interface, which it extends, is the XGMII, the 10 Gigabit Media Independent Interface. The XGMII is a 74 signal wide interface (32-bit data paths for each to transmit and receive) that may be used to attach the Ethernet MAC to its PHY. The XAUI may be used in place of, or to extend, the XGMII in chip-to-chip applications typical of most Ethernet MAC to PHY interconnects. See Figure 3.2.9.

The XAUI is a low pin count, and self-clocked serial bus that is directly evolved in the Gigabit Ethernet 1000BASE-X PHY. The XAUI interface speed is 2.5 times more than that of 1000BASE-X. By arranging four serial lanes, the 4-bit XAUI interface supports the ten-times data through the required 10 Gigabit Ethernet.

3.2.2.3. *Differences between 1 Gigabit Ethernet and 10 Gigabit Ethernet*

Figure 3.2.10 shows the main differences between 1 Gigabit Ethernet and 10 Gigabit Ethernet:

Characteristic	**1 Gigabit Ethernet (1 GbE)**	**10 Gigabit Ethernet (10 GbE)**
Physical media	Optical and cooper media	Optical media only
Distance	LANs up to 5 km	LANs to 40 km. Direct attachment to SONET/SDH equipment for WANs
PMD	Leverages Fiber Channel PMD's	Creates a new optical PMD's
MAC Protocol	Half-duplex (CSMA/CD) + full Duplex	Full-Duplex only
PCS	Re-uses 8B/10B coding	Establishes new coding schemes
Additions	Carrier extension for Half-Duplex	Throttle MAC speed

Derived from: A brief introduction to 802.3ae available at www.10GEA.org

Figure 3.2.10. Differences between 1 Gigabit Ethernet and 10 Gigabit Ethernet [**Ryan**].

3.2.3. ATM vs EPON

Below (Figure 3.2.11), there are two tables where EPON and APON are compared with respect to several points of view. The last table (Figure 3.1.24) shows the advantages of EPON versus ATM in the First Mile.

	EPON	APON
Layer 2 Protocol	Ethernet	ATM
Transport	Frame	Fixed Cell
Speed	100Mb/s, 1.25 GBE, 10 GBE	155 and 622 Mbps
Service Providers	CLECs, ELECs, DLECs, MSOs, ILECs	FSAN ILECs
Standard	IEEE, 1ª Draft by March 02	FSAN
Upstream	TDMA, Other	TDMA
Cost	Ethernet Switch	ATM Switch
Delivery	POTS, Data, VOIP, IP Video	POTS, Data

Figure 3.2.11. EPON versus APON **[Pesavento 01]**.

Advantages of Optical Ethernet (EPON) vs. ATM (APON) in the First Mile

- Low cost due to economies of scale.
- More or less 320.000.000 Ethernet Ports deployed worldwide.
- LANs are more or less 90% Ethernet today.
- Byte life begins and ends as IP/Ethernet.
- Ease of Scalability, Management.
- Cable Modems and DSL Modems have Ethernet interfaces.
- Many IP/Ethernet chipset solutions to chose from.
- Future cost curve of components (electrical, optical) steep for Ethernet.
- Ethernet is a Universal standard, no variations.
- CLECs (ELECs) can start with IP centric Networks.
- Carriers are now offering Ethernet WAN Services.
- Ethernet beat FDDI, Token Ring, Fiber Channel, ATM in LAN.
- Work force solution: Many LAN technicians comfortable with Ethernet.

Figure 3.2.12. Advantages of optical EPON vs. ATM in the First Mile **[Pesavento 01]**.

Finally Figure 3.2.13 shows a comparison about price/port depends on technology used:

Technology	Equipment Type	1996 Equipment Price/Port	1999 Equipment Price/Port	Change %
Shared Fast Ethernet	Hub	$137	$85	–39%
Switched Fast Ethernet	Switch	$785	$390	–50%
Shared FDDI	Concentrator	$835	$650	–22%
Switched FDDI	Switch	$4000	$1860	–54%
ATM 622 Mbps				
Estimate (multimode fiber)	Switch	$6600	$4800	–27%
Shared Gigabit Ethernet IEEE Goal (multimode fiber)	Hub	N.A.	$470 to $700* (2× to 3× Fast Ethernet MM)	
Switched Gigabit Ethernet IEEE goal (mulrimode fiber)	Switch	N.A.	$1070 to $1610* (2× to 3× Fast Ethernet MM)	

* Estimates based on Dell'Oro Group Info and IEEE goals.

Figure 3.2.13. Comparison of price/port EPON-ATM technologies **[Gigabit 98]**.

3.3. HFC

Hybrid Fiber Coax (HFC) access network represents the natural evolution of the existing mature tree and branch CATV networks. An HFC network provides transparent communication channels between the subscribers and the CATV hubs or head-ends. This means that no signal processing occurs between the subscriber and the hub besides the optical to electrical conversion, that eventually leads to frequency translation and amplification. The HFC network can therefore be viewed as a number of wide band shared repeaters.

The hub provides the interface between the ATM core network and the HFC distribution network (view Figure 3.3.1). At this level, communication equipment is used to transmit an analogue frequency multiplex constituted of both the TV broadcast signals (analogue or digital) and the modulated digital interactive ones.

Fiber optic nodes provide the interfaces between the fiber optic part of the trunk network and a more classical fully coaxial network. Such a fiber node is generally designed to supply between 500 and 1000 subscribers.

Over the whole HFC network, the lower frequencies are used for the return path, while the higher ones are reserved for the downstream signals.

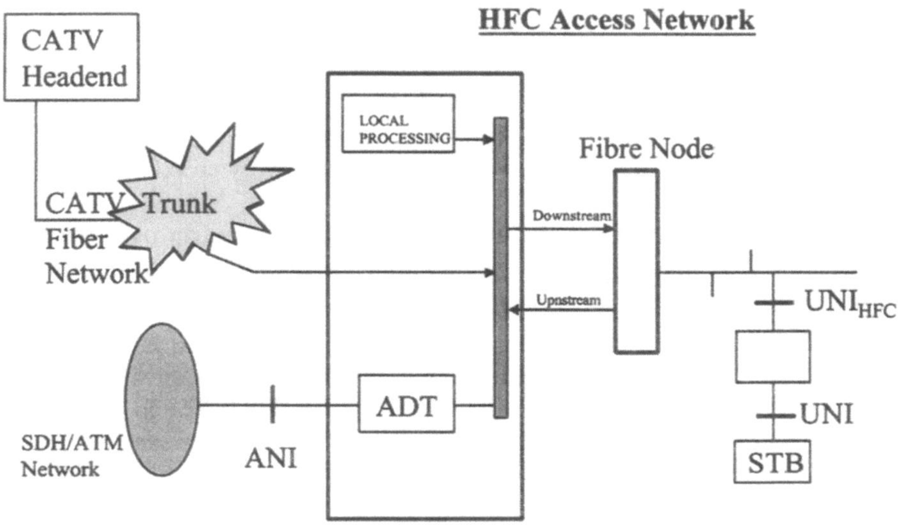

Figure 3.3.1. HFC Access Network.

Upstream frequencies range from 5 MHz to 30 or 60 MHz and downstream frequencies range from 50 or 80 MHz to 600 or 860 MHz.

The downstream frequency band is divided into two parts:

- The widest one is devoted to the broadcast services among which are analogue and digital TV, pay-TV, NVOD, FM radio, Digital radio, Broadcast MPEG2 data. *A channel spacing of 8 MHz is the standard used in the UHF frequency band.*
- A small number of these downstream 8 MHz channels will be reserved for the interactive services. Each channel carries a potential payload of 30 to 40 Mbps using a complex multi-level modulation scheme (64 QAM typical) according to the DVB standard.

For upstream direction, the frequency band will be completely devoted to the interactive digital services.

The broadcast services are by nature sent to all subscribers. This means that the capacity of the network measured as the number of channels which can be broadcasted is limited by the maximum bandwidth available.

The interactive services, on the contrary, can be introduced on a node by node basis, or even more, a reuse of the frequency spectrum can be achieved per coaxial feeder originating from a node. The capacity of the network in terms of bit rate per subscriber depends therefore on the size of the nodes, the number of coaxial feeders per node, and the actual efficiency of the modulation scheme. The node size will be limited to, as it has been mentioned before, a maximum of 500 to 1000 subscribers. Following the demand for the new interactive services, the need of more bandwidth per subscriber will increase in the future. The offer can be adapted in a flexible way by decreasing the size of the nodes, on a node by node basis.

Digital interactive services could be made available to the subscriber via either an Ethernet 10BaseT (IP over ATM) or directly by an ATM 25.6 Mbps interface.

In both cases, ATM will be used as transport mechanism over the HFC network. ATM indeed provides the appropriate flexibility both in terms of compatibility with the various types of traffic and with respect to the bandwidth management requirements.

A fully ATM approach would include the following specific functional blocks : ADT (ATM Digital Terminal) located at the hub level and ANT (ATM Network Terminal) which includes the AIU (ATM Interface Unit) and the terminal equipment STB (Set Top Box). This last device can be either a specific one or an ordinary PC.

Three interfaces are identified in [Ref 25]:

- UNI (User Network Interface), also called Home UNI (HUNI).
- ANI (ATM Node Interface).
- UNIHFC (User Network Interface for the HFC network).

The UNI (User Network Interface) is an ATM interface complying with an appropriate subset of the private or public ATM UNI defined by the ITU-T or the ATM Forum.

The ANI (ATM Node Interface) is the ATM interface between the hub and an ATM network. This interface is based on the existing ATM standards. Potential differences in the control plane are expected, namely the transport of user signalling messages to an ATM switch within the ATM network.

The UNI_{HFC} (User Network Interface for the HFC network) is an HFC-specific interface that backs all the protocol elements necessary to allow the ANT to support a standard UNI and provide connectivity across the HFC to an ATM network over ANI. The physical interface associated with UNI_{HFC} will also include other services, among others as the analogue video, that will be multiplexed onto this interface using Frequency Division Multiplexing (FDM).

In this architecture, the ATM transport protocols at a UNI_{HFC} are described as consisting of a Physical, MAC, and ATM layer. This architecture is only intended to address the transport of the ATM and not the control or management of the ATM (hence there are no management or signalling protocols). Either a separate, or a more detailed model will be necessary to address these topics.

The Physical Layer includes the modulation schemes for both the upstream and downstream channels. The TC sublayer includes the mapping of the ATM cells into the appropriate physical frames in both the downstream and upstream channels.

The MAC layer will be responsible for managing the distributed access to the upstream HFC resource across the multiple ANTs. This is a key protocol element and will directly affect the resulting ATM QoS.

The ATM protocols should not have any change in the way they operate over the HFC.

Advantages

- The HFC scheme represents the easiest evolution of the mature tree and branch CATV networks, as they have been build in a number of European countries.
- A large number of commercial products are now available and are being tested in large scale experiments including all kind of services: telephony, data communications, broadband accesses.
- Low cost, as a large part of the existing networks are reused, including the distribution network and the drop cables.
- Wide bandwidth is available.
- Modular approach as the number of nodes can be increased if the traffic increases.

Limitations

- Current lack of standard in CATV architecture: bandwidths, signal levels, return path performances.
- The quality of the system depends on the quality of the return path, which may suffer, in the existing mature networks, from ingress. Further developments on the physical layer level are still expected to improve both the quality of the networks and the robustness of the transmissions.
- Large number of the active elements (coaxial amplifiers) is subject to failure. Equipment redundancy and monitoring must be improved in the existing systems.
- The analogical nature of the signal makes it necessary to have a higher SNR than a digital signal.
- There will be the copper bottleneck when the broadband services are deployed.
- The transmitter is too complex.
- The noise is high compared to PON.
- The upstream and downstream traffic is asymmetric; the current access networks are looking for symmetrical traffic in order to be able to send and receive information at the same rate.

SUMMARY

FTTH standards still need some work to be done. Roughly half of the PON products now in the market are compliant with the Full Service Access Networks (G.983-FSAN) specifications released last year by the International Telecommunication Union (ITU). They are based on asynchronous transfer mode (ATM) technology, so-called ATM PON (APON), and it has some detractors which argue that it is limited in terms of speed and flexibility, and very expensive. Until now, only few companies, (Quantum Bridge Communications Inc. and Terawave Communications), have achieved 622 Mbit/s both upstream and downstream on PONs. Efforts are now to get new rates, such as 1244 Mb/s and 2488 Mb/s, both in the upstream and downstream direction.

A new series of specs being explored by the First Mile Study Group of the IEEE (EFM) with the objective of having faster networks, which support more users, and are cheaper to deploy. But Ethernet PON (EPON) solutions are still proprietary, and standards are far from being ready for the broad market. What is more without the approval of the ITU is that it is unlikely that large incumbent carriers would be willing to risk deploying this kind of networks.

It seems then really clear that standardization processes have still to resolve which architecture is the most suitable to exploit all the potential advantages of the optical access networks. As every manufacturer wants to sell its products, it is difficult to reach a great consensus on what is the best strategy. In spite of this and taking into account that most of the systems work with Ethernet cards, EPON networks are maybe the right choice (detractors say that it is difficult to give real time services with Ethernet . . .). At the end, some people are talking about making a new instance of access networks taking the best of both approaches...

REFERENCES

[10Gigabit01] 10 Gigabit Ethernet Alliance, '10 Gigabit Ethernet Overview', White Paper; May 2001.

[Chand 99] Naresh Chand, Peter D. Magill, Swami V. Swaminathan, Thomas H. Daugherty, 'Delivery of Digital Video and Other Multimedia Services (>1 Gb/s Bandwidth) in Passband above the 155 Mb/s Baseband Services on a FTTx full Service Access Network', Journal of Lightwave Technology, Vol. 17, No. 12, December 1999.

[Congdon] Herb Congdon, Allen Dixon and Martyn Easton, Corning Cable Systems, Hickory, NC, 'Optical Fiber Cabling for Gigabit Ethernet Applications'.

[Effenberg 01] Frank J. Effenberg et al., 'Advances in Broadband Passive Optical Networking Technologies', IEEE Communications Magazine; December 2001.

[eluminant] http://www.eluminant.com/atm3_txt.htm, 'Passive Optical Networks (PONs) Tutorial'.

[Gigabit 98] Gigabit Ethernet Alliance; 'Gigabit Ethernet: accelerating the standard for speed'; Whitepaper 1998.

[Kamino] John Kamino, RCDD, Lucent technologies, 'Gigabit Ethernet for High-Speed Networking'.

[Pesavento01] Gerry Pesavento; Alloptic, Inc., 'Ethernet Passive Optical Networks EPON', 2001.

[Ryan] Jerry Ryan (Editor); 'Building 10 Gigabit/DWDM Metro Area Networks'; The technology guide series.

[swt] 'http://www.swt.iao.fhg.de/eurorim/guidelines/siig2/architectures21.htm'.

Chapter 4

PROJECTS AND FIELD TRIALS

INTRODUCTION

Several projects and papers about optical access networks have been analysed. At the end of the chapter a summary table about projects and field trials are presented to have a fast view about their features (companies involved in the trial, split, uni or bi-directional system, scalability, cost . . .). There are about 30 deployments of PON technology worldwide and several companies are actively working to look for the best solution.

4.1. EUROPEAN UNION

4.1.1. CORDIS

4.1.1.1. *PLANET (Photonic Local Access Netwoks). ACTS-. 2000*

The SuperPon architecture supports a long range, a high splitting factor, and a large bandwidth capacity. The realized system parameters are a total splitting factor of 2048 and a span of 100 km. The span consists of a maximum feeder length of 90 km and a drop section of 10 km. The transport system supported in the SuperPON is based on asynchronous transfer mode (ATM) cells. A downstream bit rate of 2,5 Gb/s is distributed to the ONUs using time-division multiplexing (TDM). A time-division multiple access (TDMA) protocol is used to share the 311 MB/s upstream bit rate. In order to assure

J. Prat et al., Fiber-to-the-Home Technologies, 149–188.

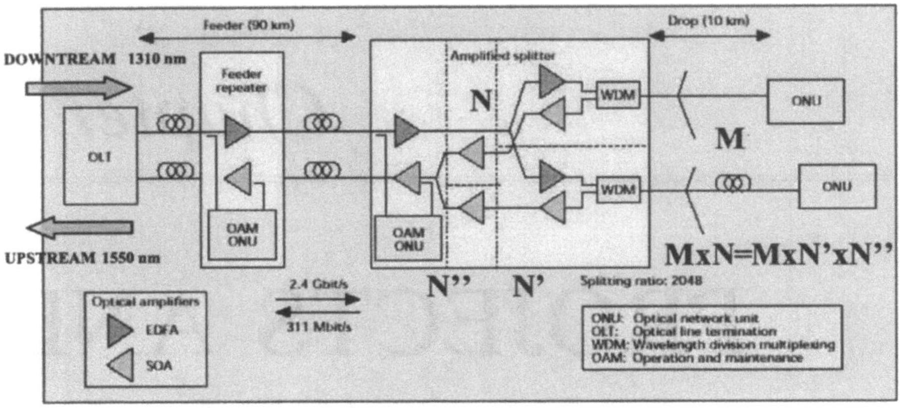

Figure 4.1.1. SuperPON architecture **[VandeVoorde 00]**.

efficient upstream transmission with minimum overhead in the upstream transmission direction, all ONUs are synchronized in time. The overall time synchronization of the ONUs, called ranging, is realized during system initialisation and its stability is constantly verified during operation.

Once all the ONUs are synchronized, the OLT distributes permits in the downstream propagation direction toward the various ONUs to prevent collision on the upstream data path. The procedure to synchronize the ONUs is similar to that used for a regular APON. The major difference for the SuperPON is the additional need to synchronize the optical amplifiers. In the upstream direction, two types of cells are supported, namely, data cells that transport ATM cells, and mini-slots that can transport data of four bytes long. The mini-slots can, for example, be used for the efficient polling of the ONUs.

The optical amplifiers are housed in optical repeaters units (ORUs). They are located in the feeder section (feeder repeater) and in the intersection between the feeder and the drop sections (amplified splitter) to compensate for the fiber and splitting ratio losses. An operation, administration and maintenance ONU (OAM-ONU) is located in each repeater location for burst-mode control of the ORUs and for operation and maintance tasks.

The capacity of the currently developed SuperPON might not be sufficient in a longer roll-out term of the system. End users might need higher capacity. Large business users might even require a direct SDH link or an all-optical link to the termination (OLT) and ATM cross-connect (ACC) equipment. These are the majors drivers to propose the WDM upgrade of the SuperPON. Figure 4.1.2 depicts the generic WDM upgrade scenario of the SuperPON with four major building blocks: the access node with the OLTs, the all-optical node, the ORUs, and the ONUs. At the all-optical node, the

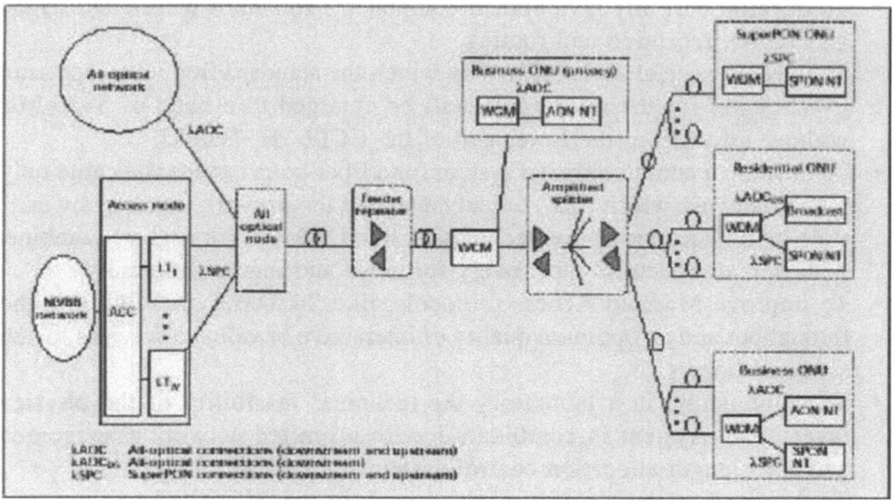

Figure 4.1.2. Scalability of SuperPON architecture [**VandeVoorde 00**].

access network interfaces with the all-optical network. Each OLT emits at a specific wavelength and addresses a separate group of ONUs. These "SuperPON wavelengths" are combined at the all-optical core network and travel along the upgraded SuperPON in order to address the different ONUs.

The ultimate goal of the evolution of the optical access network is to provide a generic optical access line to a variety of different classes and categories of customers and offer both dimensions of time and wavelength in order to support a multitude of bit rate and service requirements as shown in Figure 4.1.2.

4.1.1.2. *TOBASCO (Towards Broadband Access Systems for CATV Optical Networks). ACTS-. 1998*

TOBASCO aims to upgrade existing CATV networks with high splitting counts, and with broadband interactive services by applying High-Density Wavelength Division Multiplexing (HDWDM). TOBASCO's main objectives are:

- To develop a strategy to upgrade partly optical CATV networks for provisioning interactive broadband services to subscriber's homes.
- To develop HDWDM techniques to offer 2Mbit/s (ATM-oriented) connection capability via a partly fiber, partly coaxial-cable based CATV network.
- To develop strategies for flexible allocation of wavelengths, yielding an optimum Quality of Service.

- To develop cost-effective optical components for multi-wavelength signal generation, reception and routing.
- To develop special diplex filters by which the standard frequency spectrum (5–25 Mhz) for the return path will be enlarged to a band of 5–30 Mhz without influencing the lower end of the CCIR "B" Band I.
- To develop a communication system for a fiber-coax (or coaxial cable only) CATV-network which takes full advantage of the network topology by using state-of-the-art modulation techniques, like QAM-16 (or higher) combined with Spread Spectrum technology for noise and ingress immunity.
- To improve Medium Access protocols, like S-CDMA, yielding a higher throughput and an optimum quality of interactive broadband services to subscribers' homes.
- To demonstrate in a laboratory the technical feasibility of the physical layer of the system in combination with a limited network management and wavelength allocation control system.
- To demonstrate the viability of the system in a field trial of a commercial cable operator to show compatibility with interactive broadband communication services (ATM-based over the fiber optic cable and Ethernet-based over the coaxial cable) and with CATV services, and to perform user evaluation/acceptance with the commercial cable operator.

The next schematic (Figure 4.1.3) shows the structure of Tobasco architecture:

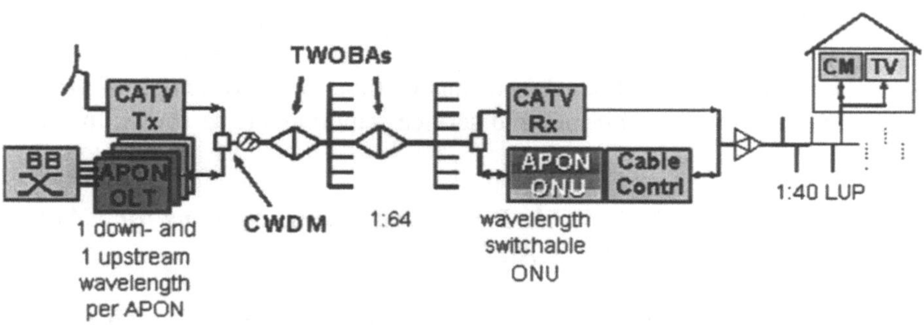

Figure 4.1.3. Structure of Tobasco architecture **[Lucent 98]** and **[Koonen 97]**.

In this architecture, the ONUs are in charge of the wavelength selection. It enables network reconfiguration without having to intrude the outside fiber plant. A different wavelength can be assigned to an ONU by switching it to that wavelength. The network operator is in control of the network configuration which can be easily modified via the network management and control system without affecting the physical topology. Thus, the logical topology can be adjusted to the traffic characteristics making very efficient use of the network.

The wavelength-switched transceiver in the ONU consists of an array of laser diodes and photodiodes, partly integrated with wavelength (de)multiplexing functions. Using arrays instead of tunable devices allows setting up a new wavelength path and the corresponding network control parameters before breaking the old one. This yields to a minimal disruption of services. Furthermore, it offers transmitter and receiver redundancy at the ONU and at the headend. In the fiber part of the network, each wavelength carries 622 Mb/s data streams within the ATM format.

Four wavelengths are used for downstream transmission of the interactive signals, and four for upstream (see Figure 4.1.4). To reduce reflection-induced crosstalk, the up- and downstream wavelengths are interleaved with a spacing of 100 Ghz, and are positioned in the 1535–1541 nm range. The launched power per wavelength is –8 dBm at headend interface A and ONU interface D. The received power at A and D is –16 dBm. The analog CATV signal is positioned in the 1550–1560 nm range, the optimum range for low-noise, and high-power erbium -doped amplifiers.

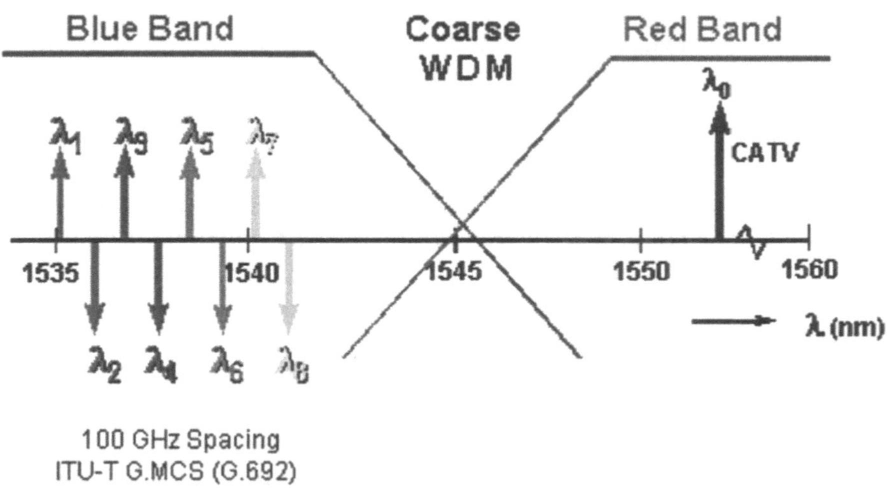

Figure 4.1.4. Wavelength distribution in Tobasco architecture **[Lucent 98]** and **[Koonen 97]**.

The bi-directional amplifier is shown in the Figure 4.1.5.

The TOBASCO project has proposed an upgrade of highly-split fiber-coax CATV networks towards high speed IP/ATM Full Services Access Networks. The proposed upgrade uses High Density Wavelength Division Multiplexing (HDWDM) techniques on the fiber and advanced CDMA/QAM techniques

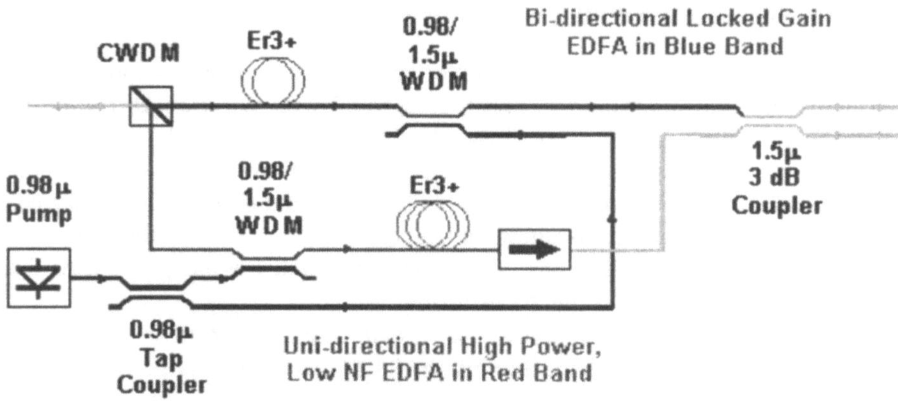

Figure 4.1.5. Configuration of bi-directional amplifier **[Lucent 98]** and **[Koonen 97]**.

on the coaxial cable part of the network. The use of wavelength switchable optical network units enables the access network to dynamically adapt itself to changing traffic conditions. The TOBASCO system will give subscribers peak access bandwidths of 32 Mbit/s downstream and 8 Mbit/s upstream over their coaxial cable network connection using a cable modem. The average bandwidth per subscriber given 30% take-rate is approximately 2 and 0,5 Mbit/s downstream and upstream, respectively.

4.1.1.3. *HARMONICS (hybrid Access Reconfigurable Multi-wave length Optical Networks for IP-based Communication Systems). IST-. 2001*

The HARMONICS system is an IP-based network composed of two main parts:

- *The Last-Mile Networks.* Different fiber-to-the-cabinet (FTTCab) and fiber-to-the-curb (FTTC) scenarios are studied where the final access is provided by a variety of technologies such as WDSL, UMTS and HFC.
- *Optical Feeder Network.* A WDM PON provides the connection between thousands of users and the backbone. The reference architecture studied in the project consists of up to a thousand ONUs. 16 ONUs share a wavelength channel, while 8 channels are multiplexed on a fiber by using passive arrayed waveguide gratings (AWGs). A total number of 8 fibers are connected to an optical cross connect (OXC) by protection rings. The OXC performs the fast optical packet switching of packets to a number of receivers at the optical line termination (OLT). By doing so, the network is capable of redirecting the capacity to different locations depending on the load. A novel *wavelength-and-time medium access control* protocol is

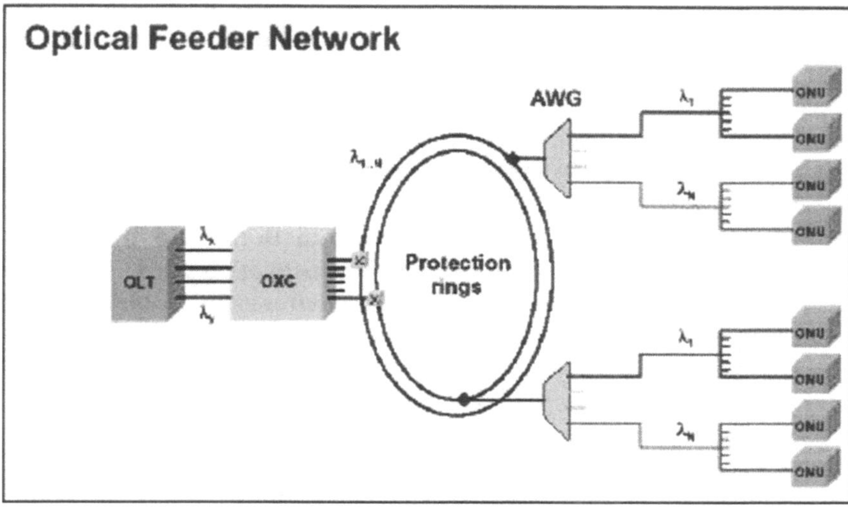

Figure 4.1.6. Structure of Harmonics optical feeder network **[Wellen 01]**.

designed to control the access of the ONUs of the different wavelength channels.

The optical network unit design is also important in HARMONICS. In this project, the use of reflective modulators is investigated. Continuous light from the central office is sent downstream and, after modulation with the data signal, reflected back by the ONU (see Figure 4.1.7). This way, the ONU remains wavelength independent and can be installed anywhere in the

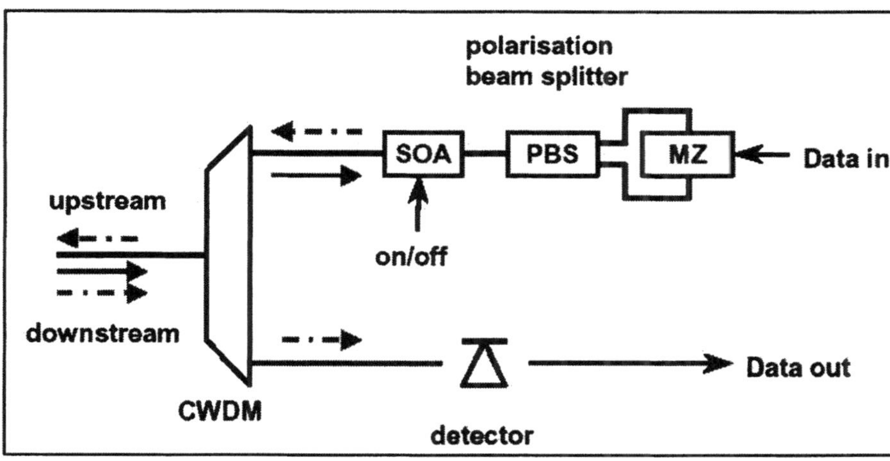

Figure 4.1.7. ONU design. Transmission and reception **[Wellen 01]**.

network. As an alternative to this, the use of tuneable lasers can be considered although up to now they have a limited output power. For proper burst-mode operation, the ONU transmitter must be switched off fully when no packets are sent to avoid excessive noise levels upstream.

At the end of the project, a field trial is planned in Berlin. T-Nova will facilitate a trial where the HARMONICS feeder network will be connected to Ethernet in-house LAN and VDSL access technologies. Friendly users will have access to the Internet, Video Services in Berlin and IP services in Darmstadt that are connected by a WDM core network testbed. The network performance will be testbed in a near-real environment on both the delivered services and the network management.

4.1.1.4. *SONATA (Switchless Optical Network for Advanced Transport Architecture)*

The customers are attached to wavelength-agile PONs that is directly connected to a single passive wavelength-routing node. Any customer wishing to communicate with another customer has to simply tune the transmitter and receiver, in the allocated time slots, to a known wavelength carrying multiplexed traffic through the PWRN between the pair of PONs to which the transmitting and receiving customers are attached. Additional capacity is provided by the wavelength conversion devices around the PWRN as traffic demand requires. These devices are also able to render the PWRN completely non-blocking, for any connection permutation.

The optimum technical solutions for a "switchless" network depend on the size of the network itself. A small scale network for 5000 terminals could be realized using direct detection receivers with optical filters and simple amplified PONs, while to cope with a national scale network for 20 million customers, more complex solutions (like coherent detection and amplifier gating in large-split "hyperPONs", should be adopted.

A small node scale could support many thousands of customers with guaranteed minimum continuous rates of 25 Mbit/s (i.e. regardless of traffic destinations). A consolidated node scale is capable of supporting hundreds of thousands of customers with up to 622 Mbit/s continuously guaranteed. A national scale network can be achieved by the use of amplifier gating in the upstream PONs to enable multiple stages of amplified splitting to be coupled into large "hyperPONs" with around 50000 way-splits. This could provide guaranteed minimum continous customer rates of 2,5 Mbit/s (regardless traffic destinations), and maximum network throughput of 200 Tbi/s (with uniform traffic destinations).

National scale switchless network features:

- 20 millions terminals (each terminal has 1 wavelength agile transmitter and 1 wavelength agile receiver);
- the maximum distance between two terminals is about 1000 km;
- each terminal is able to operate in burst mode at bit-rate of 622 Mbi/s on the optical line;
- the 20 million terminals are divided into 400 different groups;
- the 50.000 million terminals of the same group are connected to the central routing node by means of the same 2-fiber hyperPON infrastructure;
- 400 (input and output) ports of the central routing node are connected to the 400 hyperPONs;
- 400 (input and ouput) auxiliary ports of the central routing node are connected to 400 wavelength converter array in a loop-back configuration;
- each wavelength converter array includes 400 wavelength converter modules (160.000 in total).

Each terminal has 400 wavelengths to address directly all the different hyperPONs of the network and 400 auxiliary wavelengths to reach the destination terminal by means of wavelength in the case of bandwidth saturation of the wavelength directly connecting the two involved terminal hyperPONs. As in each hyperPON there are 400 + 400 wavelengths available to address all the other terminals of the network. The average available bandwidth per terminal is 800 × 622/50.000 Mbit/s = 10 Mbit/s.

Upstream traffic

Upstream traffic in each PON has to be organized according to destinations of the cells from transmitters not only of the same PON but of all the PONs of the network.

The easiest way to implement such a centralized control is to use a high power computer (network controller), co-located with the router allowing it to communicate with all the terminals by means of the same network structure used by the terminals to communicate among them. At the architectural level, two main configurations of Figure 4.1.8 can be implemented. In the first configuration, some wavelengths and PWRN ports are used to connect the network controller to the terminals: the extra complexity is minimum, while the cells exchanged for the control have to use the same cell slots used for the payload traffic. In the second, a dedicated wavelength is used between the terminals and the controller, and is added and extracted at the extremes of the fiber infrastructure: the extra complexity is high (a second non-tunable transceiver in each terminal is required), but the traffic for control is completely independent from the payload one, and no resource of this traffic has to be used.

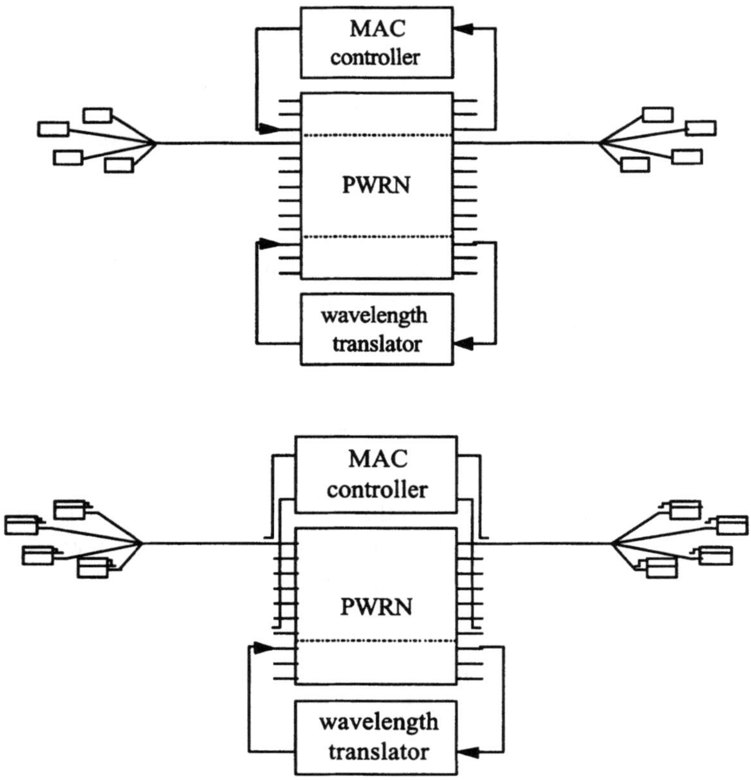

Figure 4.1.8. Reference architectures for the MAC control: (a) by means of dedicated PWRN ports; (b) with splitters and combiners.

4.1.1.5. *BONAPARTE (Broadband Optical Network using ATM PON Access facilities in Realistic Telecommunication Environments). ACTS-. 1998*

The main goal of the project was to enhance the access system prototype inherited from the RACE II project BAF and make it suitable for the realistic scenarios BONAPARTE. The objective was the deployment of the BONAPARTE ATM PON system (Network element, NE) as an Access Network (AN) in a real-life communications environment in which the system can be managed by an ATM Network Manager as part of the overall ATM Network. To accomplish the project, defined management architecture compliant with OUT-T recommendation M.3010.

The ATM PON system had to be deployed in the field connecting real users working in the section of teleteaching and telemedicine.

4.1.1.6. *BOURBON: Broadband Urban Rural Based Open Networks-. ACTS. 1999*

The project objectives was supported by the following Technology oriented objectives:

- Define and demonstrate an augmented range of integrated, scaleable telematic services based on real needs identified in the user requirements research.
- Integrate and demonstrate different access methods within the BOURBON testbed sites for SME (small and Medium Enterprise) users so that they can use telematic services and applications in the most efficient and cost effective manner.
- Provide network gateway functions between the core ATM network and other networks to demonstrate the interworking of ATM with other heterogeneous networks.
- Study network protocol issues in terms of application availability, QoS, throughput and cost effectiveness. Examine the technical feasibility of providing ATM access to SME's utilising ATM over HDSL or other transmission techniques over copper subscriber lines.

The Figure 4.1.9 depicts the proposed field trials to develop in the Bourbon project:

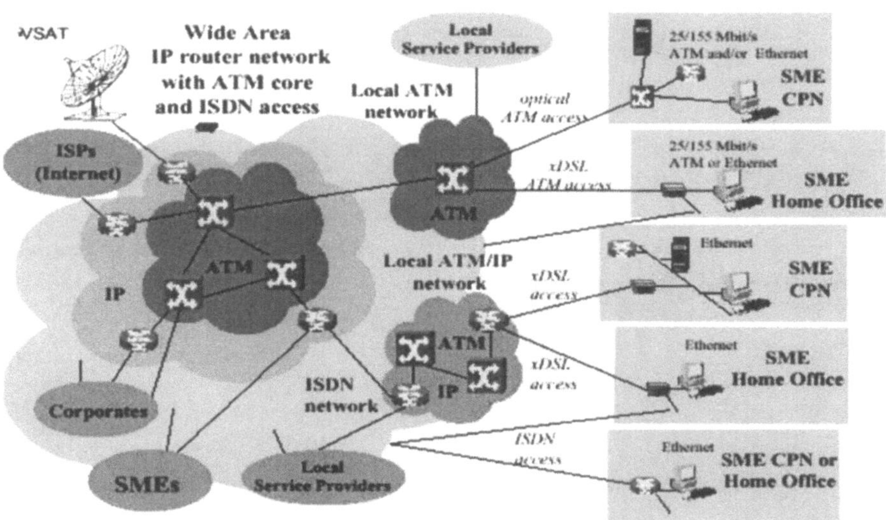

Figure 4.1.9. Proposed field trials to develop in the Bourbon project **[Engell 97]**.

4.1.1.7. *BBL: Broadbandloop Project*

The objective of the BROADBANDLOOP project is to define and test in field trials a concept for a cost-effective broadband access network which will allow fiber to migrate gracefully into the local loop as bandwidth demand increases. The project will demonstrate a graceful upgrade of PONs to high bandwidth and use high speed VDSL modems to upgrade coax and copper twisted pair networks, and to transport broadband services over the last drop into the customers' premises. The target customer group is small to medium size business enterprises and residential customers.

This system is based on a single fiber PON architecture with a splitting ratio of up to 16. The bandwidth capacity of the PON is 155 Mb/s bidirectionally in the basic version which can be gracefully extended to provide 1 Gb/s in the downstream direction and 576 Mb/s in the upstream direction. The PON consists of the Optical Line Termination (OLT), a single fiber optical distribution network, and up to 16 optical network units per link. Multiple links are connected to a single subrack system. Customers are connected directly to the ONUs or use a VDSL modem in the final drop, depending on the application.

The BBL PON integrates transport of synchronous and asynchronous traffic in the access network. The switched network continues to be a separate narrowband and asynchronous transfer mode (ATM) based broadband infrastructure. Separation and grooming of the synchronous and asynchronous traffic is performed at the OLT at the access node. In the simplest form, the OLT is

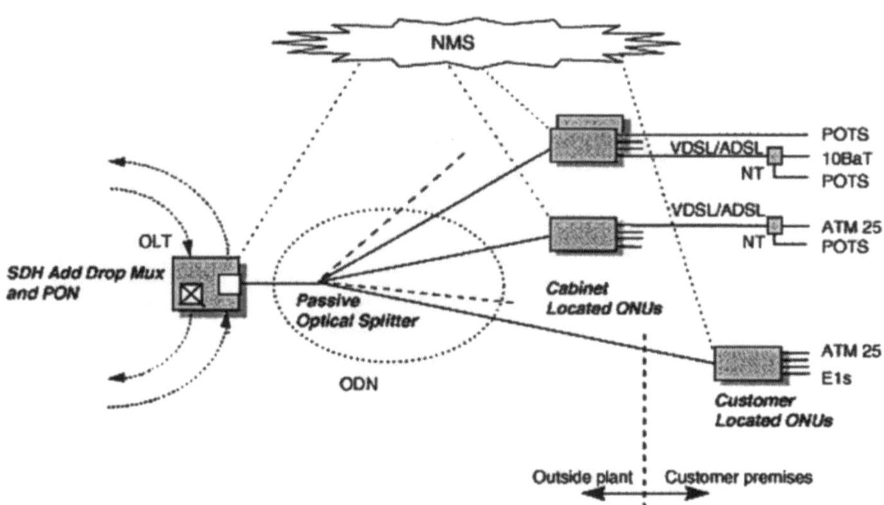

Figure 4.1.10. Detailed BBL architecture system **[Engell 97]**.

a single module forming an integrated part of the SDH network. Standard SDH virtual containers with 2 Mb/s payload (VC-12) are dropped through the PON to the ONU. The ONU provides ITU compliant with G-703/G-704 connections to a primary rate multiplexer with narrowband service interfaces such as POTS or data line cards. When cell-based broadband services are required, an ATM crossconnect module is installed in the OLT and ATM services modules are installed in the ONUs.

In BBL, subcarrier multiple access (SCMA) technology is chosen as the method to transmit from individual ONUs to OLT (multipoint-to-point) direction. With SCMA one or more electrical subcarriers are assigned to each ONU. In the OLT, the traffic from the individual ONUs is selected via electrical filters. Additionally, SCMA eliminates the ranging issues commonly found in TDMA based PON systems. This issue is particularly important when protected redundant routing must be provided. SCMA allows for fast resynchronization in case of switchover to an alternative route. The basic optical system provides a total of 9.72 Mb/s of upstream capacity on a narrowband subcarrier for each ONU. In addition, a subcarrier-based bandwidth upgrade can be demonstrated providing extra upstream capacity of 51 Mb/s on a broadband subcarrier per ONU.

The BBL PON benefits from single-fiber working. This is fast becoming a network operator preference, since it saves on optical fiber outlay and hence simplifies fiber handling. Loop lengths below 20 km are considered, and optical amplifiers will not be used in the system. Optical wavelength-division multiplexing (WDM) provides a clear separation of downstream and upstream transmissions which are carried on 1550 and 1310 nm, repectively. This enables the reuse of electrical frequency bands in both transmission directions. The required laser modulation bandwidth is therefore reduced which enables low-cost laser technology to be used.

In BBL the key optical component is an integrated optical transceiver containing a Fabry Perot laser diode, a PIN photodiode, and an optical WDM to

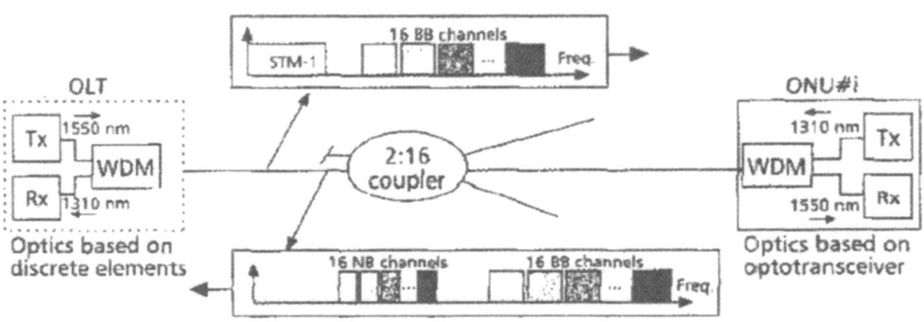

Figure 4.1.11. Detailed BBL PON distribution channels **[Engell 97]**.

separate 1550 and 1310 nm channels. Furthemore, an additional optical filter is incorporated to keep crosstalk as low as –50 dB. The ONU optotransceiver is a commercially available component developed for digital applications, enabling the BBL system to benefit from its low price.

In downstream direction, every ONU accesses on STM-1 baseband signal on a TDM basis and may have access to a broadband subcarrier channel. In the upstream direction, a narrowband subcarrier is allocated to each ONU.

4.1.1.8. *Others projects.*

The following projects were developed in the middle of the nineties:

4.1.1.8.1. *FIRST: Fiber to Subscriber Terminal-. RACE 1994*

Project R2014 FIRST deals with fiber to the residential subscriber terminal based on the developments in first-phase project R1030. The main characteristics of R2014 are:

- Tree and branch PON topology with a splitting ratio of 32 giving access for 32 customers.
- Use of two fibers, one for interactive services and one for CATV distribution.
- The ability to provide each customer with 2048 Mbits/s access using the V5 interface. The downstream transmission uses time-division multiple access (TDMA) and the upstream transmission uses subcarrier multiplexing (SCM). Investigations have shown that this combined solution offers significant advantages over the pure solutions.
- The system is designed for analog TV distribution of 40 AM-VSB channels, but will also demonstrate that analog and digital TV-signals using MPEG-1 can co-exist.
- Addresses fault, configuration, performance, and security management.

4.1.1.8.2. *BAF: Broadband Access Facilities-. RACE 1994*

This is the project R2024, and it deals with interactive broadband access facilities. The BAF system is an ATM PON (it is based on a Passive Optical Network infrastructure and the transfer mode is ATM). The main characteristics of R2024 are:

- Tree and branch PON topology with a splitting ratio of 32, allowing 32 network terminating units (B-Nus) to be connected.
- Up and downstream transmission at 622 Mb/s. Upstream transmission uses TDMA; downstream transmission uses TDM.
- The TDMA character of upstream transmission requires a ranging procedure to obtain approximately equal B-LT to B-NU/NT distances for every B-NU/NT through an adjustable electronic delay timer.

- The 1300 nm wavelength window is used, while the 1550 nm window is left available to other services (e.g. CATV).
- Maximum fiber length between OLT and ONU is 10 km.
- Routing of ATM cells, Usage Parameter Control and cell header translation are based on the Virtual Path identifier.

In R2024, collisions on the upstream combining points are avoided. The medium access control (MAC) determines for each upstream time slot which T-interface is allowed to transmit a cell. A permit for this purpose is sent in the preamble preceding each ATM cell sent in the downstream direction.

4.1.1.8.3. *COMFORT: Components for Fiber to the Home-. RACE 1994* Project R2062 deals with distributive services. The TV transmission format used is analog SCM-VSB for compatibility with existing TV sets.

The R2062 demonstrator provides these distributive services:

- More than 40 TV channels;
- FM radio broadcast in the band 88 to 108 Mhz;
- Digital radio in the band 111 to 147 Mhz.

COMFORT MATCHES the first splitting factor of 32 through a 1:2 coupler between the expensive low chirp (< 50 Mhz/mA) DFB laser with high linearity and EDFAs with + 15 dBm output power, followed by 2 splitting stages (1:2 and 2:8). The 4:4 splitter in the interactive part is used to realize a redundant architecture.

4.1.2. EURESCOM

4.1.2.1. *BOBAN (Building and Operating Broadband Access Network)*

Boban study confirms that large business customers are typically served today by optical fiber access rings using STM-1 (155 Mbit/s) connections. However, the overall penetration of fiber in the access remains low, and most of the business customers are still served via copper.

Looking at the residential customer market, it is found that digital TV and video have not taken off yet, and high speed Internet access at present is limited by the core network. Therefore providing more bandwidth in the access would not result in to a better service experience of the customers. However, the situation concerning the backbone is rapidly changing. A number of large-scale projects are underway to increase the bandwidth available. Clearly the next step is to provide more capacity in the access, especially, because demand forecasts predict an exponential growth of the traffic. Such a growth in the demand cannot be met by any other means than deploying DWDM in the metro networks.

A range of application fields of DWDM in the access is defined based on real or foreseen needs. It is important to note that some of these applications could be implemented in some forms either today using available DWDM technology or in the very near future (probably in 2001) taking advantage of the emerging more flexible network elements such as configurable OADMs.

From the operator point of view, DWDM technology is an option to introduce flexibility and add new features to their networks that represent value to their customers.

DWDM provides the following advantages:

- Simultaneous access to several service optimised networks from one or different operators.
- Virtual private networks on shared fibers through wavelength dedicated optical ADM with enhanced security.
- Cost efficient upgrade of PON systems.
- Increased network flexibility that allows rapid provisioning of bandwidth to satisfy unexpected demands and dynamic configuration and assignment of QoS.
- Higher fiber plant efficiency through fiber sharing, allowing the mixing of different technologies in the same area (different protocols and transmission formats, analogue and digital with different transmission codes).

Technology maturity is another key factor of deployment. For example, available transponders emit at pre-selected wavelengths and only fixed OADMs are available that cannot be dynamically configured. Moreover, optical cross-connects that would enable the implementation of all-optical fully configurable networks are still in the development stage.

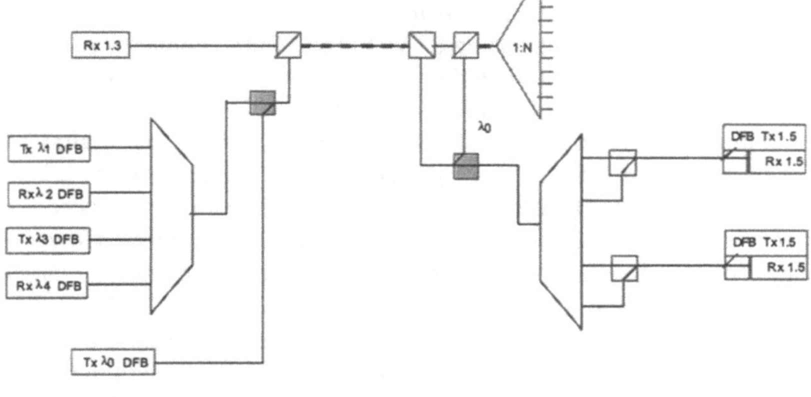

Figure 4.1.12. Upstream and downstream channels using AWGs in a single fiber PON.

Ultimately, cost will determine the use of DWDM in the access network. Here quick and inexpensive provision of services is essential. The responses to the RFI circulated gave very little information regarding hardware/system costs. One aspect, however that should be taken into account is that almost all vendors offer "metro-oriented" WDM solutions that should allow the development of WDM access network at a lower cost. The investment study revealed that today the deployment of DWDM technology in the access cannot be justified purely on a cost basis.

Commercially available DWDM access/metro systems are fully compatible with the legacy SDH systems. Furthermore, in the framework of a laboratory trial the feasibility of the PON upgrade using commercially available DWDM components has been demonstrated.

4.2. OTHERS

4.2.1. FiberVista Architecture-. 1999

FiberVista is an FTTH system that delivers Internet access via cable modems, while providing consumers with the broadcast video signals they have a demonstrated willingness-to-pay for.

FiberVista differs from de FSAN system in two important ways:

1. High-speed data services, such as Internet access, can be added in small increments as customer demand grows.
2. This system can deliver broadcast analog and digital video which is the only broad-band service to which majority of consumers currently subscribe to.

Figures 4.2.1 and 4.2.2 depict the optical power in each point of the system and the FiberVista architecture.

The modulator output is amplified by a first EDFA, split 4 ways, and then transmitted through 13 km of standard single-mode fiber. A second EDFA, at a hub location, reamplifies the downstream signal. The EDFA's have noisier figures of 6 dB. The output of the second EDFA is split into 64 ways, and then sent through drop fibers (which ranged from 1 m to 1 km in length) to ONU's.

The cable modems transmit upstream data using a single QPSK-modulated RF carrier at 35 Mhz with 1.28 Msy/s for an upstream line rate of 2.56 Mb/s. The various cable modems share the upstream RF channel using a time-division multiple access (TDMA) protocol controlled by the CMTS. The upstream RF signal goes through the same coax network as the downstream signal, but it is separated out by the diplexer in the ONU and applied to an uncooled 1,3 μm Fabry-Perot laser after passing through a bias control

	Downstream Power (1,5 μm)
Out of DFB	+16,3 dBm
Into First EDFA	+9,3 dBm
Out of first EDFA	+17,4 dBm
Into transmission Fiber	+9,4 dBm
Into second Fiber	+6,7 dBm
Out of second EDFA	+18,0 dBm
Into analog receiver	–4,6 dBm
	Upstream Power (1,5 μm)
Out of Fabry-Perot Laser	+2,4 dBm
Into transmission fiber	–20 dBm
Into upstream receiver	–28,1 dBm

Figure 4.2.1. Detailed upstream and downstream channel power [Wood 99].

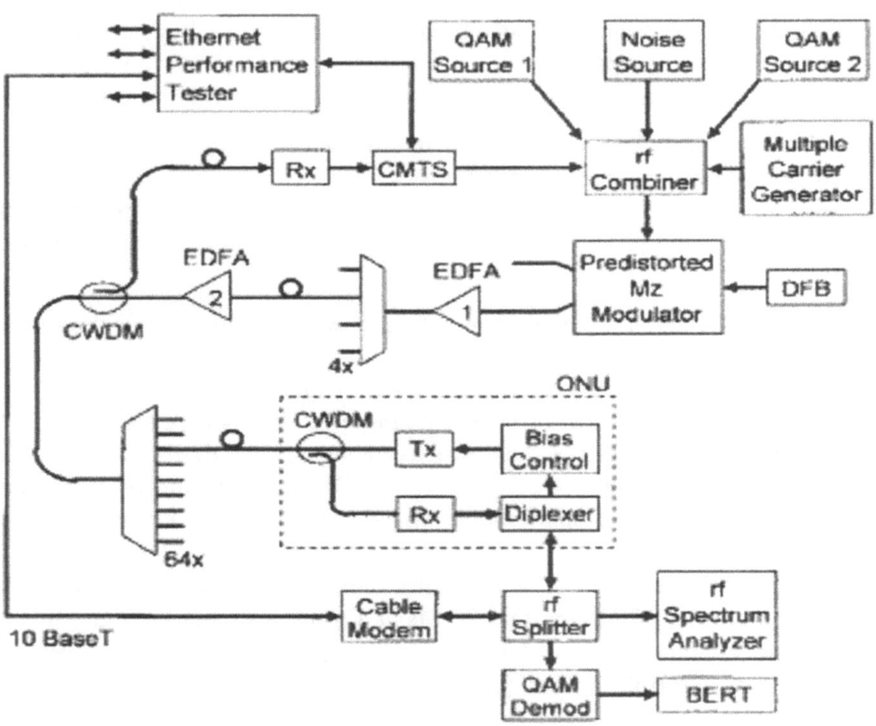

Figure 4.2.2. Fibervista architecture [Wood 99].

circuit. This bias control circuit keeps the laser below threshold until it senses an RF burst from the cable modem. This keeps all but one upstream laser off which greatly reduces the relative intensity noise (RIN), shot-noise, and optical beat interference (OBI) at the upstream receiver. The light from the upstream laser is put on the same fiber network used for downstream transmission by a coarse wavelength-division multiplexing (CWDM) coupler. After being combined with light from the other ONU's, this light is separated off by another CWDM coupler, routed through a 10-km fiber, and sent to the upstream receiver. The output of this receiver is applied to the CMTS.

Figure 4.2.1 above shows the optical power levels through the system. Using EDFA's, a reasonable optical power can be delivered to the ONU's even with the large splitting ratios.

4.2.2. RINGO architecture-. 2001

The Ringo project tries to put the complexity of the implementation of a network protocol mostly at the electronic level, but at the same time aims at achieving an efficient use of the optical bandwidth.

4.2.2.1. *RINGO architecture and protocol*

The general architecture of the RINGO network is shown in Figure 4.2.3. RINGO is based on an unidirectional fiber ring with N nodes that implement the interface between the electronic and optical domains. Each node queues at the electronic level fixed-size packets to be sent on the ring. The main features of the architecture are:

- The number of nodes N in the network is equal to the number N of used wavelengths (or "channels"); extensions to number of nodes larger than the number of wavelengths are possible.
- Each node optical interface is equipped with a tunable transmitter, and with a fixed receiver operating on a given wavelength λ_i that identifies the node itself; in order to communicate to node k, a node must tune its transmitter to send a packet on λ_k. Moreover, each node must remove the information carried by its wavelength λ_i from the ring using a fixed passive optical filter. If required, some particular nodes may receive on more than one wavelength by simply dropping them and replicating the receiver subsystem.
- Each node is able to check the status of the wavelength occupation (λ-monitoring) on a slot-by-slot basis, and is thus able to avoid collisions by a generalization of the empty-slot approach which allows to transmit packets only in those slots that are actually available.

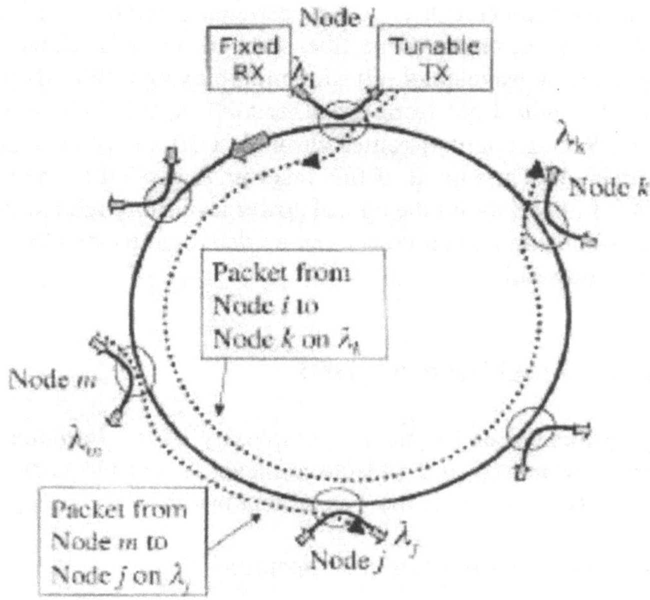

Figure 4.2.3. RINGO architecture **[Gaudino 01]**.

- The packets to be transmitted are queued using a Virtual Output Queuing (VOQ) algorithm, i.e., by storing packets waiting for ring access in separated queues, each corresponding to a different destination or to a different set of destinations.
- A packet can be simultaneously sent to multiple destinations, thus efficiently implementing multicasting by turning on in the same slot more than one wavelength of the transmitter laser array.
- Packet transmission is time-slotted and synchronized on all channels, thus a global packet synchronization is required.
- Packets have a fixed length corresponding to one time slot; the packet format adaptation, including segmentation and reassembly is left to higher (electronic) layers of the node protocol.

The proposed architecture tries to combine in an optimal way optical and electronic technologies: the aggregate bandwidth is handled in the photonic domain by working on each wavelength, while packet queueing, MAC protocol and statistical time multiplexing is handled in the electronic domain.

4.2.2.2. *RINGO node structure*

The main functions that the node must support are:

1. Amplification of the optical signals in order to compensate for the loss of the node passive elements and of the following fiber link. Amplification is obtained by placing an EDFA at the input of the node. Other alternatives are:
   - Placing the amplifier at the node output: this solution would require to launch a high power directly into the fiber, thus potentially exciting fiber nonlinear effects;
   - Placing one amplifier on each of the fibers at the output of the demux. This solution offers several advantages from the transmission performance viewpoint (reduced interaction among wavelengths, Amplified Spontaneous Emission (ASE) filtering and easier gain balancing), but seems to be definitively too expensive.
2. Demultiplexing of the WDM comb after the amplifier. The most promising devices for this purpose are the Arrayed Wave-Guide (AWG) filters.
3. Monitoring the status of wavelengths on each slot. This is done by tapping a fraction of the power on each of the fibers at the output of the demultiplexer and by sending it to a DC-coupled photodiode array. The λ-monitor requires electronics with a much smaller bandwidth than the channel bit-rate. In fact, it should detect the received average power on a packet-by-packet basis, so that the cutoff frequency can be chosen to be much lower than the bit rate, and of the order of the inverse of the packet duration.
4. Burst-mode detection of the incoming data-stream on the wavelength λ_i assigned for the node.
5. Packet generation for the local traffic. This is done on the optical side by a laser array driven by the node controller. The lasers on each wavelengths are turned on by injection current direct modulation when a packet has to be generated. The data bits are then "written" inside the packet by an external modulator. The election of a laser array for the transmitter has two different motivations:
   - To allow efficient multicast. In fact, this transmitter setup exhibits a promising features for multicast transmission since bits are written simultaneously by the external modulator on all wavelengths that are generated by the laser array in a given time slot.
   - To allow using commercial and very stable lasers on the ITU grid.

An alternative solution is based on tunable laser. The tunable laser solution is probably going to be less expensive than a laser array with a high number of channels though the integration of an array on a single chip has been demonstrated and can significantly decrease costs with respect to a discrete array solution.

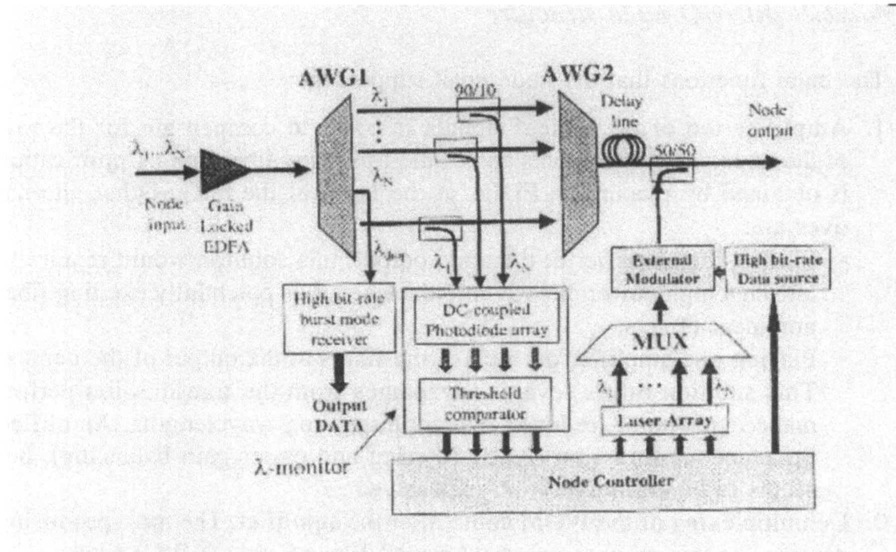

Figure 4.2.4. RINGO node structure **[Gaudino 01]**.

6. Node control. This is done on a programmable logic that implements the VOQ protocol and several other features.

4.2.3. Modified Star-Ring Architecture (MSRA)-. 2001

MSRA consists of a star network on the upper level and many concatenated ring subnets on the lower level and is proposed as subcarrier multiplexed passive optical network (SCM-PON). The upper level star network ensures high network capacity and its weakness in reliability is overcome by the concatenated ring subnets with self-healing capabilities. The self-healing function can be performed at remote nodes (RNs) by using optical switches (OS) to reconfigure the ring subnets in any link failures. Moreover, a CAT (cascade add/drop transceiver) structure is designed in the lower level ring subnets to completely eliminate the optical beat interference (OBI) problem and achieve an OBI-free optical network.

The proposed MSRA is based on the two-level star-ring architecture with special arrangements on the lower level ring subnets (see Figure 4.2.5). Downstream signals originating from the CO are distributed to the remote nodes (RNs) via the upper level star network, and they are further delivered to ONUs through the lower level ring subnets. On the other hand, upstream signals generated at each ONU are transmitted to the RN responsible for the ONU of interest via the lower level ring subnet, and then further delivered

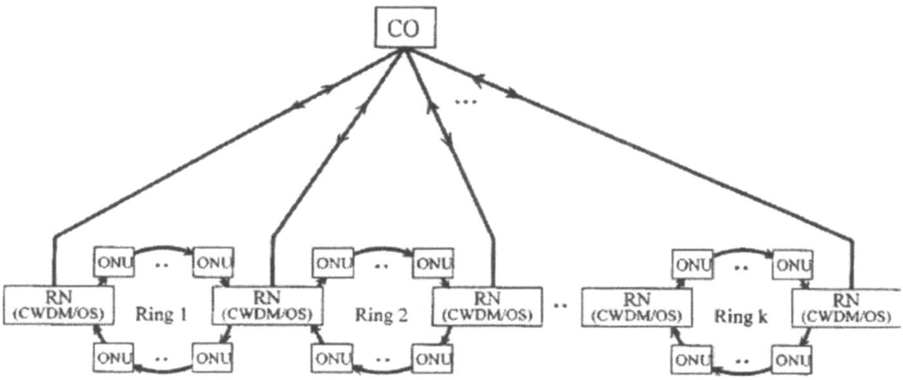

Figure 4.2.5. MSRA Architecture **[WenPiao 01]**.

to the CO using the upper star network. The 1.55-μm band is used for downstream transmission and the 1.3 μm band for upstream transmission.

Every lower level ring subnet is divided into two semirings by RNs with each consisting of several ONUs. The function of RN is to properly transfer signals between the CO and the lower ring subnet in the normal condition, and to perform self-healing function if link failure occurs in the upper level star network.

As shown in Figure 4.2.6, each RN contains two 1.55-μm/1.3 μm coarse wavelength-division multiplexers (CWDMs) and several optical switches (OS). A signalling tone is constantly sent by the CO and detected by the control circuit in the RN. The switch states are normally set as shown in (a), therefore the 1.55-μm downstream signals and the 1.3-μm upstream signals are properly forwarded by the two CWDMs. If the corresponding link in the upper level star network is broken, the signalling tone will be absent and the control circuit will reconfigure the OS functioning as (b) such that the upstream signals can bypass the RN of interest and go to the neighboring RNs. In this scenario, the network is virtually reconfigured, as shown in Figure 4.2.7, such that the affected ONUs will be assisted by the two neighboring RNs. Obviously, the downstream signals can still be transmitted from the CO to the affected ONUs.

In each ONU, there is the CAT structure to overcome the OBI problem. Figure 4.2.8 shows CAT schematic.

The downstream optical signal originated from the CO is first converted into electrical signal by PIN module. Then this signal is forwarded to an add/drop module in which the upstream signal (S_1) is added, and the message destined to the ONU_1 (C_1) is dropped. The output of the add/drop module, which contains the downstream signal from the CO and the upstream signal of the ONU_1, is transmitted to the next ONU by a 1.3 μm laser transmitter. The same operation is performed in the following ONUs in a cascade manner.

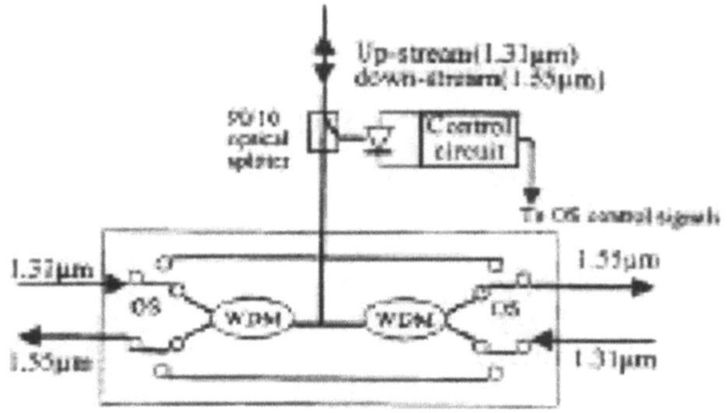

(a)

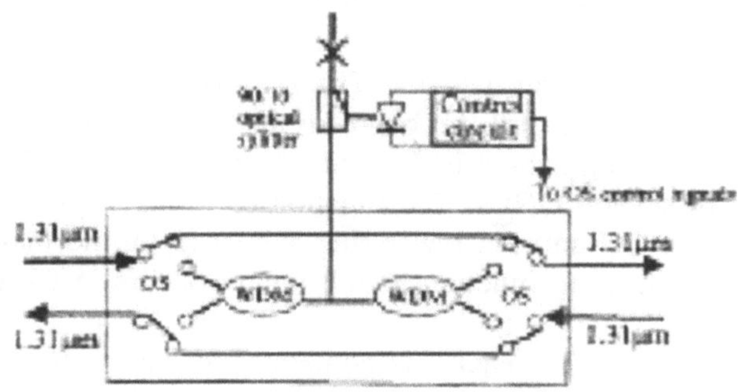

(b)

Figure 4.2.6. RN contains two 1.55-μm/1.3 μm coarse wavelength-division multiplexers (CWDMs) and several optical switches (OS) [WenPiao 01].

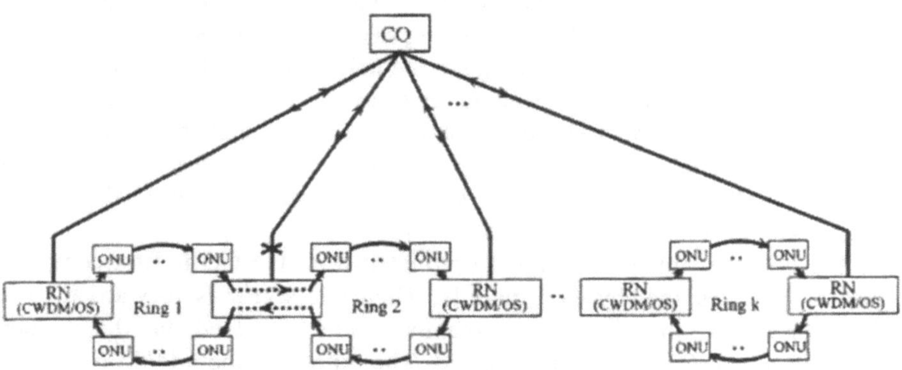

Figure 4.2.7. When a fiber cut occurs, the network is virtually reconfigured [WenPiao 01].

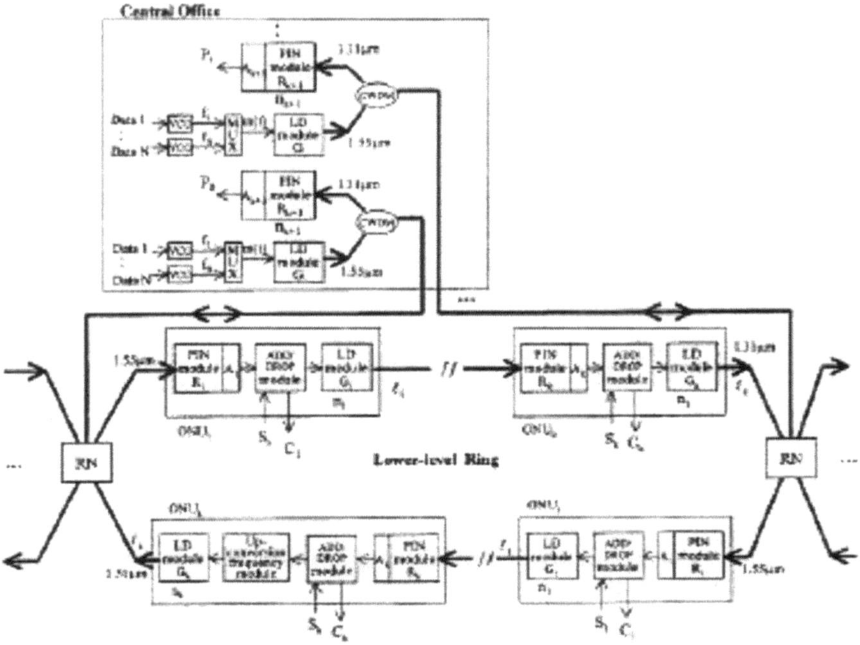

Figure 4.2.8. CAT schematic **[WenPiao 01]**.

Thus, all the downstream signals can be received by the destined ONUs and all the upstream signals can be transferred to the RN, and then to the CO afterwards. The CAT structure indeed brings two major advantages:

1. Since the upstream optical signals come from a sole laser transmitter all the time, there is no OBI problem in the lower ring subnets, and in turn the whole network is OBI free.
2. The downstream/upstream signals are regenerated at each ONU, so that low-power lasers can be used and there is no branch loss at each ONU. This is not the case for other systems in which optical amplifiers are usually needed to compensate for branch losses upstream/downstream signals.

The detail configuration of ONU is shown in Figure 4.2.9.

The electrical signals from the PIN module output include the downstream signal in the high-frequency band and the upstream signals in the low-frequency band. Using a band-pass filter (BPF), a demultiplex (DMUX), and a demodulator (DFSK), the downstream signal (digital data or multi-channel video signals) can be dropped. On the other hand, a local upstream signal is added to the PIN module output before modulating a 1.3-μm laser. The laser output is sent to the next ONU. In each ONU, a pair of OS is used to bypass the node when it fails.

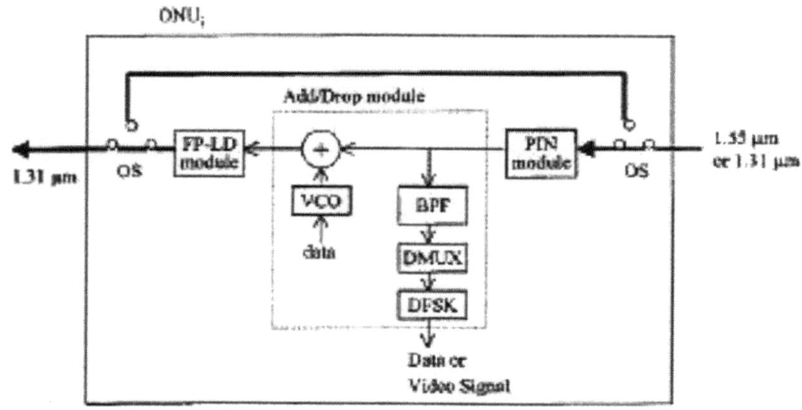

Figure 4.2.9. Detail configuration of ONU **[WenPiao 01]**.

4.2.4. The next generation optical regional access network-. DARPA 2000

In the following section, the protection and node implementation about the advanced architecture from the project Darpa is explained. The general architecture and node structure is shown in the chapter 2.

4.2.4.1. *Dual-fiber ring and optical protection switching*

Figure 4.2.10 illustrates a possible architecture for the feeder network: dual fiber ring architecture.

Each fiber is unidirectional; one carries traffic clockwise (CW) while the other carries traffic counter-clockwise (CCW). Thus, duplex connections on the network consist of two simplex connections. The design has two wavelength bands on the ring, the A band on the CW fiber and the B band on the CCW fiber. In case of a fiber cut, which can be detected by a loss of light in the fiber, traffic in the cut fiber redirected onto the second fiber going in the opposite direction around the ring. In this protection mode, fibers carry both their normal band and the protected band from the opposite fiber. After going around the ring in the protection band, the signal reaches its intented node on the other side of the fiber cut, where the signal is redirected back onto its normal band fiber. In the protection mode the signals are dropped at access nodes only when they pass them in their normal band and not in the protection band; this simplifies the protection control scheme and the access node structure.

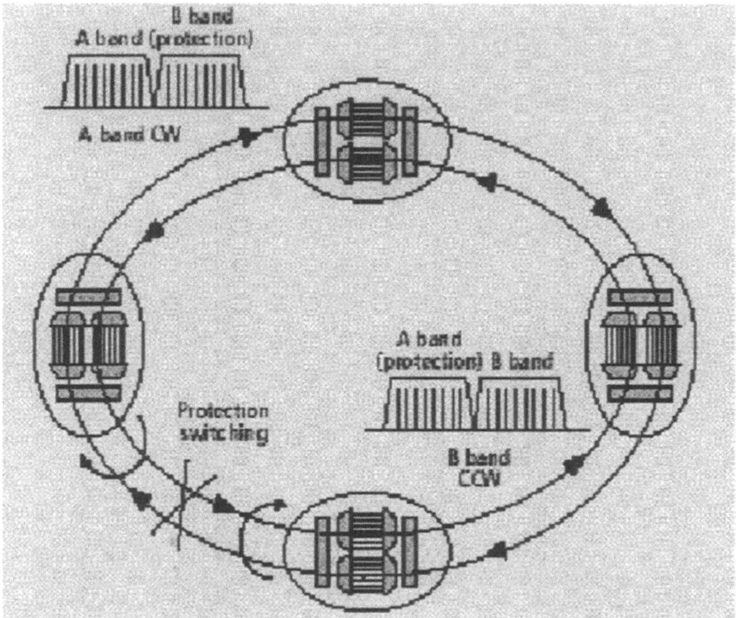

Figure 4.2.10. Dual fiber ring architecture **[Kuznetsov 00]**.

4.2.4.2. *Access node implementation*

In the Figure 4.2.11, a one possible physical implementation of the dual-fiber ring feeder access node is shown (the electronic IP router is shown, wavelength router is similar).

WDM multichannel signals from dropped feeder bands of both CW and CCW fibers are first demultiplexed into individual wavelength channels. The optical 2 × 2 switch for each channel then selects whether the channel is added/dropped from the stream or continues undisturbed back onto the feeder. Wavelength channels destined for the feeder are then remultiplexed with variable attenuators first equilibrating the channel power levels.

On the feeder drop side, the dropped feeder wavelength channels can go directly to the distribution network or be terminated in the IP router with a set of 1 × N optical switches plying the role of a tunable channel selector. The IP router drop output ports connect to the distribution network using the distribution wavelengths. On the feeder add side, signals from the distribution network can be added directly to the feeder if they are on the feeder wavelengths. Signals on the distribution wavelengths (dashed lines) are electronically terminated in the IP router where they are electrically processed and routed through the feeder using the routers output ports tunable over the

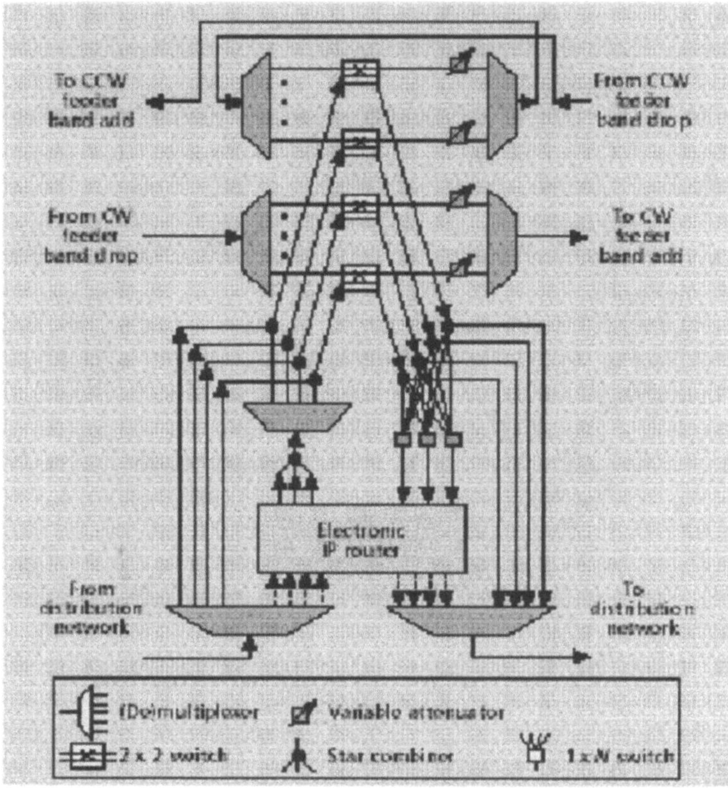

Figure 4.2.11. Possible physical implementation of the dual-fiber ring feeder access node **[Kuznetsov 00]**.

feeder wavelengths. Optical amplifiers can be used in order to compensate for propagation loss between access nodes and the various splitting and excess losses inside the access node.

Several schemes for implementing the access node wavelength router/transponder are shown in Figure 4.2.12.

In all cases, the use of N × M signal switches allows many inputs signals to time share a smaller number of output ports, while wavelength routing is done by routing the output ports' wavelengths. Control signalling is used for signal switching and output tuning in these wavelength routers. In the two router/transponder configurations, the signals undergo optical-electrical-optical conversion with either electrical or optical switching of the signals; output transmitters are wavelength-tunable. It is easier to make optical-electrical-optical transpondering for specific bit rate signals, but bit-rate-independent transponders are also possible. In the all-optical wavelength router, the signals are switched optically; bit-rate-independent tunable all-optical wavelength con-

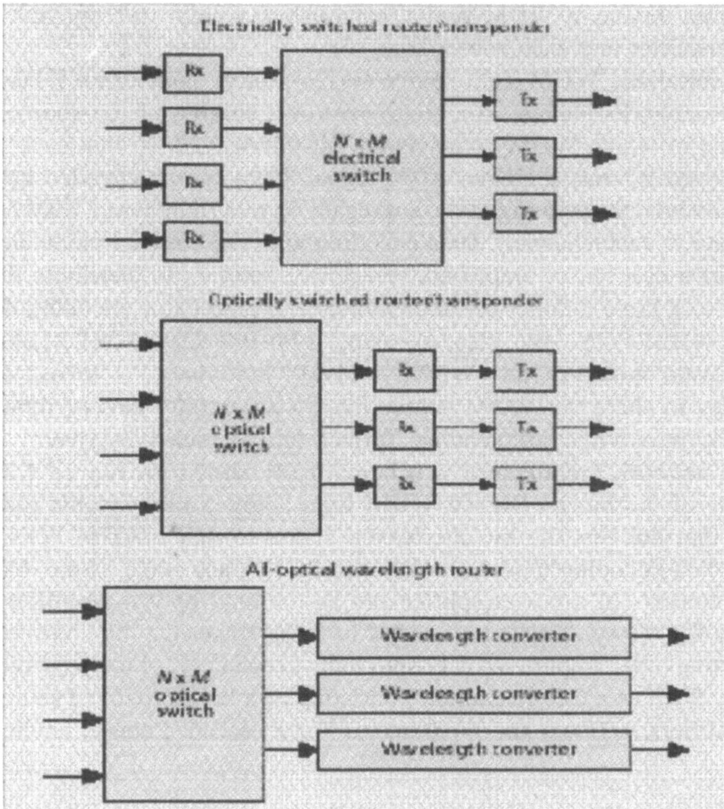

Figure 4.2.12. Several schemes for implementing the access node wavelength router/transponder **[Kuznetsov 00]**.

verters are used in the output ports. However, all-optical wavelength converters are research type devices at present.

4.2.5. Others projects and experimental projects

This chapter explains several ideas about PON architectures.

4.2.5.1. *Cascaded WDM passive optical network with a highly shared source. 1997*

Wavelength Division Multiplexed WDM PON's, in which each transmitted wavelength carries data for a single subscriber, are attractive because the branching device, a WDM router, establishes a point-to-point link between the transmitter and each subscriber.

A centralized WDM light source serves a large number of WDM passive optical networks leading to a significant cost sharing of the source. We use a source in which WDM channels are bit-interleaved, so that each time-slot is occupied by only a single WDM signal. Data can be encoded by a single TDM modulator operating at the aggregate bit rate (number of WDM channels × bit rate in each channel). Data encoding with the external modulator allows to insert a cascade of amplifiers and power splitters to distribute the WDM carrier to a large number of modulators, each modulator encoding data for a single WDM-PON. The idea, by using a bit-interleaved WDM source and an external mixed-mode "WDM + TDM" modulator, is to use amplifier cascades to share the WDM source among potentially tens of thousands of network users while still sending distinct data to each subscriber.

The network architecture is schematically shown in Figure 4.2.13. and consists of the bit-interleaved WDM light source, a cascadable distribution fabric that delivers the bit-interleaved signal to many WDM-PON's, and a single data-encoding modulator for each PON. Each stage of the distribution fabric consists of a power splitter and an optional optical amplifier to compensate for the splitting loss. Several of these stages can be concatenated by connecting other amplifier/power-split units to each port of the preceding power splitter. Such as a CO transmitter consisting of a bit-interleaved source, banks of amplifiers, splitters and modulators. For k cascaded stages having M-fold

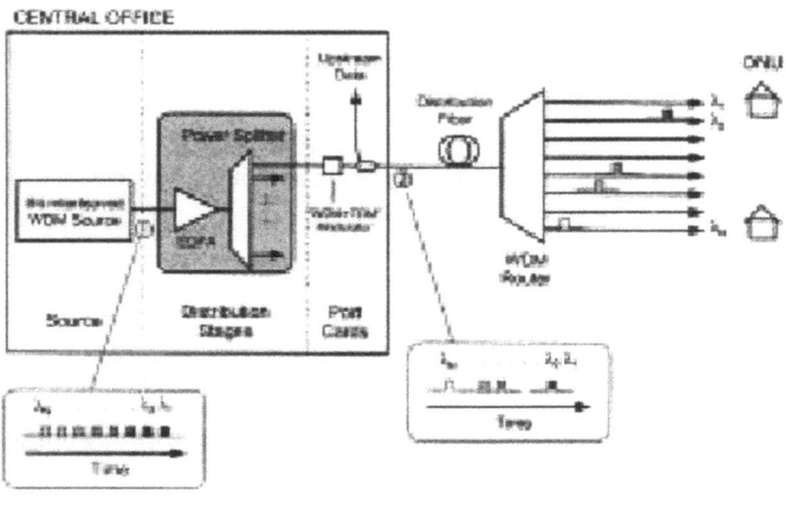

Figure 4.2.13. Network architecture cascaded WDM passive optical network with a highly shared source **[Stark 97]**.

power split and N WDM channels, a single light source serves $N*M^k$ subscribers.

This WDM distribution network addresses only the downstream portion of the network. The upstream portion can be implemented in a number of ways, like for example by TDMA or SCMA using coarse WDM at 1300 nm. A coarse WDM splitter then separates upstream and downstream traffic at the CO.

4.2.5.2. *A WDM access system architecture based on spectral slicing of an amplified led and delay-line multiplexing and encoding of eight wavelength channels for 64 subscribers-. 1997*

In this architecture, the pulsed output from an amplified LED is spectrally sliced by a WDM router and multiplexed with short fiber delay lines into individually addressable wavelength channels that can be modulated with an external modulator.

The proposed WDM access architecture is shown schematically above

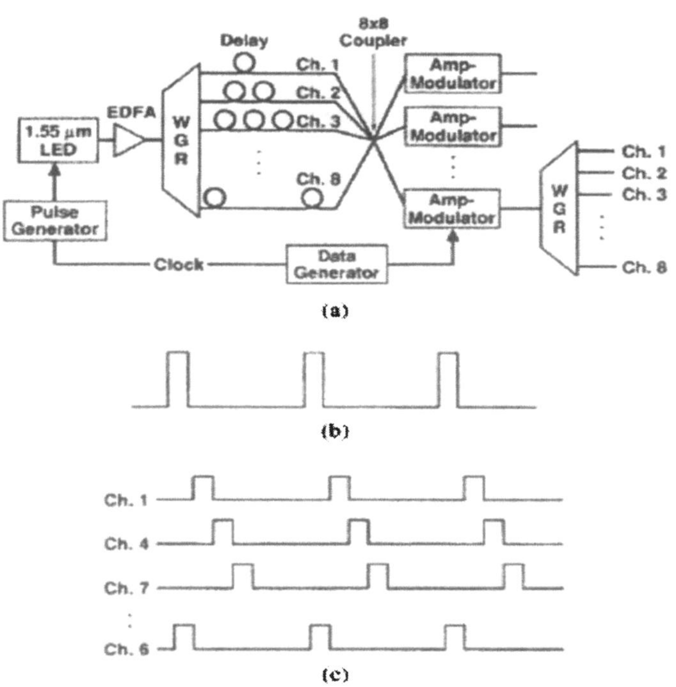

Figure 4.2.14. (a) The WDM PON architecture; (b) The conceptual pulsed output from the broad-spectrum LED; (c) The delayed multiple-wavelength output from the WGR **[Liou 97]**.

(Figure 4.2.14). The output from a 1.5-μm LED is amplified by an EDFA, followed by spectral slicing using an eight-channel silica WGR. The channel spacing of the WGR is 200 Ghz (1,6 nm) and the –3 dB width of the pass-bands is about 120 Ghz. To deliver 50-Mb/s data each to eight channels, the LED is directly pulsed with 2.5-ns width and 20-ns repeating period, as shown conceptually in (b). The inverse of the LED pulse width equals to the channel number times the data rate per channel. Fiber delay lines of different lengths are used at each of the output ports of the WGR to delay one pulse width per channel. The delayed outputs are the multiplexed with a star coupler into a single-mode fiber, resulting in a TDM sequence of WDM channels. Fig (c) shows conceptually the delayed WDM channels, that are reordered in sequence of channel number 1, 4, 7, 2, 5, 8, 3 and 6, before they are combined by the star coupler. The use of this sequence may be desirable to separate the arrival times of adjacent wavelengths channels at the demultiplexing receiver to reduce crosstalk due to spectral overlapping which is most serious for adjacent channels. A WDM multiplexer can also be used, instead of star coupler, for combining the delayed WDM channels. An 8 × 8 star coupler, however, can split the output to multiple PON's without additional loss (an 8 × 8 coupler has the same insertion loss as an 8 × 1 coupler). An external modulator is then used at each of the coupler output ports to modulate the wavelength encoded pulses. The modulator operates at 400 Mb/s (8 × channel bit rate). At a remote node, the transmitted signals are WDM demultiplexed and routed by another WGR to separate subscribers, where the signals are received at 50 Mb/s.

4.2.5.3. *"Rite-Net" Architecture-. 1994*

"Rite-Net" is the acronym for "Remote Interrogation of Terminal Equipment", and the basic implementation is shown below (Figure 4.2.15).

The system's channels are set by a wavelength-division multiplexing router (WDM/R) at the RN. At the CO, a tunable laser, whose modes match those

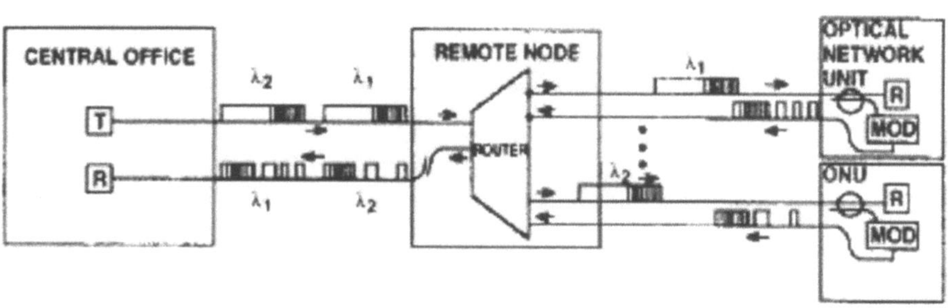

Figure 4.2.15. Rite-Net Architecture **[Frogo 94]**.

of the WDM/R, is a wavelength-stepped through the ONU channels and modulated with packet data. Upon arriving at the WDM/R, each downstream packet is routed (according to wavelength) to the appropiate ONU. This device also multiplexes all upstream signals onto a single fiber to the CO.

At the ONU, the light is split by a passive tap with a portion of the light detected by a receiver. The remainder is "looped-back" towards the CO through a modulator. This light forms a carrier, or "optical chalkboard" of sorts upon which the user writes as it passes by the modulator. The up and downstream signals can be separated at the CO by time partitioning, wavelength, subcarrier frequency, format, modulation depth, gating, coding, etc.

It is likely that multiplexed upstream packets will "collide" due to differing RN-to-ONU path lengths. Solutions to this problem include upstream WDM at the CO or packet sequencing as a function of optical path length.

The RITE-Net architecture takes advantage of each layer of multiplexing to increase functionality while decreasing costs: packet TDN at the CO allows single laser to be shared among N users; routed WDM provides network security, power advantage and fiber gain.

4.2.5.4. *Wavelength-tolerant optical access architecture*

In this approach, a passive two-stage AWG embodiment is considered as shown in Figure 4.2.16.

The free spectral ranges (FSR's) of the AWG's are tailored so that the

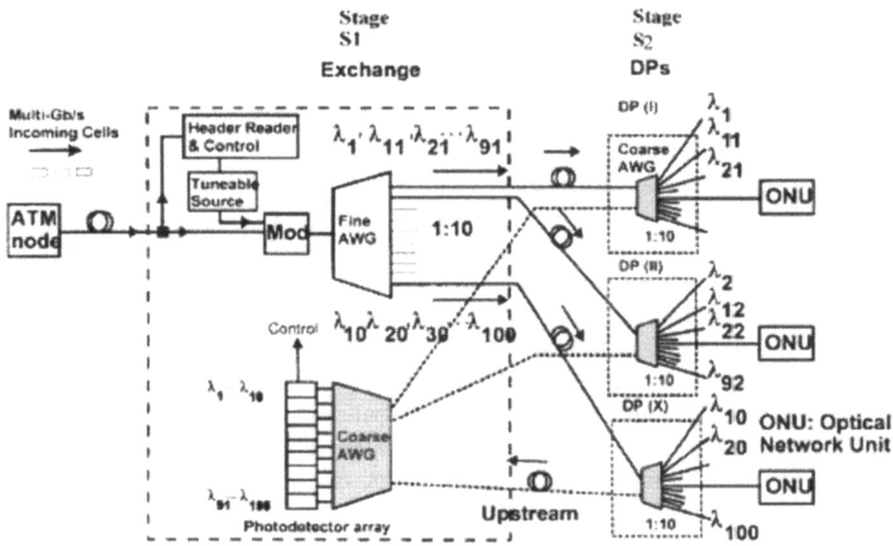

Figure 4.2.16. Network Architecture for two-stage **[Parker 94]**.

FSR of the fine AWG at stage 1 is equal to the passbands width of the coarse AWG at stage 2. First-stage downstream routing at S_1 is carried out within the exchange by exploiting the space/wavelength routing equivalence of AWG's. At the distribution point (DP), second stage (S_2) routing is provided by a coarse passband AWG. In this architecture, ATM cells enter the exchange serially where the header address information is extracted. Using a tunable source, the cells are modulated onto one of 100 wavelengths. The wavelength range of the tunable source is much larger than the FSR of the fine AWG. Consequently, packets of various wavelengths separated by multiples of the FSR emerge from each output port of the fine AWG (10 × in this case). The relatively large wavelength separation between packets allows the use of coarse AWG's at the DP. These route the packets according to wavelength to optical network unit (ONU's) situated at the customer premises, each ONU receiving just one wavelength.

For the upstream traffic, the periodic properties of the AWG at the DP are used to separate the upstream and the downstream packets, thus, avoiding the need for a circulator. Advantage is taken of the reciprocal nature between the input and output ports of $N \times N$ AWG's, as well as their cyclic-shifting properties. The upstream wavelengths are assigned such that each ONU transmits at the wavelength of its neighboring ONU. This results in all the upstream traffic being multiplexed onto a single, separate upstream port different from that used for the downstream traffic. The upstream traffic is demultiplexed at the exchange using a single coarse AWG with an array of ten photodetectors positioned at its output ports. By time-sequential polling of the DP's, collision are avoided between ONU's of different DP's transmitting upstream to the same photodiode. At the photodetector array, each ONU is uniquely defined by a space-time coordinate, thus allowing easy establishment of incoming upstream packet source addresses.

4.2.5.5. *Coherence multiplexed access network*

A simple bi-directional PON is proposed based on CM (Coherence Multiplexed) techniques using the half-duplex communication procedure and a novel configuration: hybrid transceiver units.

Figure 2.3.22 shows the configuration of the proposed PON. Each ONU consists of a directly modulated light source (LS), one optical tap, two PDs, two 3-dB couplers, and an appropiate optical delay line. This configuration reduces the number of optical components used. The OLT simply consists of almost the same components as the ONU. A bi-directional optical amplifier (BOA) may be employed in the OLT.

Figure 2.3.23 (below) shows time charts of the proposed system and the TDMA scheme in the conventional ATM-PON. Users can independently send or receive signals using the half-duplex communication procedure. The ratio of up-to downstream bandwidth can be changed as demanded by each user.

The proposed concept also suits a CM system using phase-shift keying as shown in the Figure 2.3.23 (above). The bi-directional CM access network features fewer taps and photodiodes in the transceiver units, but requires external phase modulator (PHs) System performance is expected to be worse because unbalanced receivers are used.

4.3. FIELD TRIALS

Figure 4.3.1 depicts the main field trials that are being developed these days (more detail in section 2.19). As it's shown, both active and passive solutions are considered as solutions for FTTH. Figure 4.3.2 contents PON vendors. Several companies take the ATM as communication protocol; the other ones take Ethernet.

SUMMARY

There are currently several projects focused on optical access networks and their components. New broadband services are reclaiming more bandwidth per user. The PON solution is a key technology to meet the challenge. Apart from the technical view about scalability, upgradibility . . . the main objective for any company is the COST. Obviously, companies are looking for a tradeoff between cost and performance. Projects as Ringo, SuperPon Architecture and Harmonics have architecture models that are able to be deployed as solution. In all cases, the distance, bandwidth per user and splitting factor are very important parameters to be taken into account; there is also a general idea to maintain the ONU as simple as possible and to complicate the OLT which will be an element to share by all users (share the value among users).

Trial	Partner	Splitting	Protocol	Others	References
Palo Alto	Marconi	Active Star	GEthernet	OLT-Switch: SM fiber; Switch to each home: two MM fibers	www.pafiber.net
Grant County	CISCO, World Wide Packets	Active Star	GEthernet	OLT-Switch: SM fiber; Switch to each home: two MM fibers	
Nebraska	BellSouth; Lucent, OKI	PON	Subcarriers FSK	Technology: Optical Solutions modulation	www.americasnetwork.com
Kanazawa	NTT, Matsushita	Active	Ethernet	Maximum speed: 10 Mb/s	www.ntt.co.jp
Bredandsbolag	Telia, NTL, Framlab	Active	Ethernet	FFTBuilding	www.bredband.com
Ca*net3	Bell Canada, Cisco, Nortel, Newbrigde, JDS, Cambrian Systems	APON	ATM, Gigabit Ethernet		www.canet3.net
Fastweb	Pirelli				
Dubai Marina	Marconi, EMAAR			From video to smart house applications	
LINA.NET					www.lina.net
Hankyu Nissei New Town	NTT Telecom Engineering Kansai			10 Mbits	www.japan-magazine.com/2000/February/news.htm

Figure 4.3.1. Main field trial table.

PON Vendor	Product	Technology	Department Status	References
Alcatel		ATM	In trials	www.alcatel.com
Alloptic	GigaForce™	Ethernet	One customer plus trials	www.alloptic.com
Fujitsu	FMX Series	ATM		www.fujitsu-telecom.fr
Iamba Technologies		N/A	In development	www.iamba.com
Marconi	DeepFiber	Ethernet	Two customers plus trials	www.marconi.com
NEC Eluminant Technologies	FiberSLAM	ATM	In trials	www.eluminant.com
OKI Network Technologies		ATM	In trials	www.oki.com
OnePath Network	iPath	Ethernet	In trials	www.onepathnet.com
Optical Solutions	FTTH FiberPath	RF-Bsed, ATM	23 customers plus trials	www.opticalsolutions.com
Paceon		ATM	One customer plus trials	www.paccon.com
Quantum Bridge	DWS	Ethernet		www.quantumbrigde.com
Communications Inc.		ATM	Six customers plus trials	
Salira Optical Systems Inc.	Salira Access	Ethernet	In development	www.salira.com
Teraware Communications	TeraPon	ATM	Six customers plus trials	www.terawave.com
Wave7 Optics Inc.		Ethernet	In trials	www.wave7optics.com

Figure 4.3.2. PON Vendor table.

REFERENCES

[Caponio] N. Caponio and et al.; CSELT, BT, ETH-Z, Instruments SA, Thomson-CSF; 'SONATA specifications'.

[Engell 97] Niels Engell Andersen (DSC Communications A/S), Paulo M. N. Nordeste (Portugal Telecom CEET), 'Broadbandloop: A full service access network for residential and small business users'; IEEE Communication Magazine, December 1997.

[European] The European Community's RACE projects demonstrate fiber-based broadband infrastructures for local access. 'Field Trials for Fiber Access in the EC'.

[Final 98] Final Report; 'Vertical'; December 1998.

[Frigo 94] N. J. Frigo, P. P.Iannone, P. D. Magill, T. E. Darcie, M. M. Downs, B. N. Desai, U. Koren, T. L. Koch, C. Dragone, H. M. Presby and G. E. Bodeep; 'A Wavelength-Division Multiplexed Passive Optical Network with Cost-Shared Components'; IEEE Photonics Technology Letters, Vol. 6, No. 11, November 1994.

[Gaudino 01] R. Gaudino, A. Carena, V. Ferrero, A. Pozzi, V. de Feo, P. Gigante, F. Neri, P. Poggiolini; Dipartamento di Elettronica, Politecnico di Torino; 'RINGO: a Demonstrator of WDM Optical Packet Network on a Ring Topology'; February 16 2001.

[Kani 01] Jun-ichi Kani, Katsumi Iwatsuki, Noboru Takachio, Nobuo Fujii; 'A Simple Broad-band Coherence Multiplexed Optical Access Network and Its Scalability'; Journal of Lightwave Technology, Vol. 19, No. 4, April 2001.

[Koonen 97] Ton Koonen, Theo Muys, and Chris van der Plaats, Lucent Technologies; Sonia M. Heemstra de Groot, H. J. H. N. Kenter, Ignas G. M. M. Niemegeers, and Frans N. C. Slothouber, ctit, university of Twente; 'TOBASCO: An innovative Approach for Upgrading CATV Fiber-Coax Networks for Broadband Interactive Services' IEEE Communications Magazine, April 1997.

[Kuznetsov00] Mark Kuznetsov, Nan M. Froberg, Scott R. Henion and Hemonth G. Rao (MIT Lincoln Laboratory); Jeff Korn (Axsun Technologies); Kristin A. Rauschenbach (Photonex Corp.); Eytan H. Modiano and Vincent W. S. Chan (Massachussetts Institute of Technology; 'A Next-Generation Optical Regional Access Network'; IEEE Communications Magazine, January 2000.

[Labtujere 93] Ad C. Labrujere, M. Oskar van Deventer, Oscar J. Koning, Johan P. Bekooij, Antonius Han Hiong Tan, Gert Roelofsen, Martin K. de Lange, Jean-Paul Boly, J. Anita H. W. Berendschot-Aarts, Cyriel P. Spruijt, Maaten F. L. van Nielen, Rob F. M. van den Brink, Kees M. de Blok, A. Kees van Bochove; 'COSNET-A Coherent Optical Subscriber

Network'; Journal of Lightwave Technology, Vol 11, No 5/6, May/June 1993.

[Lake 99] LAKE Communications; 'Bourbon: Broadband Urban Rural Based Open Networks; 31-08-1999.

[Liou 97] K.-Y. Liou, U. Koren, E.C. Burrows, J. L. Zyskind and K. Dreyer; 'A WDM Access System Architecture Based on Spectral Slicing of an Amplified LED and Delay-Line Multplexing and Encoding of Eight Wavelength Channels for 64 Subscribers' IEEE Photonics Technology Letters, Vol. 9, No. 4, April 1997.

[Lucent 98] Partners: Lucent Technologies, University of Twente, Bosch Telecom, Corning Europe, Optoelectronics Research Centre, Royal PTT Nederland, KPN Research, Norcontel Ireland, Nerderlande Philips Bedrijven, Uniphase Netherlands BV, Mechelen Technologies, Interuniversity Microelectronics Centre, Burger Breedband Net; 'Final Report AC028 TOBASCO', November 1998.

[Parker 98] Michael C. Parker, Farshid Farjady, Stuart D. Walker; 'Wavelength-Tolerant Optical Access Architectures Featuring N-dimensional Addressing and Cascaded Arrayed Waveguide Gratings'; Journal of Lightwave Technology, Vol. 16, No. 12, December 1998.

[Prat 01] Josep Prat i Gomà, Pere Emili Balaguer i Rodrigo, Joan M. Gené i Bernaus, 'Optical CSMA/CD strategy for an Ethernet Access Network'; NOC 2001, UPC.

[Stark 97] Jason B. Stark, Martin C. Nuss, Wayne H. Knox, Steven T. Cundiff, Luc Boivin, Steven G. Grubb, Don Tipton, David DiGiovanni, Uziel Koren and Kevin Dreyer; 'Cascaded WDM Passive Optical Network with a Highly Shared Source'; IEEE Photonics Technology Letters, Vol. 9, No. 8, August 1997.

[Tan 98] Anton H. H. Tan; 'SUPER PON- A Fiber to the Home Cable Network for CATV and POTS/ISDN/VOD as Economical as a Coaxial Cable Network'; Journal of Lightwave Technonolgy, Vol. 15, No. 2, February 1997.

[VandeVoorde00] I. Van de Voorde and C.M. Martin (Alcatel Corporate Research Center); J. Vandewege and X.Z. Qiu (IMEC-INTEC); 'The SuperPON Demonstrator: An Exploration of Possible Evolution Paths for Optical Access Networks'; IEEE Communications Magazine; February 2000.

[Wellen 01] Jeroen Wellen; Bell Labs Advanced Technologies; Lucent Technologies; 'HARMONICS' Whitepaper; March 29, 2001. Project Partners: Portugal Telecom, Corning SA, T-NOVA, Intracom, KPN Research, Mason, IMEC, University of Limerick.

[WenPiao 01] Wen-Piao Lin, Ming-Seng Kao, Sien Chi; 'The Modified Star-Ring Architecture for High-Capacity Subcarrier Multiplexed Passive Optical

Networks'; Journal of Lightwave technology, Vol. 19, No.. 1, January 2001.

[Wood 99] Thomas H. Wood, Gordon C. Wilson, Robert D. Feldman and J. Anthony Stiles, 'FiberVista: A Cost-Effective Fiber-to-the-Home (FTTH) System Providing Broad-Band Data Over Cable Modems Along with Analog and Digital Video'; IEEE Photonics Technology Letters, Vol. 11, No 4, April 1999.

Chapter 5

COMPONENTS

INTRODUCTION

In this chapter an extended review of main optical components of the optical access networks is done. Availability of optical components is the most determining factor for a feasible solution. Furthermore, the cost of those components is probably the most limiting factor in FTTH environment.

In section 5.1, a study about different fibers, fiber cables and cabling systems is presented. Fiber is a basic component and dealing with it is unavoidable.

The most important component is the transceiver. In section 5.2, a review of the state-of-art in optical transceivers for access systems is developed. Integrated transceiver is the main point but independent points about sources and receivers are included.

5.1. FIBRE CABLES AND CABLING SYSTEMS

5.1.1. Introduction

Network planners and end users are increasingly concerned about the impact that higher data rates and more bandwidth intensive applications will have on their current cabling infrastructure, and optical fiber cable is emerging as the medium that offers them the *performance, reliability and longevity* that they seek. Many of the obstacles that have discouraged end users from choosing optical fiber have also been overcome: lower prices for fiber cable, new products that support fiber-to-the-desk applications, the development of lower

J. Prat et al., Fiber-to-the-Home Technologies, 189–302.

cost electronic components, the availability of trained installers, and now the opportunity to take advantage of centralized cabling architectures, all position optical fiber cable as the best long term cabling solution – and that is available today.

Advantages of fiber optic cables in a network:

- Wide bandwidth
- Low loss
- Electromagnetic immunity
- Security
- Light weight
- Small size
- Safety
- Less expensive than copper
- More reliable than copper
- Easy to upgrade to higher speed and performance

Although it is clear that fiber is the best solution for the access network (the only one that will support the exponential growing of data traffic) there are some open issues in addition to whether it is already the FTTH era. Probably, the *most important question at this moment is what kind of fiber* is the appropriate for implementing FTTH. While some companies are promoting multi-mode fiber, others say that single-mode is possible right now (f.i.:Pirelli has introduced 850nm single-mode fiber **[pirelli]** called Fine Light). There are even people who propose hybrid cables as the best solution (multi-mode for the near term and single-mode for the future). Another important issue is that the current most employed SMF is the ITU-T Rec.G.655 as it has the best performance with 1550 nm light. This kind of fiber (usually known as NZ-DSF) is not specified as single-mode for 1300 nm light. On the other hand, 1300 nm devices are being promoted (due to their lower price) as the right solution for the optical access . . . so "old" (G.652) single-mode fiber would be the appropriate. Anyway, this is an important open issue and it is not clear what strategy would be the best.

5.1.2. Optical fiber

Nowadays, revolution in telecommunications environment is widely leaded by optical communication systems because they are most suitable solution for the accelerated growth of data traffic, thanks to the invention of optical fiber. To reach the current point, a lot of research works have been carried out through the last century. The first attempt to transmit information by means of light was in 1880, when Alexander Graham Bell patented an optical telephone system, which he called the Photophone. He sought to send signals through the air, but he was unsuccessful due to the atmosphere that limited the light transmission.

Before, in the 1840s, Swiss physicist Daniel Collodon and French physicist Jacques Babinet showed that light could be guided along jets of water, and to the late 19th century John Tyndall demonstrated that light could be conducted through a curved stream of water by means of the phenomenon of total internal reflection of light in a material surrounded by other materials with lower refractive index. By developing this theory one stage further, inventors realized that bent quartz rods could carry light and patented them as dental illuminators.

Other techniques to transmit data by means of light were developed at the beginning of the 20th century. For instance, John Logie Baird in England and Clarence W. Hansell in the United States used arrays of hollow pipes or transparent rods to transmit images for television or facsimile systems, but Heinrich Lamm, a medical student in Munich, was the first person who transmitted images of inaccessible parts of the body through a bundle of unclad fibers, but the fibers transmitted images poorly.

In 1950s, Abraham van Heel of the technical University of Delft in Holland made the most important innovation on covering the unclad fiber with a transparent cladding of lower refractive index. However, glass-clad fiber attenuation of about one dB per meter was much too high for communications.

In 1964, Dr. Charles K. Kao identified that to achieve the lowest light signal loss in optical fibers; he had to achieve glass of the highest purity. Although Dr. C. K. Kao's observations over experimental optical waveguide developed by Standard Telecommunications Laboratories had a loss of about 1000 dB per Km, but he forecast that fiber loss could be reduced below 20 dB per Km. Only in 1970, the forecast of the specification given by Kao necessary for long-range communication had a loss below 20 dB per Km.

At the Corning Glass, Robert Maurer, Donald Keck and Peter Schultz reached losses below 20 dB/Km in single mode fiber at 633 nm. They deposited purified materials from the vapor phase adding carefully controlled levels of dopants to make the refractive index of the core slightly higher than that of the cladding. It was not commercially available until 1975.

In the early 1970s, due to the low bandwidth of step-index multimode fiber, researchers started to develop gradient-index multi-mode fibers. The first-generation systems used as fibers. Losses of about 2 dB/Km allowed to transmit light at 850 nm. several kilometers without repeaters until a new InGaAsP laser appeared which emitted at 1300 nm where fiber loss was 0.5 dB/Km. Afterwards, in the early 1980s, single mode fiber was used with 1300 nm. sources.

However, the last generation of single-mode systems operate at 1550 nm. where fiber loss is 0.2 to 0.3 dB/Km in applications in submarine cables and serving large number of users. Optical fiber is expected to arrive to the home this decade.

In a simple way, fiber optic (Figure 5.1.1) cables are made from a very clear glass, much more transparent than window glass. As it has been explained before, fiber technology uses lightwave(from Lasers, VCSEL's, Led) to transmit signal whereas, other medias uses electromagnetic signal or radio frequency signal. The performance and the clarity of the glass permit the distribution of lightwave impulses through hundred of kilometers.

Coding is easy. It uses simple On/OFF keying. Light is perceptible when the receptor sense photons. In chapter 2, other coding schemes have been developed.

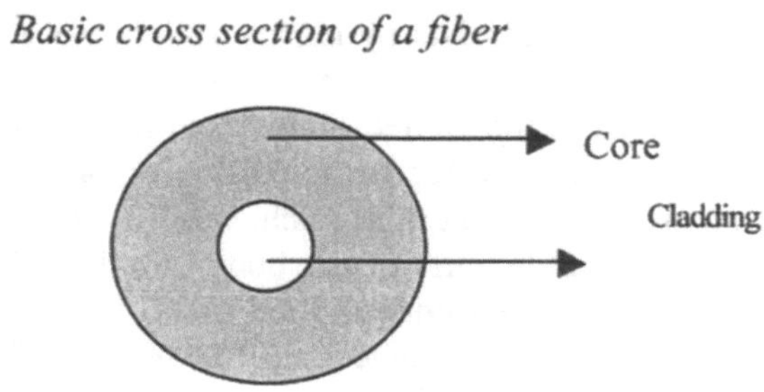

Figure 5.1.1. Basic cross section of a fiber.

Light can be guided down the fiber optic cable if it enters at less than the critical angle. This angle is fixed by the indices of refraction of the core (n_1) and cladding (n_2) and is given by the formula:

$$\alpha_c = \text{arc cosine } (n_2/n_1).$$

The critical angle is measured from the cylindrical axis of the core. By way of example, if $n_1 = 1.44$ and $n_2 = 1.43$ then a quick computation will show that the critical angle is 8.5 degrees, a fairly small angle.

Of course, it must be noted that a light ray enters the core from the air outside. The refractive index of the air must be taken into account in order to assure that a light ray in the core will be at an angle less than the critical angle. This can be done fairly simply. The following basic rule then applies. Suppose a light ray enters the core from the air at an angle less than an entity called the external acceptance angle – α_{ext} It will be guided down the core. Here

$$\alpha_{ext} = \text{arc sin } [(n_1/n_0) \sin (\alpha_c)]$$

with n_0 being the index of refraction of air. This angle is, likewise, measured from the cylindrical axis of the core. In the example above, a computation shows it to be 12.4 degrees – again a fairly small angle.

5.1.3. Types of fiber

Currently, as described, fiber is categorized into two groups: single-mode and multi-mode fiber, nevertheless, there exists another kind of fiber like the POF. The mode refers to the path a photon takes in going from one side of the fiber to the other.

5.1.3.1. *Single-Mode Fibers (SMF)*

In this type of fiber (Figure 5.1.2), all photons take the same path along the fiber. This is because the center of the fiber is very narrow. According to the ANSI the fiber has 8.5 micron core with 125-micron cladding. This factor reduces the signal loss and the ISI (intersymbol interference). Therefore, with single-mode fiber used primarily in telecommunications networks, fibers are characterized by long distance links and high capacity

It is the most reliable medium, but it has an important drawback for the fact that, at this moment, it is not clear that SFF connectors perform well with SMF (at least, it is not easy to reliably connect two SMFs to the field) **[OFC97]**. As fabrication processes are improving quickly, it will be possible to have enough concentricity in fibers as to connect them aligning their claddings.

Advantages of using SMF are well known:

- Its bandwidth is enough to support, not only the current traffic but also the future one.
- Greater distance: fiber can be run several kilometers.
- Low attenuation: Light signals meet little resistance, so data can travel faster

It has also some drawbacks that can be reduced to one:

- The core diameter is very small so, problems to access every home (connectivity) will arise.
- The narrow size of the core causes that other elements of the system, such

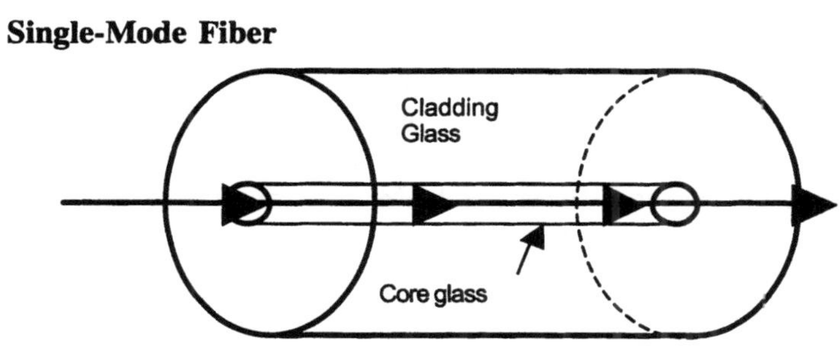

Figure 5.1.2. Section of a Single Mode.

as transmitters, connectors and transceivers would have to operate at a slower tolerance, so their cost would then be higher.

The next table will show the principal features and characteristics of the single-mode fiber described by the *Standard ITU*.

- ITU-T G.652: Standard single mode fiber. General applications fiber suitable for a lot of uses
- ITU-T G.653: Dispersion Shift Fiber. Fiber optimized for transmission in the third window (1550 nm), recommendable in very high speed and long distance applications.
- ITU-T G.655: Freelight Fiber TM. Fiber designed for DWDM applications. It is characterized by very low dispersion at 1550 nm and a high effective area, which prevents the non-linear effects of high speed in this type of transmission, offering improved service in comparison to the previous fibers.

Standard ITU-T	**Window**	**Maximum attenuation (dB/Km)**	**Average attenuation (dB/Km)**
G.652	1.310 nm	**0,38**	**0,36**
G.652	1.550 nm	**0,25**	**0,23**
G.653	1.550 nm	**0,25**	**0,23**
G.655	1.550 nm	**0,25**	**0,23**

5.1.3.2. *Multi-Mode Fibers (MMF)*

The strong point (with respect to SMF) is its easy connection. On the other hand, some difficulties may arise from the fact that its lower bandwidth (of the order of 500 MHz·Km) can be a problem in some networks. New developments have been done and MMF with 2000 MHz·Km are available today **[corning]**, so they can not be discarded. Another advantage is that, at this time, 850nm emitters are cheaper than 1300 or 1550 nm.

These fibers are often used in local area networks (buildings, campus, etc.), usually of short distance (< 3–4 km) and low capacity (< 150 Mbit/s).

In multimode fiber (Figure 5.1.3), photons careens off the fiber wall as it goes from one end to the other, thereby defining a path. Different photons will probably describe different path. In fact, the number of paths is functions of the core diameter with wider diameter.

The below table shows the principal features and characteristics of the multimode fiber described by the *Standard ITU*.

Standard ITU-T G.651	Window	Max. attenuation (dB/Km)
62,5/125	850 nm	3,1
62,5/125	1.300 nm	0,25
50/125	850 nm	3,0
50/125	1.300 nm	0,8

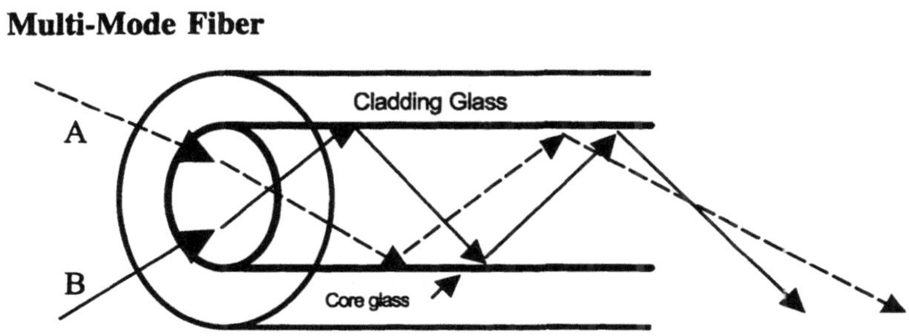

- - - - - - -	Photon A	Will get to other end before A
————	Photon B	Has a higher angle of incidence; Has a longer path

Figure 5.1.3. Section of a Multi Mode Fiber.

Currently, due to the great demand for increasing bandwidth, it has prompted the release of the 802.3z standard (IEEE) for Gigabit Ethernet over optical fiber. 1000Base-Lx requires the use of launch conditioning patch cord to allow the proper propagation of the laser VCSEL light along the multimode fiber.

The Mode Conditioning Patch Cord is designed for long wave (-LX) multimode applications of Gigabit Ethernet. It is compliant with this IEEE 802.3z application standard. This patch cord consists of connectors on each end of a cable assembly with a single-mode fiber offset to a multimode fiber connection point in between.

The need for this patch cord is due to the single-mode launch nature of the -LX or long-wave (1300 nm) transceiver modules used for Gigabit Ethernet. These modules have to operate for both single-mode and multimode fibers. Launching a single-mode laser into the center of a multimode fiber can cause multiple signals that confuse the receiver at the other end of the fiber. These multiple signals, caused by Differential Mode Delay (DMD) effects, severely limit the cable distance lengths for operating Gigabit Ethernet. A mode conditioning patch cord eliminates these multiple signals by allowing the single-mode launch to be offset away from the center of a multimode fiber.

This offset point creates a launch that is similar to typical multimode LED launches.

5.1.3.3. *Plastic Optical Fiber (POF)*

POF is similar to glass fiber¡ and consists of a core surrounded by cladding. POF can deliver data at the speed of up to three gigabits per second, and has the characteristics similar to copper and glass(the two other methods of physically linking sources to transfer and communicate data). The distance for POF is limited to about 50 meters (about 150 feet), but it is a simple way of getting high bandwidth distributed into the home with easy connectivity. It works with visible light and current designs have losses as high as 1.5 dB/m. This suggests that it only could be appropriate for short (maybe "in home") transmission.

The main advantages of POF over glass fiber are lower installation and maintenance costs, which could run 50 percent less than glass, and the need for less technical expertise to develop and maintain it. Its ability to withstand a severe bend radius of up to 20 mm with no change in transmission also allows for easier installation through walls. It is a distinct advantage in the growing home network market.

Features and advantages over silica optical fibers:

- 3 to 10 times larger in diameter.
- 2 to 3 times larger in numerical aperture (hence it does not require high precision of connectors).
- Hard to break by bending.
- High resistance to impact and to vibration.
- Easy to cut and to treat fiber ends (easy to install).
- Low total cost.

Improvements needed:

- To reduce the high attenuation loss.
- To improve heat resistance.

Although it seems very improbable that POF could be a solution for the access network, there are some issues that have to be considered:

- IEEE 802.3 committee included it as an acceptable material to use in short lengths in ATM (asynchronous transfer mode) networks.
- IEEE 1394 Standard. Enables simple, low-cost, high-bandwidth connections between consumer electronics **[Pofeska]**.
- Lucent & Asahi planned to develop a special (Perfluorinated) low loss POF (it is planned to be in the market next year).
- News from WDMSolutions 29/6/01. Nanoptics-Samyang joint venture to manufacture GRIN POF.

The proliferation of DVD drives in the consumer electronics market has resulted into a large supply of relatively cheap light sources at the 650 window. With these considerations in mind and taking into account what plastic fiber has as best loss value around 650 nm, it seems reasonable to think that POF could be the right choice for the last section of the network. Anyway, as POF is not appropriate for distances greater than 100s meters, its use would imply the use of active elements (i.e. converters) at the premises. The POF industry continues to pursue development of a viable fiber which is transparent in the same region as silica glass. Recent work with fluorinated plastics to achieve this breakthrough has been encouraging **[LR 00]**.

5.1.3.4. *FineLight™: Special fiber for FTTH*

To avoid the high costs of the ONU, currently, a new kind of fiber has been developed by Pirelli. It is suitable to be used under FTTH deployments. As principal characteristic of the Fine Light is that the ONU can use low-cost-laser-sources avoiding the effects of working with the first or second window in a G.652 fiber.

Pirelli Telecom Cables and Systems has expanded its lineup of industry-leading optical fibers with the addition of FineLight™ Giga fiber and FineLight™ Base fiber, creating a distinctive new family of products for fiber-to-the-business, premises and LAN applications. FineLight™ Base and FineLight™ Giga optical fibers are single-mode fibers with the capability of operation at the 850 nanometer window. As a consequence, system providers can build a network with a fiber optic cable that meets today's requirements for multi-mode operation, and which can be upgraded, on the same fiber, to single-mode operation as tomorrow's demands dictate.

The 850 nm operation uses standard low-cost VCSEL (Vertical Cavity Surface Emitting Laser) devices, while the upgrade can be accomplished simply by switching to single-mode equipment. "FineLight™ fibers offer the first real opportunity to bring all three operating windows within the capability of a single optical fiber. This allows the customer to benefit from any system cost saving that may be offered in the future, independent of the wavelength utilized and without the need to upgrade the cable. FineLight™ fibers meet high-capacity requirements with a transmission capability of 100 Mbit/s up to 5 km for FineLight™ Base fiber and 1 Gbit/s up to 1km for FineLight™ Giga fiber. FineLight™ fibers can operate in all three standard operating windows – 850 nm, which is the accepted operating window for multi-mode operation, and 1310 nm and 1550 nm, the accepted operating windows for single-mode operation.

FineLight™ fibers meet existing universal standards, including the ITU-T standard G.652, and standard SC, ST and FC style connectors. Attenuation for both fibers is = 2.2 dB/km at 850 nm. Bandwidth performance specifications are = 600 MHz/km for FineLight™ Base fiber and =1,000 MHz/km for FineLight™ Giga fiber.

The table below describes the specifications of the FineLight™:

Dimensional specifications	
Glass geometry	
Cladding diameter	125.0 ± 1.0 μm[a]
Cladding non circularity error	< 2.0%
MFD/cladding concentricity error	< 0.8 μm
Coating geometry	
Outer coating diameter	245 ± 10μm
Coating-cladding concentricity	< 12.5 μm
Typical MFD (@ 1310 nm)	9.2 μm
Bandwidth @ 850 nm	> 0.6 GHz·km
Attenuation coefficients	
@ 850	< 2.20 dB/km
@ 1310 nm	< 0.40 dB/km
@ 1550 nm	< 0.30 dB/km
Dispersion coefficients	
@ 1310 nm (absolute value)	< 3.5 ps/(nm·km)
@ 1550 nm	< 20.0 ps/(nm·km)

[a] μm: micrometers = 1 × 10 exp-6; nm: nanometer = 1 × 10 exp-9.

5.1.4. Impairments

Although fiber is a good stuff, it also has some limitations. It is subjected to attenuation under certain circumstances.

Some of these impairments are:

Dispersion: It causes light to spread as it travels down the fiber and limits bandwidth. In the fiber, photons that bounce will arrive at the end later than photons that travel straight. The affect of dispersion is loss of temporal integrity. To ensure that pulses do not overlap, the bit rate must be the adequate

Attenuation: It refers to the loss of optical power along the fiber. The loss is measured in dB/Km and not all the links have the same attenuation, it depends on the fiber age, construction, installation care, splices, etc. . . . The attenuation ranges are between the 300 dB/Km for plastic fibers to 0,2 dB/Km for 1550 nm SM fiber.

Bending: Bending the fiber causes light escape from the core. The tolerable bend in a fiber is function of the refractive index of the core and surrounding cladding.

Handling problems: It will become particularly acute for fiber to the Home. For example the ends of the fiber must be perfectly aligned and as clean as possible.

5.1.5. Protecting the fibers: Cable designs

Optical fibers are quite fragile, they are made of glass. But optical fibers are a lot stronger than fine glassware, and, once they are enclosed in protective cables, they can be even tougher than copper communication cables.

To achieve this strength, all optical fibers are designed in three concentric layers. At the center is the core. It is the central piece of glass that carries the light. Surrounding the core is the cladding, a second layer of glass, which keeps the light from escaping the core. And around both of these lies the buffer, an outer layer of plastic, which provides the real protection and strength.

There are two basic cable designs: Tight buffers and Loose tubes: Tight Buffers Cables and Loose Tubes Cables.

5.1.5.1. *Tight Buffers Cables (TBC)*

TBC are primarily used inside buildings. It consist of a thin layer of plastic extruded over the individual fiber, much like the insulation placed on copper wires (except a lot smaller). This design is suited for jumper cables, which connect outside plant cables to terminal equipment, and/or linking various devices in a premises network. It provides a rugged cable structure to protect individual fibers during handling, routing and connections.

Single-fiber tight-buffered cables are used as pigtails, patch cords and jumpers to terminate directly into opto-electronic transmitters, receivers and other active and passive components.

Figure 5.1.4 shows a cross section in a Tight-Buffered Cable.

Tight-buffer cables offer the flexibility, direct connectability and design versatility necessary to satisfy the diverse requirements existing in high performance fiber optic applications.

Tight-buffer fiber optic cables incorporate the following attributes that are most important in network applications:

- *Excellent fiber protection:* superior moisture and mechanical protection provided by multiple fiber buffers and advanced jacket design.
- *Easy handling:* dense fiber packaging for smaller cable diameter, tight bend radius and easier pulling.
- *Ease of termination:* direct termination of fibers reduces mess and expense

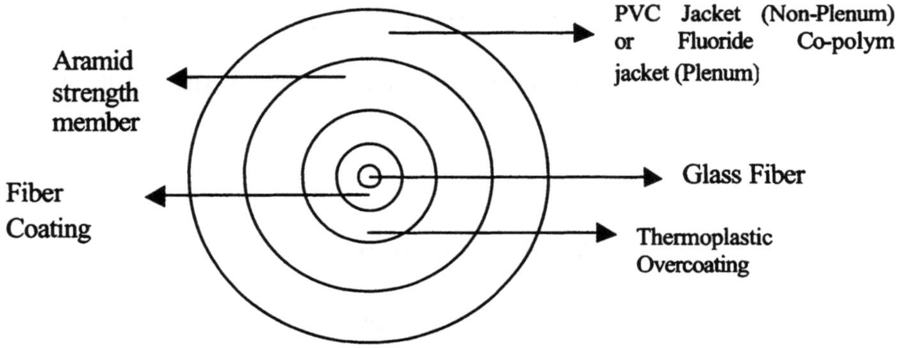

Figure 5.1.4. Cross section in a Tight-Buffered.

of installation by eliminating steps and materials required. Plus, a built-in ripcord speeds the stripping process. Flame-retardant and UL-listed.

- *Indoor/outdoor versatility:* superior moisture resistance, UV resistance, material durability and extended temperature range make the cables suitable for outdoor runs.
- *Higher survivability standards:* based on military technology for survival under mechanical and environmental stress.

Water blocking without messy gel is available in most cable designs.

5.1.5.2. *Loose Tubes Cables (LTC)*

Meanwhile, LTC are used in most outdoor cables. In a loose-tube setup, one or more fibers (usually up to 12) are installed inside of a plastic tube that resembles a drinking straw, and the assembly is then bound into the cable. Frequently, these tubes are filled with a water-block gel (usually and descriptively called "icky-pick" on the job site), which prevents water from entering the cable.

The loose-tube assembly provides particularly good protection for the fibers, but such cables are messy to terminate and, because the fibers have a tendency to fall out of the vertical cables, they should never be used as risers.

The cable core, typically surrounded by aramid yarm, is the primary tensile strength member. The outer polyethylene jacket is extruded over the core. If armoring is required, corrugated steel taped is formed around a single jacket cable with an additional jacket extruded over the armor

With larger cables, strength members may be made of a steel rod, steel wires, or fiberglass/epoxy rods. Most strength members, however, consist of a layer of Kevlar yarn in the cable.

Figure 5.5 shows a cross section in a Loose Tubes Cables:

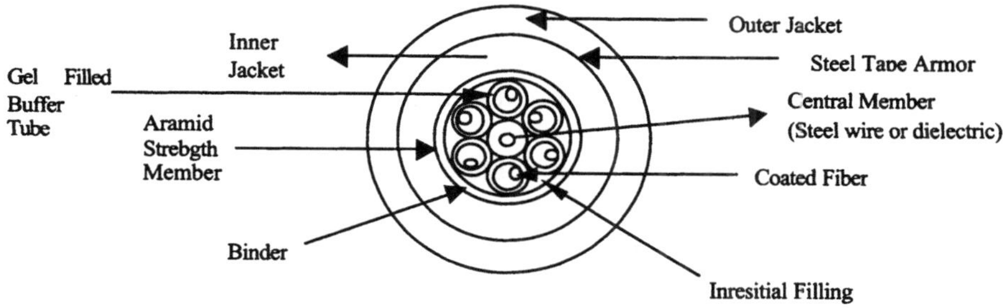

Figure 5.1.5. Cross section of a Loose Tubes Cables.

5.1.6. Types of fiber cable

Internet usage and new mobile services are creating an enormous demand for bandwidth. However the weakest link in the chain is the "last mile" which provides access to the subscriber. It is here where bottlenecks occur mainly due to low speed transmission.

We have different variants of cables, according to the different modes of installation depending of their use: Indoor, Indoor/Outdoor, Outdoor Cable, aerial

5.1.6.1. *Some common cables*

Beyond these basic configurations, fiber optic cables can take many forms. Here is a look at some of the most common:

Simplex and zip cord cables consist of one or two fibers that have been tight-buffered and Kevlar-reinforced and jacketed. Such cables are used mostly for patch-cord and backplane applications.

Tightpack cables, also known as distribution cables, are made up of several tight-buffered fibers bundled under the same jacket with Kevlar reinforcement. These cables are small in size, but because their fibers are not individually reinforced, they need to be terminated inside a patch panel or junction box. They are generally used for short, dry conduit runs as well as riser and plenum applications.

Breakout cables are made of several simplex units, cabled together to produce a strong, rugged design suitable for conduit runs and riser and plenum applications. Because each fiber is individually reinforced, this type of cable allows for a strong termination to connectors and can be brought directly to a computer back plane. Breakout cables are larger and more expensive than tightpack cables.

Loose-tube cables are considered ideal for outside plant trunking applications. Depending on the actual construction, they can be used in conduits, strung overhead, or buried directly into the ground.

Composite cables are those that contain a number of copper conductors (jacketed and sheathed depending on the application) in the same cable assembly as optical fibers. Be careful of the terminology there. Prior to 1993 NEC, such cables were known as *hybrid cables*, but today that term typically refers to cables that contain only optical fibers – but both multi-mode and single-mode fibers.

Those are different types of cables most commonly used, but there is another kind of division. Cables are designed for being used with or without ducts, so, another way to define them is the following.

Duct Cables may be either pulled into position, which requires a cable of higher tensile strength, or blown into the duct using compressed air. In this case, the cable glides into place on an airflow which imposes less stress on the cable. In any case, the cables should feature low dimensions and weight and should be allowed for small bending radii. Cables for installation in ducts usually only need a single PE[1] sheath as outer protection. As loose tube designs, Companies offers two special variants for duct cables according to the different modes of installation: while the version for pulling installation shows a maximum tensile force of $2 \times W$ (where W denotes the kilometric weight of the cable), the variant for blowing installation shows a value of $1 \times W$, together with a slightly reduced jacket thickness. Therefore, it is possible to select a suitable product already from a standard cable range. It should be noted that armored cable designs may also be laid in ducts – especially in cases where a rodent protection is desired.

Direct Buried cables are the favored alternative if no ducts are available for the laying of optical cable. In loose soil, a tractor and vibrating plow buries and covers the cable at about 80cm below the soil, while for extremely hard ground, a special saw is used to cut a small 10 cm trench into which a duct is placed, or a cable directly buried. Due to forces acting perpendicularly to the cable axis, this installation method is only recommended for armored designs which have greater resistance against crush and impact. Moreover, protection against rodents is a strong requirement for cables which are directly buried. An standard solution for these requirements is an armor of corrugated steel tape which strongly increases the values for crush and impact

---

[1] PE means Polyethylene. The cable with this kind of sheath has got a good resistance to the mechanic stress.

and at the same time serves as a perfect rodent protection. Cables with steel tape armoring are available as single and double jacket versions where the latter features an additional PE sheath beneath the armoring layer.

In addition there are Aerial cables, but are not used in the access network. Currently each manufacturer has its particular family of cables.

Lately, some factories are working hard to provide Ribbon Cable fiber. Ribbon, are cables in which many fibers are embedded in a plastic material in parallel forming a flat ribbon-like structure. Although there is no definite standard for this type of cable, the advantages will become a good solution for the access networks. It is expected to be one of the most important and used cable for fiber to the home.

Some benefits of cable Ribbon are:

- Low weight, easy handling and shipping
- Allows time and labor saving during fiber splicing
- Optimum utilization of ducts
- Permanent and easy fiber identification
- Ease of installation, identification and Routing
- Reduces installation costs and extends pull distance
- Excellent mass fusion-splicing yields
- Compact and robust design for high fiber counts
- Suitable for access applications

5.1.7. Cabling systems

Air Blown Fiber, Push & Pull Installation Fiber, Blowing Cable (method to get cable into conduit using air) and Sewer cabling are methods used to install fiber in a deployment. Some of them are discussed.

5.1.7.1. *Push & Pull Installation Fiber*

This system is when fiber cable is pushed or pulled manually or mechanically along the duct. This method is used for short distances when it is done manually.

Outside plant personnel are very familiar with the pulling method of installing cable: Thread a line through the duct, attach the line to the cable, and pull or tug the cable through the duct. The force needed to pull the cable usually comes from a capstan or hand-over-hand pulling of the rope. This force is needed to overcome the cable's frictional resistance to movement. Length of installation is limited by the maximum force allowed on the cable.

Fiber optic cables should be installed according to the manufacturer's specifications. The following guidelines apply:

- Cable that runs without splices are preferred. No splices should exhibit an insertion loss greater than 0,5 dB. The integrity of cable should be maintained for all splices.
- Length markers should be imprinted on the cable jacket at reasonable intervals.
- The maximum pull force on cable installation should not exceed the manufacturer's specifications. All strain gauge or dynamometer-type device should be used to monitor pull force during installation.
- The minimum bend radius of the cable should not exceed the manufacturer's specifications.
- Cable ties or protective devices should be used to prevent chafing.
- Avoid sharp bends and corners.
- Provide additional crush/mechanical protection in high-risk environments.
- Secure the fiber optic cable for existing supports or large cables wherever possible.
- Protect connectors when installing preconnectorized cable.
- Do not deform the cable jacket, especially when using cable fasteners or ties.

5.1.7.2. *Air-Blown Fiber*

Air-Blown Fiber, or ABF (as it is called), is a fiber optic cabling and installation method that utilizes a compressed nitrogen gas source to propel a thin fiber optic bundle, without stress, through a previously installed duct. Because the fiber can easily be blown out and reused, ABF technology provides exceptional flexibility in dynamic environments. It is the newest technique to deploy fiber.

A particular advantage of ABF is the ability to provide point-to-point fiber connections easily. A very important issue is that an ABF infrastructure is needed and it must be intelligently designed as point-to-point requirements change. The actual destination and origin of the fiber can be changed to meet these news requirements. This is accomplished by rerouting the tube cable through the reconfiguration of the tube cable's pathway at selected node points (tube distribution unit).

A critical advantage this system has is its ability to install fiber in a strain-free manner. This provides two advantages. The first is guaranteed Usable Fiber. No strain, no possible damage. The second advantage is reusability. Since no strain is placed on the fiber during insertion or removal, it is available for reuse. An additional feature enhancing reliability is the ease in which point-to-point runs are created. Point-to-point installation minimizes field splicing, increasing reliability and the quality if the link.

As indicated under reconfigurability and reliability, air-blown fiber cable may be reused saving the costs of new media in appropriate circumstances.

One of the major benefits of ABF is in emergency restoration. This system takes less time for restoration.

Blowing requires good planning and preparation, especially Microducts and duct preparation. All of the units require seals sized to the cable to prevent leakage from the pressure chamber around the entering cable. Properly sized seals should be obtained before starting the job.

Blown fiber avoids the two most common causes of failure in conventional fiber optic networks. There is no stress on optical fibers during installation. This avoids the danger of damaging fibers when pulling conventional cables through conduit. All end-to-end links installation are single fiber runs eliminating fiber splice points, thus decreasing attenuation and increasing system integrity.

As a conclusion, great advantages of new technique in Blowing fiber in FTTH are:

- *Extraordinary Design FLEXIBILITY*
  - Innovative fiber technology can be cost-effective added to any network architecture.
  - Add, remove, relocate, reroute, reconfigure, repair or upgrade easily and quickly Push-lock Microduct connectors simplify building a network infrastructure.

- *ADAPTABLE to Any Environment*
  - Install optical fibers only when as needed as the network evolves and changes.
  - Minimal workplace disruption during future network modifications and upgrades.
  - No need to change Microduct infrastructure if later a decision to change fiber type is taken.

- *"Pay As You Grow" DEFERRED INVESTMENT*
  - Defer cost of a deploying all fiber capacity upfront, spreading investment over time.
  - Saves time and labor costs over conventional fiber cable as network evolves.
  - Installations provide future growth capabilities at a fraction of upfront fiber costs.

- *FASTER Disaster Recovery*
  - Failed network nodes can be quickly repaired and restored or reconfigured.
  - Microduct tubing can be easily spliced, and fiber can be replaced quickly.
  - Disaster recovery time and cost dramatically reduced vs. conventional methods.

- Improved RELIABILITY
– Inherently better protection of fiber against damage during installation.
– Optical fibers are blown rather pulled, with zero tensile stress on fibers.
– Fiber splice points eliminated, lowering attenuation, increasing system performance and integrity.

5.1.7.3. *Air-Assisted Fiber*

Previously, a system to blow fiber optic bundle has been defined, now a system to blow the fiber cable will be presented.

Air-assisted fiber optic installation was developed and explored in the Netherlands as well as the United Kingdom in the 1980's. While fiber-blown methods are used to install individual fibers in tubes in LAN's, Air-assisted is focused on the outside plant installation, for multi-fiber cable in duct.

Air-assisted installation must overcome the same frictional force to move cable, but it does this in a very different way. The force in air blowing first comes from a mechanical device which pushes the cable; and second, from the force of moving air on the cable jacket, or alternatively, the force of air on a piston, missile, or carrier at the front end of the cable.

These two variations of air cable assisted installation are diagrammed in the Figure 5.1.6

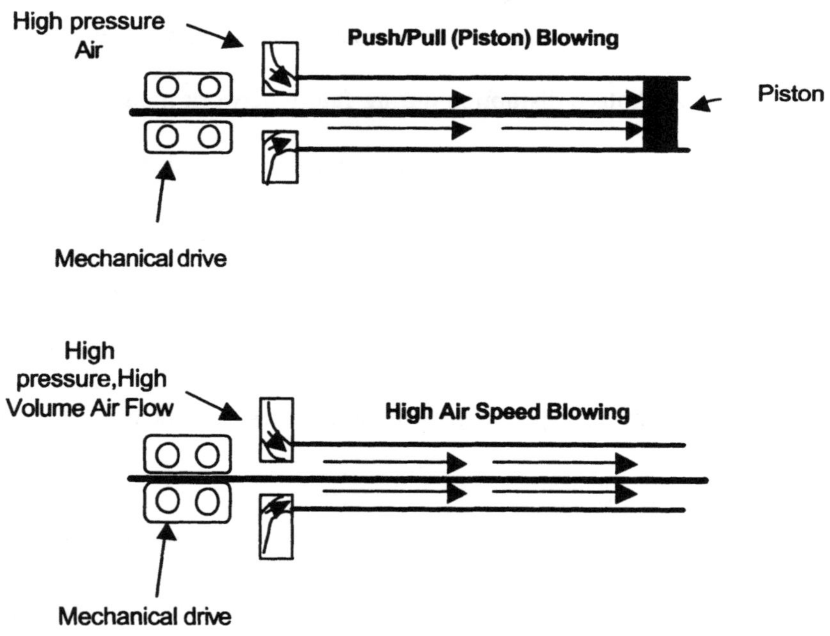

Figure 5.1.6. Air assisted techniques.

While the pushing device is common in both methods, the purpose of the airflow through the conduit is different. In the high air speed blowing (HASB) method, the conduit is wide-open, and high air volume (300–600 cfm[2]) blows through it. This "wind" pushes on the cable and advances it forward at any speed the pusher will support. There is no pulling force on the front end of the cable, only a mechanical pushing force at the back and an air drag force distributed along its length.

The piston (push/pull) method attaches a missile or carrier to the front of the cable. The missile is pushed through the conduit by the air pressure force, and the missile pulls on the cable, thus the term push/pull installation. Because the conduit is blocked, or at least partially blocked, by the piston, this method does not require as much air flow as the high air speed method, and flows of 200 to 300 cfm are typical during push/pull piston installation.

5.1.7.3.1. *Fiber Optic Pulling Vs. Cable Blowing*

Taking into account that the first technology presented needs a particular infrastructure to deliver fiber to each user, and that is a new market solution, now, a comparison between the classical methods of Pulling and the Cable Blowing (they use the same infrastructure) will be studied.

While all types of cable can be blown into duct, blowing is particularly useful for placing outside plant fiber optic cable. Fiber optic cable is lightweight and quite flexible. Long uninterrupted lengths of installed cable are desired. Both factors play with the strength of the blowing methods.

When a cable goes through a conduit bend, the pulling force on the cable actually pulls the cable into the bend. An exponential frictional force is created, that is, many times greater than what would occur with just the weight of the cable. Thus, an accumulation of bend angles greatly reduces possible pull length. But cross-country fiber optic pulls do not have a lot of bends, or do they?

Continuous flexible duct introduces "hidden bends." These are inherent undulations or waves from reel memory as sketched here.

While these waves can be minimized, they can not completely be eliminated. Typical displacements of a few inches to a foot increase pulling tension by factors of 3 to 200. Reference presents theoretical detail on the duct wave effect and its magnitude.

Maximizing distance in fiber optic pulling means minimizing duct undulations. Thick walled inner duct with a diameter of 2 inches or more can be placed "straighter" than 1 inch or 1.25 inch duct, but at a cost price. Multicell ducts keep inner ducts straight with a rigid exterior, again with a cost and labor add-on. Plowed duct is straighter than trenched duct. In fact, pulling

---

[2] Cfm: Cubic Feet per Minute.

cable into open trenched inner duct is not practical. The amount of undulation simply makes pull lengths too short.

Duct bends, including undulations, do not limit blowing processes, especially HASB, as much as they do traditional cable pulling. In HASB, the cable is not pulled into the bend by a large pulling force, but rather pushed through the bend by a lower, local air pushing force.

In push/pull blowing, the pulling from the piston is partially offset by the pushing force from the mechanical pusher. This somewhat offsets the effects of bends. A more detailed analysis of the bend performance differences in HASB versus push/pull can be found in **[Grif 93]**.

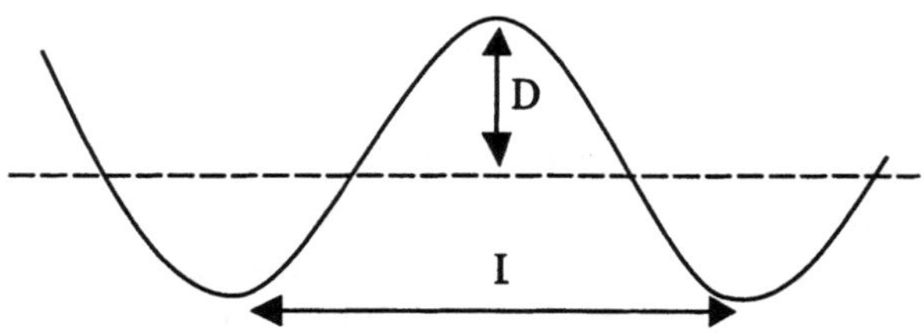

Figure 5.1.7. Signal shape.

Blowing installation of fiber optic cable has proven to be both practical and economical. The technology is not only gaining popularity in the USA and Europe, but throughout the world. The process is quite different from pulling, and requires a focus on different factors than pulling. An installer can quickly learn and use air-assist methods by attending to these factors.

Blowing processes generally install cable into duct at a higher speed than pulling. Economics can be favorable especially on large jobs where the cost of equipment can be amortized. Field crews report as much as 50,000 feet of cable installed per day. General cost factors are presented below:

	PULLING	BLOWING
Equipment	1–12 M $	15–45 M $
Speed	75–200 fpm[a]	200–350 fpm
Length	1500–3500 ft	3000–8000 ft
Pull Line	Required	Not Required

[a] fpm: Feet per meter.

5.1.7.4. *Sewer cabling*

A new method of cabling is emerging with strength enough to be established as best way for Last Mile cabling. Currently, some cities are reaping the benefits of there being more fiber in the sewer systems.

This system uses a robot to access which manoeuvres through sewers to install high-capacity fiber optic networks. It will provide thousands of businesses with high-speed, last mile connectivity to metropolitan-area network services providers around the city.

Businesses, ISPs and other telecommunications service providers will use the sewer networks to support voice, video and data services.

Highly populated cities seem more interested in this approach to updating telecommunications infrastructures because it is less invasive than traditional fiber optic deployments, which involve digging up city streets.

Utilization of existing sewer pipes/conduits confers many advantages, means less costs in terms of CapEx with a fast deployment versus other methods. Cities have sewer networks with exit to every street, so to make use of this advantage means to reduce drastically the infrastructure costs. In addition to this method are:

- Avoids the need for unsightly, costly and disruptive excavation work.
- Eliminates the costly and time consuming process to procure permits or licenses required by conventional methods.
- Ensure the timely completion of projects by using a method unaffected by weather.

The net result is faster, cheaper, more reliable technology to extend power and benefits of high speed service access through a metropolitan area. With its high productivity and efficiency, this technology is unmatched in the industry, bringing large financial savings and making it the most desirable solution on the market today.

Other advantages are:

- Strong and simple
- Eliminate excavation noise and pollution
- Eliminate excavation disruptions
- Quick, efficient, and cost-effective
- Non-damaging to pipes
- Non-obtrusive to sewers
- Low-cost materials
- Corrosion resistant
- Continuously measures gas concentration
- Changeable magazines

5.2. TRANSCEIVERS

5.2.1. Introduction

In this chapter we deal with the optical transceivers as the most important component in optical access systems. The availability, cost and integration level of this components are the main determining factors.

Optical sources and receivers are included as independent points in order to complete the study for future integrated transceivers solutions.

5.2.2. Integrated transceivers

5.2.2.1. *Introduction*

The semiconductor industry was born 40 years ago, while modern commercial optical component industry has been in existence for less than five years. So it is in its infancy. If compare with the revolution that happened in the traditional semiconductor environment, it seems clear that the evolutionary path of the optical component industry will have its turn in the next few years.

Figure 5.2.1 shows the evolution of the semiconductor industry. In the 1960s, when the semiconductor industry was in its infancy, it was characterized by vertical integration in two significant ways: first, chip design, fabrication, and second packaging functions were all performed within the same company;

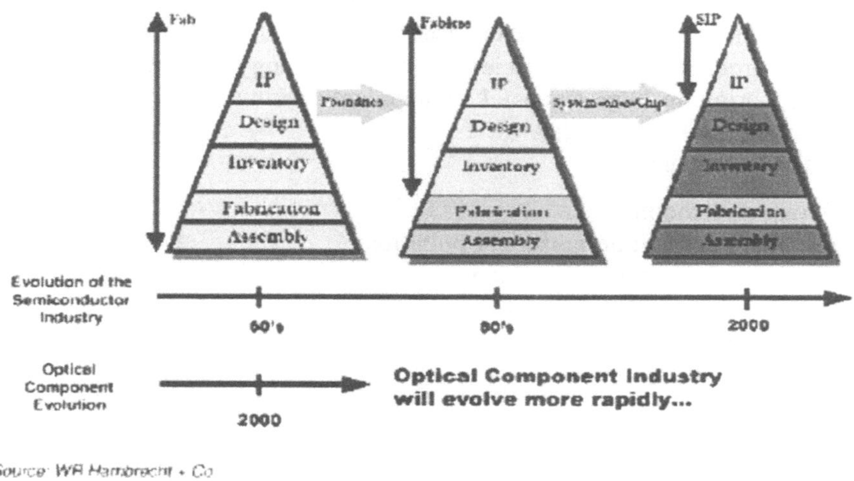

Figure 5.2.1. Evolution of the semiconductor industry compared with optical component industry (from **[HAM]**).

similarly, companies produced both the component-level chips and the box-level end products. The 1980s saw the emergence of foundries and packaging houses that provided dedicated fabrication and packaging services for fabless chip companies; furthermore, systems companies began to outsource their chip needs to merchant semiconductor companies. Today, the semiconductor industry has evolved a horizontal structure, with independent semiconductor intellectual property (SIP) providers, fabless chip design houses, foundries, and packaging/assembly houses.

Today's optical component industry is similar to the semiconductor industry in the 1960s, and the following four trends can be foreseen:

- *Outsourcing.* Today a significant portion of the optical component market is comprised of the internal component divisions of optical system companies. It is expected that in the near future optical systems companies will increasingly outsource their optical component needs to optical component suppliers. The expected spin-off of the component divisions of Lucent and Nortel underscores this trend.
- *Horizontal markets.* The optical component companies tend to concentrate on design, manufacturing, packaging and assembly, and testing functions. This is due to several factors: 1) the optical manufacturing process tends to be difficult and labor-intensive with very poor yields; 2) the optical manufacturing process is a significant differentiating factor and it provides barriers to entry as manufacturing processes are dependent on a company's unique intellectual property. However, as manufacturing volume increases, yield improves with automation, and more manufacturing standards develop across the industry, the optical component industry gradually evolves toward a horizontal structure.
- *Optical manufacturing equipment industry.* Just as the semiconductor capital equipment industry provides manufacturing equipment for semiconductor companies, an optical manufacturing equipment industry will emerge that provides the tools to manufacture, package, assemble, and test optical components.
- *Device integration.* The optical component market will move toward device integration. Initially optical component makers will focus on modules/subsystems solutions to package multiple components into a single box (current transceivers); subsequently, it is foreseen a shift toward hybrid integration to package multiple components into a single device; finally, it is expected that multiple optical components will be manufactured on the same substrate to produce single-chip solutions with the eventual goal of producing an "optical system-on-a-chip." Optical integration will have a profound impact on costs in the industry **[CIR]**.

5.2.2.2. *Characteristics*

How they work. Optical Transceivers contain both an optical transmitter and an optical receiver allowing two-way communications between two end points. The transmit module combines source lasers, modulators, wavelength lockers, and couplers with multiple passive components in order to convert incoming electrical signals to the optical domain. The receiver module, which combines a photodetector and other components, converts optical signals that have been sent through the fiber network to electrical signals.

Technology trends. While 850 nm VCSELs have been widely used in transceivers for datacom applications, currently transmitters and receivers for long-haul optical communications applications are based on edge-emitting lasers. Going forward, as long-wavelength VCSELs become commercially available over the next few years, it is expected that transmitters and receivers used in long-haul communications will also shift toward VCSELs as the source technology.

Vendors. JDS Uniphase, SDL Inc., Corning, Nortel, and Lucent are leading vendors. Emerging vendors include LightLogic, E2O Communications, and Cielo Communications.

A typical layout for a transceiver is shown in Figure 5.2.2 **[MIT]**. The optical modules used in an optical access network must be compact and low cost. Active research has therefore been directed at developing an optical transceiver module with hybrid integration of a laser diode, a receiver photo-diode, and wavelength division multiplexing (WDM) components on a silicon substrate. One design issue was how to suppress cross-talk between the transmitter and the receiver because the receiver should detect very weak signals whereas the transmitter should output powerful signals in the same package. Another was to meet the need for low cost. A description of the structure follows.

5.2.2.2.1. *Structure of an optical transceiver*

On a silicon substrate, this module integrates a 1.3 μm-band laser diode, a receiver photodiode, a polymer-waveguide chip, single-mode fiber (SMF), and multi-mode fiber (MMF) using passive alignment technology to build a hybrid integrated optical transceiver module. A transfer-mold package offers good mass productivity and low-cost packaging. At the same time, an SFF-type optical adapter is generally developed for a standard optical interface reducing cost by eliminating the fiber for connector adaptation. The polymer-waveguide chip is implemented for the ease of mass production and low cost. The chip has the WDM filter which reflects 1300 nm light and transmits 1550 nm light.

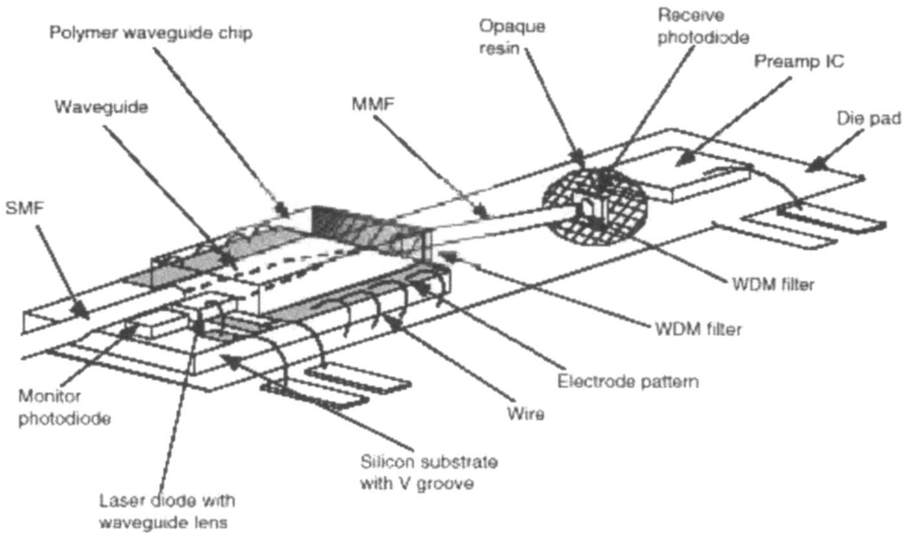

Figure 5.2.2. Typical configuration of an optical transceiver.

As noted above, a key feature of the optical transceiver module is the reduction of cross-talk between the transmitter and the receiver. Cross-talk is caused by optical and electrical coupling between the transmitter and receiver of the module. The following measures can be taken to prevent optical coupling between the laser diode and photodiode:

- A high-isolation WDM filter is applied to the edge of the polymer wave guide chip.
- Multi-mode fiber is inserted between the polymer waveguide chip and photodiode.
- The photodiode is surrounded by opaque resin.
- A WDM filter is formed on the end of the MMF.

The below measures can be taken to reduce electrical coupling between the transmitter and the receiver of the module:

- The laser diode is electrically isolated from the die pad with ground potential.
- An electrode pattern is formed on the silicon substrate between the transmitter and the receiver and is connected to the ground.
- The receiver is kept apart from the transmitter by inserting multi-mode fiber between the polymer waveguide chip and the photo diode.

By reducing optical and electrical cross-talk as described above, excellent receiver characteristics can be achieved.

Title	Source	Year	Authors
Manufacturing Technique of SFF Transceiver	Electronic Components and Technology Conf.	1999	Sumitomo Electric Ind. Ltd.
A 1.25 Gbit/s GaAs OEIC for Gigabit Ethernet	0-7803-5585-7/99 IEEE	1999	Industrial Technology Research Institute, Taiwan
2 Gbit/s Small Form Factor Fiber-Optic Transceiver for Single Mode Optical Fiber	IEEE TRANS. ON ADV. PACKAGING, VOL. 23, NO. 2	2000	Optoelectronic System Laboratory, Hitachi Cable Ltd.
Agilent Technologies' Singlemode Small Form Factor (SFF) Module Incorporates Micro-machined Silicon, Automated Passive Alignment, and Non-Hermetic Packaging to Enable the Next Generation of Low-Cost Fiber Optic Transceivers	IEEE TRANSACTIONS ON ADVANCED PACKAGING, VOL. 23, NO. 2	2000	Agilent Technologies, Suffolk, UK
A compact, low-cost WDM transceiver for the LAN	Electronic Components and Technology Conf.	2000	Agilent Laboratories, CA
A novel low-cost small-form-factor transceiver module	Electronic Components and Technology Conf.	2000	IBM Corp.
Single Mode Fiber MT-RJ SFF Transceiver Module using Optical Sub Assembly with a New Shielded Silicon Optical Bench	Electronic Components and Technology Conf.	2000	The Furukawa Electric Co.
Characterizing Multimode Fibers and Sources for Gigabit Ethernet Applications	0-7803-5947-X/00 IEEE	2000	Corning Inc.

Uniform and high-performance eight-channel bent waveguide SOA array for hybrid PICs	IEEE Photonics Technology Letters, Vol. 13, No. 5	2001	NEC Corporation
Design of Sampled Grating DBR Lasers with Integrated Semiconductor Optical Amplifiers	IEEE Photonics Technology Letters Vol. 12, No. 7	2000	Agility Comm., University of California, Santa Barbara
InGaAsP/InP scaleable, photonic cross connects using optically amplified supressed modal interference switch arrays	Integrated Photonics Research Conf	1998	University of California, Santa Barbara
Widely tunable integrated filter/receiver with apodized grating-assisted co-directional coupler	SPIE Photonics West	1998	University of California, Santa Barbara
Widely tunable Sampled Grating DBR Laser with Integrated Electroabsorption Modulator	IEEE Photonics Technology Letters Vol. 11, No. 6	1999	University of California, Santa Barbara
Different papers about VCSELs	SPIE Special Issue "VCSELs in Information Systems"	2001	Many
A Monolithically Integrated 1-Gb/s Optical Receiver in 1-m CMOS Technology	IEEE Photonics Technology Letters, Vol. 13, No. 7	2001	Technical University Vienna, Alcatel SEL
Efficient WDM Cross-Connect using widely tunable FBG	CLEO	2000	University of Southern California
Improvement of Cost-effective Optical Module for Optical Access System	Electronic Components and Technology Conf.	1999	Rohm Co., Ltd., Japan
A cost effective ceramic package for LD/PD module in subscriber system	Electronic Components and Technology Conf.	1998	KYOCERA Corp., Japan

Figure 5.2.3. Proposals to implement cheap transceivers.

The package accounts for 60 to 80% of current manufacturing expenses in optoelectronic component assembly **[PAL]**. Advances in packaging automation are critical to reducing costs in the optical networks. Automation provides the following benefits:

- Yield improvements.
- Throughput increases.
- Productivity increases.
- Reduced time to scalable volume.

A lot of effort has been lately spent in two main fields: integration and automation. These are one of the key issues for the success of the optical access networks as it is not possible to spend so much money for every user. There are some interesting IEEE references which deal with different proposals to implement cheap transceivers.

Company	Product	Web	Comments
AMP	ESCON-MM-MTRJ Transceiver	www.amp.com	200Mbit/s
Blaze	Afterburner	www.BlazeNP.com	10Gbit/s 1Gbit/s
Cognet Microsystems	VCSEL Drivers	www.cognetmicro.com	10Gbit/s
E20	GBIC, Paralel-LC, 1 × 9 GbE	www.e2oinc.com	Many products
Emcore	VCSEL CWDM	www.emcore.com	10Gbit/s
Gore	nLighteen	www.gore.com	Tx+Rx, VCSEL, GbE
Honeywell	HFT-248x-521	www.Honeywell.com	MTRJ SFF 1.25 Gbps OFE
HPLabs	1300nm mt-rj × 4 Transceiver	www.hpl.hp.com	Uncooled DFBs
IBM	SFFFamily	www.ibm.com	Many products
JDS		www.jdsunph.com	Many products
Opt. Comm.Prod	DTR-1250-LC		Transceivers, VCSEL,GbE
Stratos	GBIC	www.stratoslightwave.com	

Figure 5.2.4. Several commercial transceivers.

Finally, some commercial transceivers (mainly involved in the Gigabit Ethernet arena) are shown in the next table that have been found in the web (other related products are given in section 5.2.3.5 VCSELs).

5.2.3. Light sources

5.2.3.1. *Introduction*

One of the most important contributions in communications engineering is the invention of lasers. Two people are responsible for this revolution. In the 1940s and early 1950s, Charles Townes and Arthur Schawlow were deeply interested in the field of microwave spectroscopy. What they wanted was to develop a device to help them study molecular structures using microwave radiation of short wavelengths, but Townes realized how he could use stimulated emission from molecules and atoms to amplify waves which. This led to his invention of the MASER (Microwave Amplification by Stimulated Emission of Radiation). The idea of Schawlow was to arrange a pair of mirrors one at each end of a region of excited atoms or molecules. In the fall of 1957, they began working out the principles of a device that could provide these shorter wavelengths.

In May 1960, an american physicist Theodore Maiman built the first laser to produce a pulse of coherent light using synthetic ruby as the laser medium. The first continuously operating laser suitable for communications was achieved a few months later in July 1960. In 1975, engineers at Laser Diode Labs developed the first commercial continuous-wave semiconductor laser operating at room temperature. This opened the door to the use of lasers to transmit optically encoded telephone conversations over fiber-optic cables.

Since then, the semiconductor laser has been widely studied to achieve better benefits in the use of optical fiber. Several works of telecommunications companies in the world have been dedicated to the research of the laser structure to reach these aims launching into the market the Fabry-Perot (F-P) laser and the Distributed Feedback Bragg (DFB) laser. Also, the light emitting diode (LED) was developed which is used as a light source in communications of short reach.

In the last decade when WDM technology was developed, the laser had to evolve much more to be able to transmit several wavelengths by the same fiber, then the tunable laser appeared to generate different wavelengths. This allows to obtain greater simplification of the systems and to save costs in inventory devices and equipment maintenance.

Currently, the last technology in light sources for optical access telecommunications is the Vertical Cavity Surface Emitting Laser (VCSEL) which is also now developed for the WDM systems. VCSELs are right now the best low cost solution for access as they present the better cost-performance relation.

In this section, the different types of light sources and their applications in optical telecommunications are reviewed.

5.2.3.2. *Light emitting diode (LED)*

There are two fundamental processes that can explain light emission in a semiconductor pn junction: Spontaneous emission and stimulated emission. Light emitting diode (LED) is a pn-junction that emits light by spontaneous emission process while laser mainly emits light by stimulated emission.

Spontaneous emission occurs when electron-hole pairs of a LED are recombined, then, a photon is emitted and light is produced (radioactive recombination). This light is incoherent with a relatively wide spectral width of 30–60 nm approximately. Although the emitted light is propagated in random directions, LEDs can be coupled to optical fiber to transfer information to short distances. Thus, LED is used as an inexpensive source to inject light into the fiber with low coupling loss. However, it cannot couple enough light into a single-mode fiber to facilitate long-distance communications due to an inefficient concentration of enough power into a very small core. Figure 5.2.5 shows two structures of LED used as source in optical networks.

The materials used at the junction determine the wavelength of the emitted light. Since LEDs of Silicon (Si – 1100 nm) and Germanium (Ge – 1810 nm) are not efficient producers of light, other compound semiconductors are used which generate specific colors. Aluminum indium gallium phosphide (AlInGaP) is used to develop long visible wavelength for yellow, amber and red LEDs. For shorter wavelength LEDs as green and blue, the material is indium gallium nitride (InGaN). For applications in optical communications, the companies use LEDs with wavelength center in 1310/1550 nm.

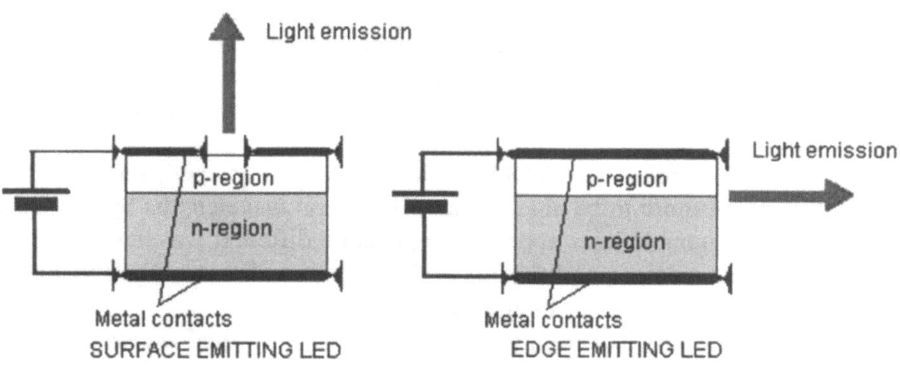

Figure 5.2.5. Light emitting diode Structures.

Optical and electrical features

The most important typical parameters of LEDs for communications optic systems are summarized in the following table (Figure 5.2.6):

Parameter	Value		Unit
Wavelength	1310	1550	nm
Spectral Width	min: 30 max: 80	min: 50 max: 110	nm
Operating Current	40		mA
Output Power	7 to 45 (singlemode) 20 to75 (multimode)	7 to 20 20 to 30	micro-W
Rise Time/Fall Time	1.5 to 3		ns
Forward Voltage	1.5		Volts
Operating Temperature	-20 to +75		°C
LED reverse voltage	2.5		volts

Figure 5.2.6. Parameters of light emitting diode (LED).

Should LEDs have pigtailed Mini-housing, the output power can become 20 to 200/20 to 40 micro-W for 1310/1550 nm singlemode operation and 75 to 300/40 to 100 micro-W for multimode operation.

Applications

Early 1990s, Philips built a LED driver for fiber optic communications that was tested over a fiber optic link of about 0.5 meter long to 50 and 100 Mbit/s (NRZ) data rate. A current source was driven to the anode of an LED when the input data was a logic HIGH and switched to ground when the input data was a logic LOW. Due to the fact that LED is a non-linear device with non-uniform impedance and non-linear transfer characteristics, it was harder to turn off than to turn on. In order to improve the optical performace in the LED, pre-biasing and pre-charging techniques were used. It was noted that some LEDs had higher power output, but lower data rates and others had lower power but higher data rates [www.semiconductors.philips.com].

LEDs have a typical power ranging from –20 to –15 dBm. These have application only for short transmission distances. Currently, LEDs are used in LANs and access up to about 2 Km and up to 200 Mbit/s. Some companies, as E2O Communications, manufacture transceiver with low cost sources as LEDs for the use in Fast Ethernet applications. This complies with specifications for IEEE-802.3z Fast Ethernet at 125 Mbps and with Multi-Source Agreement (MSA) operating with 50 micro-m and 62.5 micro-m multimode optical fibers.

In the last years, researchers have been investigating the use of LEDs in Wavelength Division Multiplexing (WDM) techniques for FTTH since the use of wavelength tunable Distributed Feedback Bragg-reflector (DBR) lasers or arrays of DFB are not affordable solutions for local access applications.

Crawford Hill Laboratory of Lucent technologies in USA has demonstrated a 24-channel WDM transmitter for local access networks with 200 GHz channel spacing by applying a novel loop-back spectral slicing. The experimental setup in Figure 5.2.7 shows how a WDM router in a wavelength comb of WDM channels spectrally slices the broad emission spectrum from a 1.55 μm edge-emitting LED. In this form, a relatively simple WDM source with eight wavelength channels for the access systems is obtained by combining spectral slicing of a LED with optical timing delay and using the bit-interleaved data modulation **[Liou 98]**

An 8-channel WDM system for local access network had been reported previously by the same group **[Liou 97]**. They also reported a system in 1995 that combined an edge emitting LED and a semiconductor single-traverse mode amplifier to be used as a broadband 1.3 μm source to achieve both high power and high modulation bandwidth. By using broad linewidth LED source, optical beat interference (OBI) can be avoided.

One single waveguide grating router (WGR) is used with the LED reducing the optical insertion loss and component cost for the access transmitter.

An amplifier-modulator amplifies and pulses the broadband LED output producing a train of 0.83 ns pulse repeating at 50 MHz. The direct modulation speed of the LED, about 600 Mb/s, is limited by the spontaneous recombination lifetime. The pulse output is further amplified by an Er-doped fiber amplifier (EDFA) followed by a WGR. By inserting fiber delay lines between the output ports and input ports of the WGR as shown in Figure 5.2.7, a pulse train of all the WDM channels can be generated.

Figure 5.2.8 shows the spectrally sliced output power spectra of 24 WDM channels and the output pulse train of the WGR. Applying complementary elec-

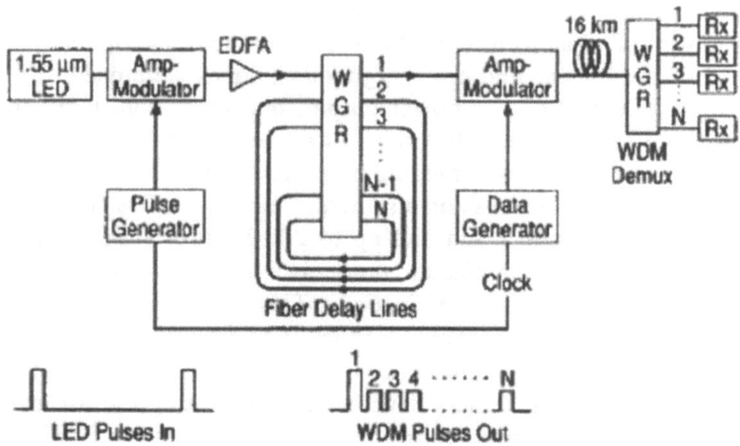

Fig. 5.2.7. Experimental setup 24-channel WDM transmitter for access.

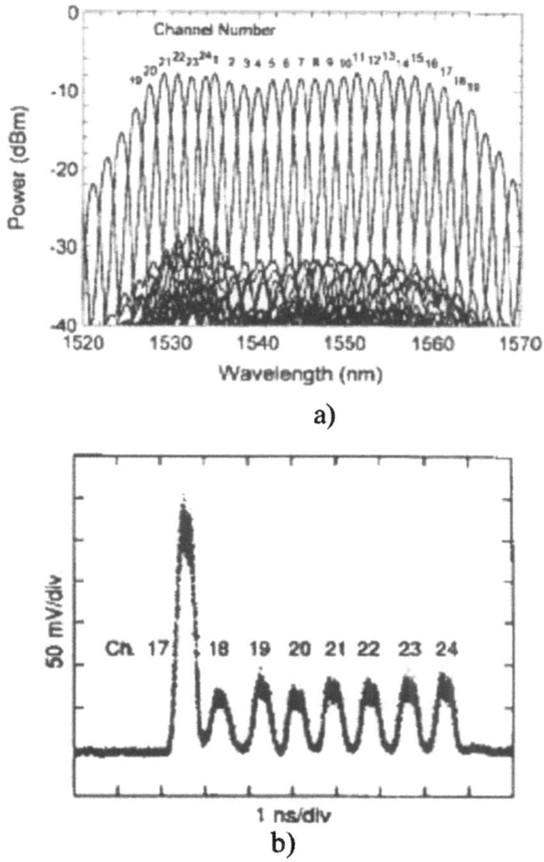

Figure 5.2.8. a). Output power spectra of 24 WDM channel. b). WDM pulse train from an 8 × 24 WGR.

trical pulses to the data modulator located after the spectral slicing, WGR can equalize the pulse train.

The Department of Electrical Engineering of Korea Advanced Institute of Science Technology has implemented PONs for CDMA-based microcellular communication systems with optical fiber using low-cost LEDs instead of wavelength-selected DFB lasers for upstream signals links. Thanks that the high power and narrow linewidth are not required **[Kim 01]**. In addition, the broad linewidth of LEDs suppresses the OBI noise to an acceptable level. In the same way, LEDs also have been used in **[Jung 98]** to provide the upstream channels in a WDM PON architecture proposed by this same institute, as shown in Figure 5.2.9 A directly modulated LED was used at each subscriber's site to transmit a 155 Mbits/s upstream channel. The output power was –5.5 dBm and 3-dB bandwidth was 60 nm.

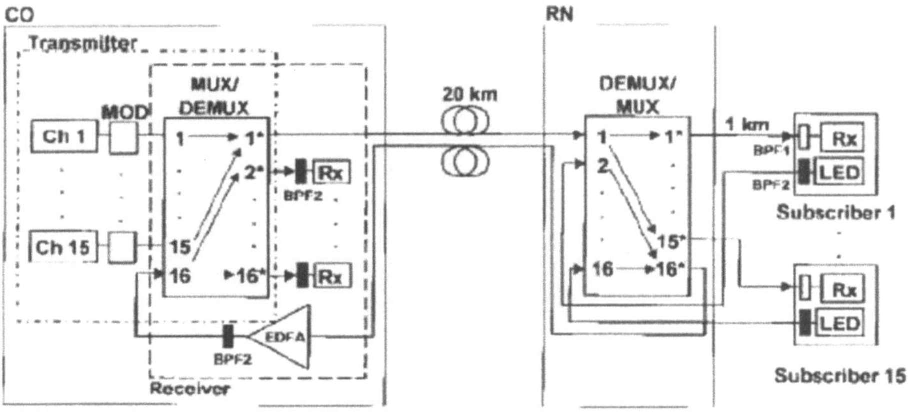

Figure 5.2.9. Experimental setup WDM PON architecture.

It was necessary to reduce the spectral width of LEDs by using a bandpass filter centered at 1540 nm to avoid in-band crosstalk at the subscriber's receiver and reduce the effect of dispersion. The The output power of LED was also reduce to about –16 dBm. Therefore an EDFA was placed in at the central office compensating also the slicing, insertion and fiber losses. In this system, the downstream and upstream channels are in different wavelengths at about 1550 and 1540 nm, respectively, as shown in Figure 5.2.10.

In conclusion, though LEDs produce a broad spectral light and low outpower, they produce a spread beam difficult to focus on an optical fiber, and have some advantages such as low cost, no-effect OBI due to broad spectra. These are creatively utilized by the researchers to built new low-cost access architectures using broadband transmitter and spectrum-sliced sources.

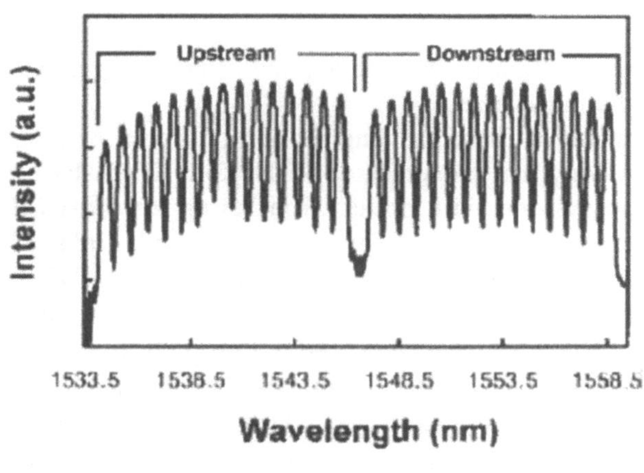

Figure 5.2.10. Downstream and Upstream Channels Spectrum.

5.2.3.3. *Laser Diode (LD)*

Currently, the source most used and developed extensively after 1970 is the semiconductor laser because it presents several advantages over LEDs and other kind of lasers, such as: high efficiency, narrow angular spread of the output beam, narrow spectral width of the emitted light that allows operation at high bit rates to be modulated directly at high frequencies, compact size, and the narrow beam of light of the semiconductor diode laser that allows to provide high concentration of light energy to be coupled to a single-mode fiber.

Semiconductor diode laser mainly emits light by means of a stimulated emission process which is produced when an atom in an upper energy level is stimulated in phase by an incoming photon of a specific energy. The excited atom may return to the normal state (ground state) with the consequent release of a photon. The excited photon matches the incident photon not only in energy or frequency, but also in phase and direction of the propagation.

The semiconductor diode laser is formed from heavily doped pn-junctions deposited between two mirrors based on modified light-emitting diode structure as shown in Figure 5.2.11. To achieve the laser action, high concentration of electron-hole pairs available need to be ensured for recombination. This is achieved by high doping concentrations across junction and long spontaneous lifetime materials. By applying a voltage through the junction p-n, the recombination of electron-hole pairs is produced and light energy is released. Light that is spontaneously generated is reflected back and forth causing stimulated emission. This is the fundamental principle of the operation of light amplification by stimulated emission of radiation (LASER) when the substance (lasing medium) is put between.

The length of the cavity formed by mirrors determines the frequency of

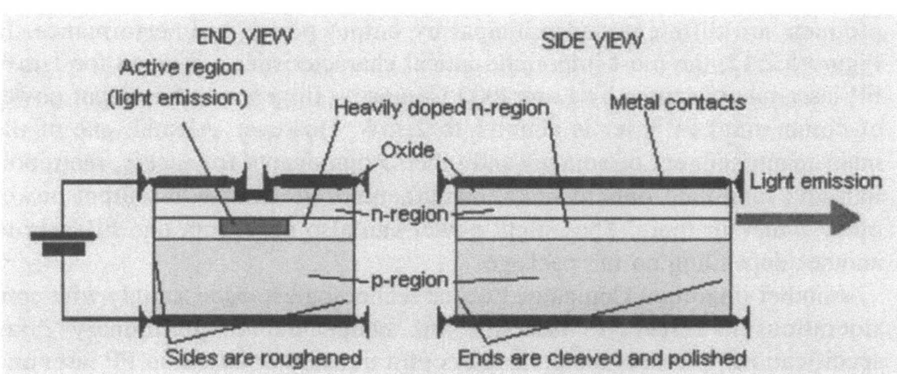

Figure 5.2.11. Laser diode Structure.

emitted light. Wavelength for which an integer multiple of the wavelength is equal to twice the cavity length could be amplified **[Borella 97]**.

The range (633 nm to 1.55 μm) is available in a laser diode that depends on material or structure. Some materials are AlGaInP, GaAlAs, and InGaAsP.

5.2.3.3.1. *Fabry-Perot Laser*

Fabry-Perot (FP) laser is an edge-emitting laser that is formed by an optical resonant cavity called Fabry-Perot resonator. The optical resonator consists of a low-loss dielectric optical waveguide terminated by cleaved facet mirrors. One mirror is totally reflective and the other is partially reflective. This structure allows optical feedback such that the stimulated beam is made to pass backwards and forwards several times. The cavity amplifies only the wavelengths that accomplish the condition of resonant attenuating the other wavelengths. The resonances of such an optical cavity are equally spaced in frequency v by $\Delta v = c/2nL$, where c is the speed of light, n is the group index of refraction of the waveguide, and L is the chip length between facets. Due to its multi-mode longitudinal resonance principle, the output of a typical FP laser is a small number of modes spaced in wavelength by about 1 nm. This can generate slight fluctuations that sometimes cause FP laser to vary among wavelengths, which contribute to the appearing of chromatic dispersion over the signal.

Optical and electrical features

The center wavelength is given as a nominal value at 25 °C with typically a tolerance of 10 nm. typically. This wavelength varies with temperature between 0.2 nm–0.5 nm/°C depending on the diode. Additionally, when temperature varies, FP laser produces the mode-hopping where the wavelength changes by a discrete value increasing the partition noise. This is a disadvantage for some applications where wavelength accuracy is necessary.

There are a lot of companies that manufacture semiconductor laser whose products are differentiated principal by output power and performance. In Figure 5.2.12, the most important optical characteristics are given for 1 mW FP laser manufactures by Laser2000 Company. In general, the output power of commercial FP laser is about 1 to 2 mW. However, Alcatel, one of the main manufacturers of sources and others components for access, metropolitan and long-haul optical links, manufactures FP laser with output power up to 7 mW or more. The output power can also vary from one FP laser to another depending on the package.

Another important Company, Lucent Technologies, made a study with considerations in respect to lasersfor link budget and in preliminary 2 km specifications, presents the parameters of a transmitter based on FP laser and standard single mode fiber (See Figure 5.2.13).

Parameter	Typical	Units
Wavelength	1270 to 1330	nm
Spectral with	2	nm
Threshold Current	25	mA
Operating Current	45	mA
Output Power	1	mW
Rise/fall Time	0.5	ns
Forward Voltage	1.2	V
Tracking Error	±1	dB
Monitor Current	0.05	mA
Monitor Dark Current	0.3	μA
Monitor Capacitance	10	pF

Figure 5.2.12. Typical values of Fabry-Perot (FP) laser.

Parameter	Min	Max	Units
Signal Speed		11.25	GBd
Average Power	-5	0	DBm
Wavelength	1290	1330	Nm
RMS Spectral Width		2.5	Nm
Extinction Ratio	7		DB
10%-90% Rise/fall time		35	Ps
Relative Intensity Noise		-125	dB/Hz

Figure 5.2.13. Transmitter parameters for preliminary 2 Km specifications.

FP lasers also are manufacture for applications in 1.5 μm wavelength. However, if narrow linewidth, high side mode suppression,and stable wavelength vs. temperature are required, then a DFB laser may be more suitable.

Applications

Near the chromatic dispersion zero of conventional single-mode fiber, the spectral width of FP laser can permit transmission to distances of 40 Km at the speed of 1.7 Gbit/s. However, the wavelength center of laser operation goes out of the wavelength tolerance at the speed of 2.5 Gbit/s and higher.

Lower fiber loss at 1.5 μm with values below 0.2 dB/Km allows loss-limited to up 100 Km or more with FP laser to span. It was thought to offer coarse wavelength division multiplexing (CWDM) with channels at both 1.3 μm and 1.5 μm simultaneously on one fiber, but the high dispersion for conventional fiber at 1.5 μm disallowed the use of FP laser.

FP laser offers: sufficient benefits in LANs, access and systems that use one wavelength per fiber, however, it is not suitable for long distances, coherent reception or wavelength division multiplexing due to the broad linewidth. However, Its multi-mode feature and major output power than LED has allowed

FP lasers to be used in some wavelength-division multiplexed access and local-area networks as a multiwavelength source of amplified spontaneous emission suitable for spectrum-sliced. The potential transmission distance of a PON is increased due to the fact that uncooled and unisolated lasers have over ten times the output power of LEDs and, moreover, they are less expensive. A wavelength-grating router (WGR) at the remote node would spectrally slice the broadband light, so the light from each ONU would reach the central office. To achieve this, the mode spacing of the WGR and the laser must be compatible and the spectrum of FP laser must be broadened to insure that light will be transmitted though each port of the WGR and the noise and distortion must be acceptable. AT&T laboratories-research in **[Woodward 98]** broaden the optical spectrum driving the laser into clipping. That is, the modulation depth is increased, so that the laser is driven below threshold. Figure 5.2.14 shows the typical optical spectrum of a commercial FP laser and the spectrum with a 2 GHz clipping-tone present.

It is thought that by accessing to several customers with WDM or time-division-multiple access (TDMA) would require the use of another WGR followed by separate receivers for each channel, then the system would became more expensive, thereby subcarrier-multiplexing (SCM) is used to provide access to multiple customers. The experimental setup is shown in Figure 5.2.15 where 10 Mbit/s was demonstrated over 18 Km of fiber.

Department of Electrical and electronic Engineering, University of Melbourne, Australia, in **[Holloway 97]** also used a FP laser with EDFAs to implement a multiwavelength source for access network of which scheme is shown in Figure 5.2.16.

The FP laser generates a comb of ASE peaks uniformly separated in frequency and of approximately uniform intensity over the EDFA. The obtained

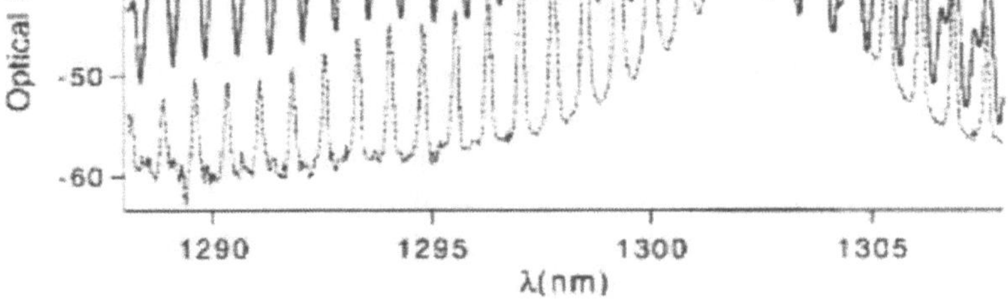

Figure 5.2.14. Typical optical spectrum of a FP laser and into clipping.

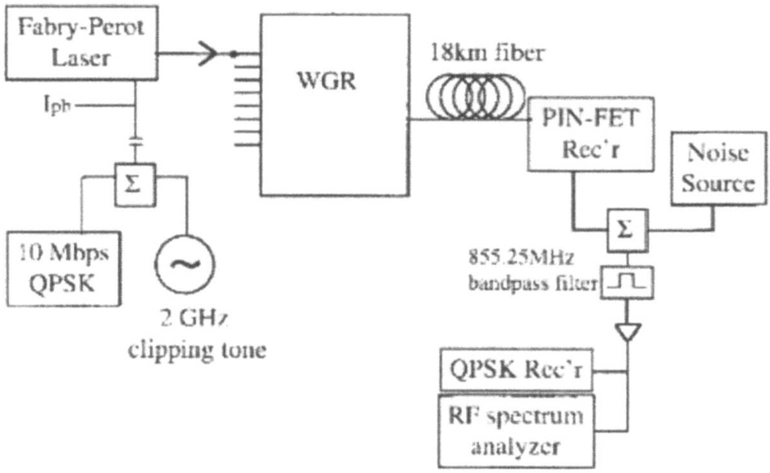

Figure 5.2.15. Experimental setup of SCM system at 10 Mb/s.

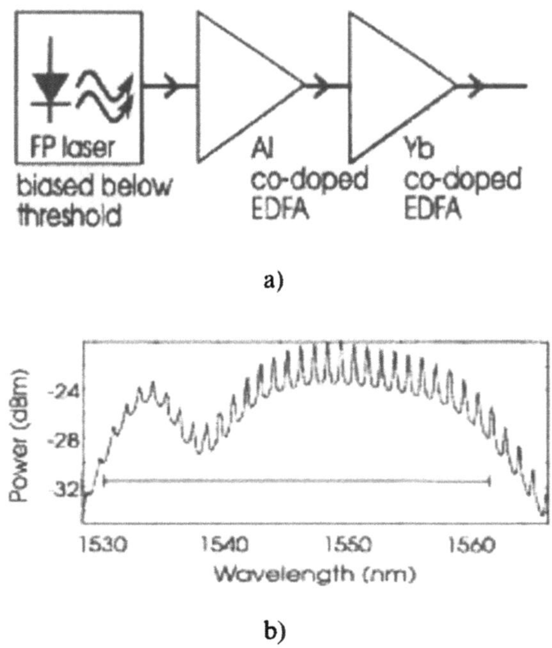

Figure 5.2.16. (a) Schematic diagram of the multiwavelength source. (b) Measured optical spectrum.

optical spectrum is shown in Figure 5.2.16. The length of the laser cavity determines the channel spacing. A longer cavity produces closer channel spacing. The channel bandwidth and center wavelength of the source can be independently tuned by changing the laser bias and temperature, respectively.

This multiwavelength ASE has several attractive features as channel wavelength tuning using a single parameter, uniformity of channel powers over a broad-spectral range, and adequate contrast between channels for wavelength-locking to external filter.

In conclusion, FP laser is perfectly suited in optical links requiring long term stable power, and stable wavelength when a proper temperature control is used. They can be used in applications at 1.25 Gbit/s upstream link over 10 Km whether FP laser has bandwidth (RMS) less than 2.8 nm. However, the broad optical spectrum of FP laser allows to be used for spectrally sliced PONs.

5.2.3.3.2. *DFB Laser*

DFB lasers are more advanced structures than FP lasers and exhibit ideal source characteristic for high bit-rate systems. Output power and wavelength stability, accuracy and fiber compatibility, therefore are inherent to the lasers. Their narrow linewidth and high-power make them ideal sources to be used in WDM networks, though, they present a higher cost for access WDM techniques due to their greater complexity.

DFB laser structures replace the cleaved facet mirror in FP laser with an optical feedback from a corrugated waveguide grating or periodic grating structure above the active layer, as illustrated in Figure 5.2.17. The light undergoes a multitude of sub-reflections through this grating structure from each corrugation period. Only a accumulative net reflections near the Bragg wavelength λ_B, given by $\lambda_B = 2n\Lambda_g$, are phased, where n is the phase refractive index of the waveguide mode and Λ_g is the spatial period of the corrugation. Thus, DFB laser provide a highly single longitudinal mode operation with a suppression rate of others modes of 30 dB or more.

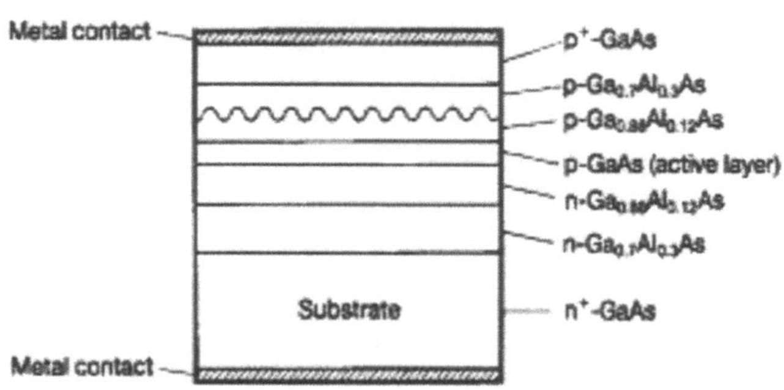

Figure 5.2.17. DFB laser structure.

Optical and electrical features

DFB laser is the most developed light source among semiconductor diode used for single-wavelength and WDM systems. However, DFB lasers are not extensively used for access because this would be a high cost for subscribers. Typical characteristics for a commercial DFB laser are given in Figure 5.2.18.

Parameter	Typical	Units
Wavelength	1530 to 1570	nm
Wavelength/Temp. Coef.	0.08	nm/°C
Spectral with	50	MHz
Threshold Current	20	mA
Operating Current	40	mA
Output Power	1 to 50	mW
Rise Time	0.3	ns
Fall time	0.4	ns
Forward Voltage	1.2	V
Thermal resistance	10	KΩ
Cooler Current	0.6	A
Cooler Voltage	1.6	V
Monitor Current	0.05	mA
Monitor Dark Current	0.3	µA
Monitor Capacitance	10	pF

Figure 5.2.18. Typical values of distributed feedback (DFB) laser.

Commercial DFB laser can achieve transmission in conventional fiber over distances in excess of 200 Km at 2.5 Gbit/s with negligible dispersion penalty. High yield of DFB lasers makes them unaffordable to be used as subscriber light source. However, some reductions in performance of DFB laser can trade especially against cost that in some cases can become up to 1000 times more expensive than a basic FP laser. On the other hand, the high performance and the great advantages make DFB lasers worth the extra money especially when they are placed in the central office.

5.2.3.4. *Tunable laser*

Actually, it is desirable to have a wide wavelength tuning range as well as fast and accurate access to the wavelength of interest in an optical network that uses multi-wavelength. Therefore, wavelength tunable single-mode semiconductor lasers are interesting components in the growing field of fiber optics. They are finding usefulness in a variety of applications in network reconfiguration, dynamic bandwidth allocation and wavelength protection, as much WDM as for optical access network **[Modiano 99]**. Although tunable lasers are still a relatively young technology, on a longer term, they could even replace fixed-wavelength lasers not only providing significant inventory savings, but also simplifying equipment operation and maintenance.

Currently, several structures for tunable sources such as Grating Coupled Sample Reflector (GCSR) and Super Structure Grating Distributed Bragg Reflector (SSG-DBR), MEMS technology, remain under intensive investigation.

5.2.3.4.1. *The Distributed Bragg Reflector (DBR) laser*

Distributed Bragg Reflector (DBR) laser is also based on the same corrugated structure as the simplest DFB laser. The difference between DFB and DBR laser is that the grating structures in DBR laser is outside of active region to avoid wavelength variations by thermal effects when injecting current in this region.

Figure 5.2.19 shows the structure of a three-section DBR laser. The three sections are the active, phase, and Bragg section, with each section electrically isolated from the other sections and current is injected through the electrodes of each section. The optical cavity includes all three sections making an entire unit.

The active section is highly doped and provides the gain necessary for lasing, whereas the phase and Bragg sections are passive and lightly doped. The tuning processes rely on changing the refractive index through injecting current into the passive regions. Coarse tuning is achieved by means of Bragg wavelength control, varying the applied current to Bragg region, and the same way fine tuning is achieve varying the current in phase section. Experimental results of coarse and fine tuning carried out with a commercial DBR, which were obtained in the Optical laboratories of Politechnical University of Catalonia, are shown in Figure 5.2.20 **[Souza]**.

Agere Systems, USA, has deployed a practical wavelength selectable source formed of five sections as shown in Figure 5.2.21. This has been developed in a highly integrated InP chip comprising of a gain section, a DBR laser coupled to a semiconductor optical amplifier (SOA), power monitor and

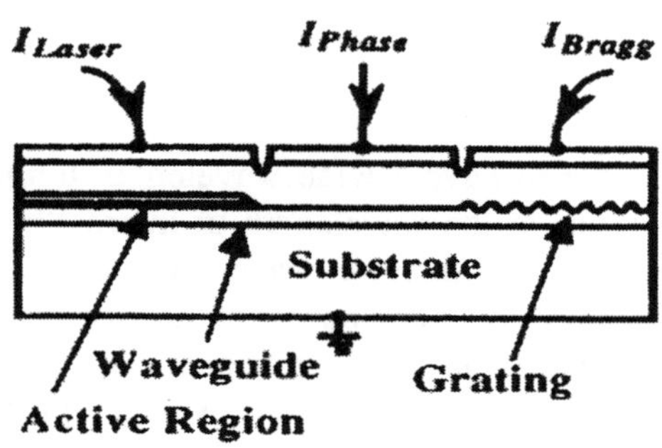

Figure 5.2.19. The structure of a three-section DBR laser.

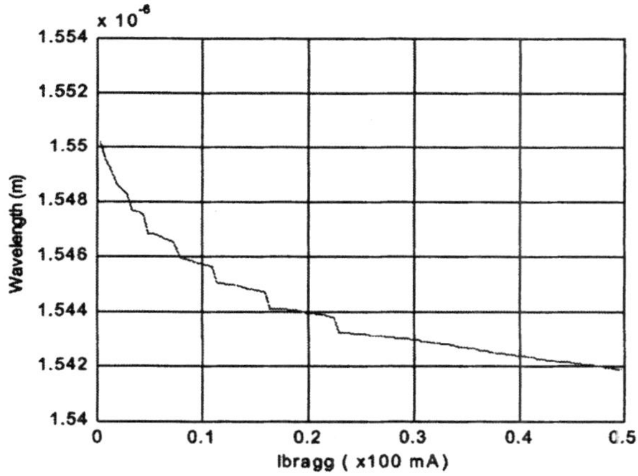

a)

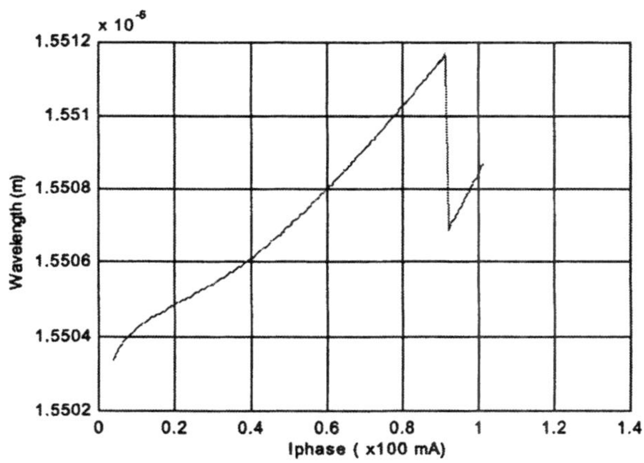

b)

Figure 5.2.20. Tuning of DBR laser. a) Coarse tuning. b) Fine tuning.

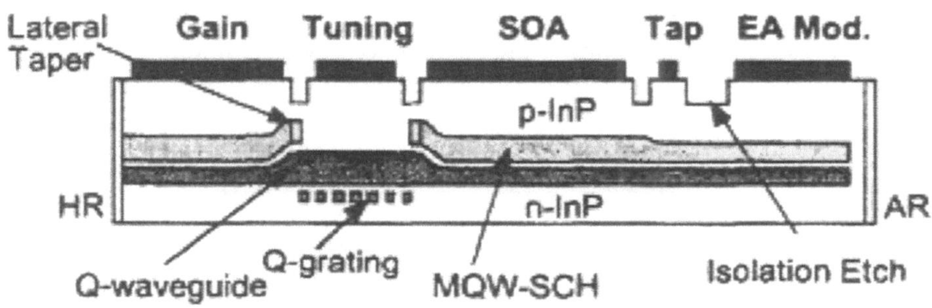

Figure 5.2.21. Schematic representation of the EA-DBR laser.

(optional) electro-absorption (EA) modulator. EA-DBR version is capable of error-free standard-fiber transmission at 2.5 Gb/s over 680 km, and 10 Gb/s over 80 km on each of twenty-five 50 GHz-spaced channels. A CW version, which is used with an external modulator, has also been developed producing 20 mW fiber coupled output across thirty-two 50 GHz-spaced channels **[Ackerman 01]**.

5.2.3.4.2. *The Grating Coupler Sample reflector (GCSR) laser*

Using a Vertical Grating Assisted Co-directional Coupler with Sampled Grating Reflector (GCSR) Laser it extends the tunability. In essence, the GCSR laser functions as multiple DBR lasers combined into one. The GCSR laser is a four-section semiconductor laser which shares several similarities with the DBR laser as can be showed in Figure 5.2.22.

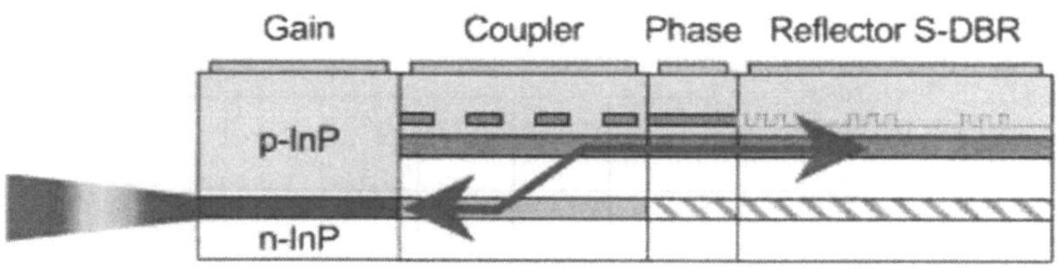

Figure 5.2.22. Schematic cross-section of GCSR laser.

Unlike a conventional DBR laser, the GCSR laser incorporates a sampled reflector grating which generates a comb of reflection peaks spaced a few nm. The tunable vertical-cavity coupler acts as a filter that selects a wavelength. When current is injected into the sampled grating section, each peak shift towards a shorter wavelength due to the refraction index of waveguide changes.

5.2.3.4.3. *The Super Structure Grating (SSG-DBR) laser*

A SSG-DBR laser is illustrated in Figure 5.2.23. A SSG-DBR laser consists of two Bragg grating reflectors at each end, a laser gain section, and a phase tuning region that allows for narrow continuous wavelength tuning. The front Bragg grating reflector is chirped with a spatial grating frequency varying from

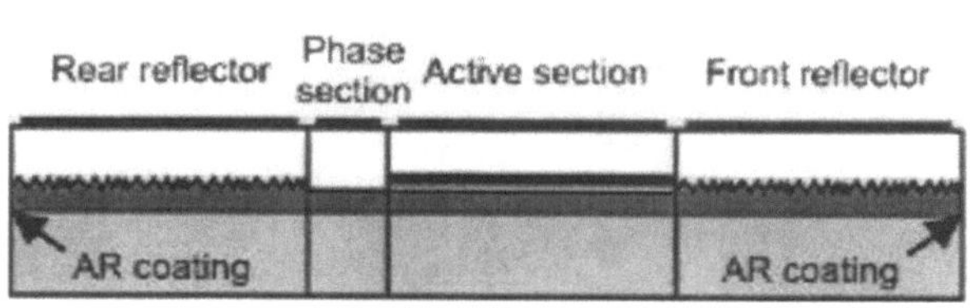

Figure 5.2.23. Schematic cross-section of an SSG-DBR laser.

Λ_a to Λ_b producing multiple spectral reflection peaks with a mode spacing $\Delta\lambda_s$ **[Rigole 95]**. The rear grating is constructed in a similar manner except with a slight offset in the grating pitch spacing Λ_s as is shown in **[Biernacki 99]**.

These lasers exhibit an inherent fast tuning ability about ns speeds via three wavelength-control currents injected into the rear reflector front reflector, and phase control regions of the laser. In addition, the laser has a gain region that controls the output power. For stable single-mode operation, a peak of each reflection comb and a longitudinal mode has to be aligned. The SSG reflectors can be tuned by current injection. A typical tuning characteristic made by the Department of Information technology, University of Gent, Belgium **[Sarlet 00]** is shown in Figure 5.2.24.

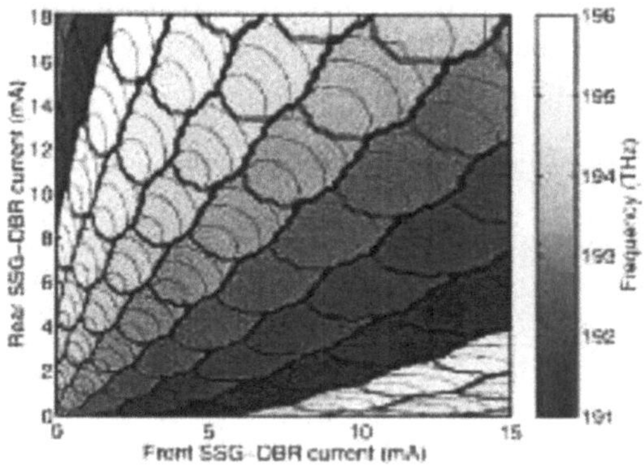

Figure 5.2.24. Tuning characteristic of a typical SSG-DBR laser.

The concept behind the chirped grating can be understood by noting that a grating with period Λ will only reflect wavelengths that satisfy the Bragg condition $\lambda = 2n\Lambda$. Therefore, if the grating period changes from λ_a to λ_b (shown in Figure 5.2.25(a)), all wavelengths $\lambda_a > \lambda > \lambda_b$ will be reflected provided that the condition $\lambda_a = 2n\Lambda_a$ and $\lambda_b = 2n\Lambda_b$ is satisfied. During fabrication, the gratings are made such that the grating period is actually step-wise discrete. Thus, the Bragg condition is actually met in a discrete manner making the lasers quasi continuously tunable. Single-mode oscillation occurs by making the rear SSG with slightly different mode spacing than the front section. In this manner, laser oscillation will take place only when both reflection peaks of the rear and front gratings align with each other.

Coarse tuning of the laser is achieved by injecting current into the Bragg reflector regions. Current injection decreases the index of refraction and shifts the reflection peaks to a slightly shorter wavelength. If one of the reflectors is tuned, different pairs of reflector peaks will successively overlap and the lasing frequency will jump by approximately the peak spacing. Therefore, the laser output wavelength can now be tuned by shifting both reflection peak spectra until the desired overlap containing the wavelength of interest is achieved. This process is shown visually in Figure 5.2.25(b).

Coarse wavelength tuning produces a mode-hopped spacing of about 5 nm for the front section and similarly about 5.5 nm for the rear section. Medium tuning is obtained by shifting both reflectors simultaneously. Fine-tuning of the laser wavelength is achieved by changing the longitudinal mode (optical cavity length) of the DBR laser through injection of the phase control current and has a range of about 0.4 nm. It is important to note that there is periodic tuning (i.e., a repeating sawtooth structure) in each of the three currents such that there is multitude of available currents that can provide the same correct lasing wavelength. Consequently, by selecting an appropriate combination of front DBR, rear DBR and phase current, the laser can be tuned to any frequency within a range of a few THz (quasi continuous tuning).

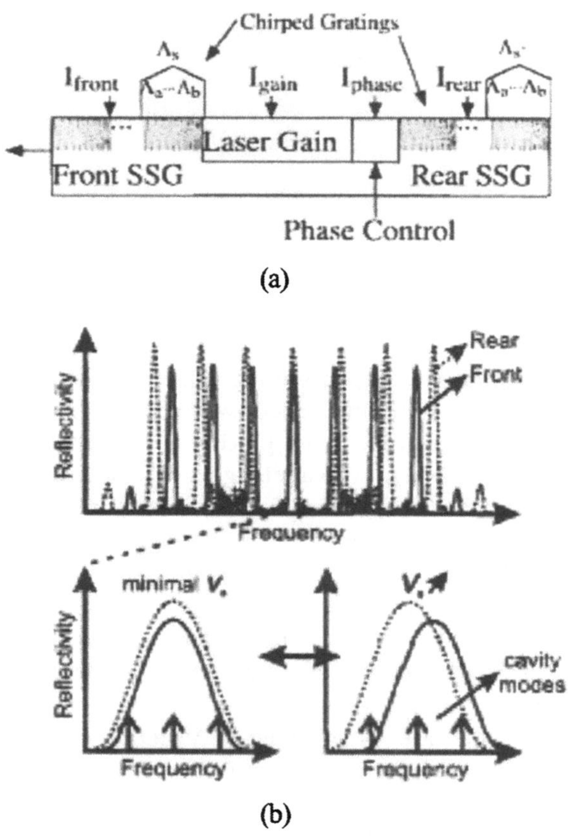

Figure 5.2.25. Reflectivity of front and rear SSG reflector.

5.2.3.4.4. *Laser with silicon MEMS*

Recently, another experimental technology tunable laser with a silicon MEMS electrostatic rotary actuator that is shown in Figure 5.2.26, has been demonstrated. An experimental tunable laser has been manufacture by Iolon, Inc. USA that matches the performance of a tunable external cavity diode laser (ECDL) **[Berger 01]**.

5.2.3.4.5. *Wavelength characterization*

Essential for all the above applications is that the laser can be tuned to a particular channel very accurately and reliably, which is largely dependent on the electronics and algorithms that are used to control the tunable laser. Typical requirements are frequency accuracy better than ±10% of the channel spacing, and a guaranteed side-mode suppression ratio (SMSR) of at least 30 dB.

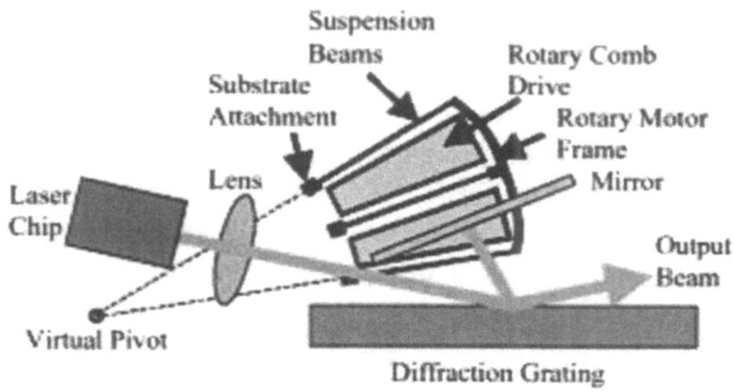

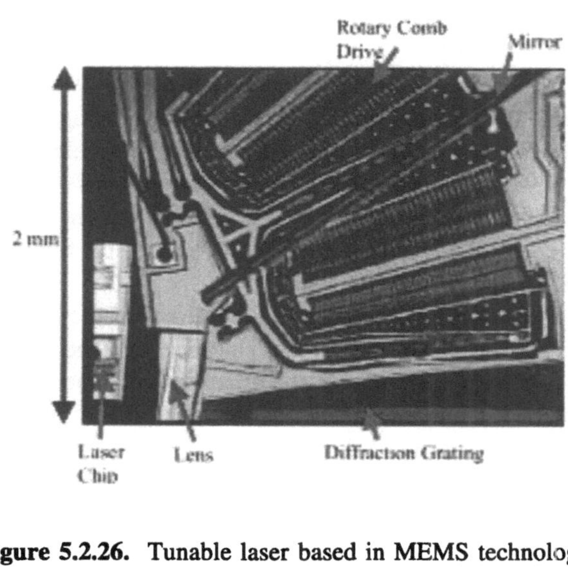

Figure 5.2.26. Tunable laser based in MEMS technology.

The tuning range of DBR lasers is limited by the maximum refractive index change that can be achieved through current injection ($\Delta\lambda/\lambda = \Delta n/n$). Typical tuning ranges are on the order of 5-1O nm, although devices with a tuning range of more than 15 nm have been demonstrated. The SCGR and SSG-GCR lasers have demonstrated quasi continuous tuning ranges of 40–60 nm.

Two aspects must be taken into account to control such devices: First of all one needs a look-up table which tells what currents should be applied to the different sections to get a certain output power and frequency (with high SMSR). The current sources driving the laser will typically be controlled by a microprocessor, which reads the appropriate data from the look-up table (stored in an EEPROM) when output power and/or frequency should be changed. On the other hand, one typically also wants some sort of feedback control which ensures long-term stability and frequency accuracy. It is crucial that the characterization time (or look-up table generation time) is reduced as much as possible, in order to minimize cost.

5.2.3.4.6. *Devices features*

In the last years, various technologies have been developed in tunable sources. Among them, distributed Bragg reflector (DBR) tunable laser of three sections, tunable laser of four sections as grating coupler sample reflector (GCSR) and super structure grating DBR (SSG-DBR), vertical cavity surface emitting laser (VCSELs) and tunable laser with MEMS technology.

A comparison among most important characteristics of various kinds of tunable lasers is made in Figure 5.2.27. Some of the tunable lasers that appears are still prototypes and others already exist in the optical communication market. The information of this table has been gotten in the web sites of each one of these companies (Iolon Inc., Altitun, Agility, Bandwidth 9, New Focus, Nortel Networks).

In most of them, the tuning is usually achieved by tuning a dispersive intracavity element such as a grating or a distributed Bragg reflector; through the laser diode (LD) gain spectrum with a Bragg wavelength control and a phase control. It was understood that the insertion of small DFB sections inside the cavity attenuates some modes while being transparent to others. Inserting these sections periodically will confer more control on the mode selectivity, filter bandwidth and spacing.

Some of the physical characteristics of lasers that may affect system performance are laser linewidth, frequency stability, and the number of longitudinal modes. The linewidth affects the spacing of channels and also affects the amount of dispersion that occurs when the light is propagating along a fiber. The spreading of a pulse due to dispersion will limit the maximum bit rate. Typical requirements are frequency accuracy better than (10% of the channel spacing, and a guaranteed side-mode suppression ratio (SMSR) of at least 30 dB.

Fabricant	Iolon Inc.	Altitun	Agility	Bandwidth 9	Lucent	New Focus	Nortel Networks	Agere
Name	Apollo	NYW55/50	3040 CW	Metroflex	CW20P	TLS 420C	LCW5	D2587P
Technology	MEMS	GCSR	SGDBR	VCSEL	CW-DBR		DFB	DFB
wavelength Tuning Range	32 nm C-band	10 nm / 40 nm	1525-1565 nm	full C and L-	1528.77-1610 nm	1528-1565 nm	6.55-9.02 nm	1528-1610 nm
No. Channels		20, / 100	100		20		20	
channel spacing			50 GHz	50 GHz			50 GHz	50 GHz
output power	10 mW	10 dBm / 3 dBm	4 mW	500 microW	20 mW	20 mW	10-16 dBm	20 mW
output power variation	0.25 dB	2 dB max			0.5 dB			
spectral linewidth	2 MHz	25 MHz	Narrow				20 MHz	2 - 10 MHz
wavelength stability	1.25 GHz	3 GHz		2.5 GHz	2.5 GHz		0.25 GHz (locked)	(+/-)2.5 GHz
Side Mode Sup. Ratio (SMSR)	50 dB	50 dB / 35 dB	30 dB min		32 dB		30 dB	35 - 45 dB
Extinction ratio		20 dB		10 dB			20 dB	20 dB
Relative intensity Noise	(-145 dB/Hz)						(-140 dB/Hz)	(-135 dB/Hz)
Optical isolation	30 dB							30 dB
Data rates				2.5 Gbps				
Tuning and locking speed		1 msec		<100 microsec				
fiber	1 m	1 m			core 8 um (1 m)			1.5 m
Back reflactance	(-50 dB)	(-50 dB)						
Optical conector		FC/PC						ST plug
Key allignment		Slow axis						
Guide Key width of connector		1.98 to 2.02 mm						
storage temperature	(-40 to 85) °C	(-40 to 85) °C			(-40 to 85) °C		(-40 to 85) °C	
operating temperature	0 to 70 °C	0 to 70 °C			0 to 70 °C		0 to 70 °C	
Total power dissipation	7.5 W				10 W			
Non operational shock	500 G							
DC supply voltage	5.0 V	5.0 V			5.5 V max			
DC supply voltage(neg)		300 mA 1.5 A			(-6.0 V) max			
DC supply current	400 mA				500 mA max			
DC supply current(neg)					250 mA			
Comments	High speed RS232 serial communication interface	anywave lasers are tuned via current injection They do not require temperature-tuning Contain no moving parts	temperature and wavelength control included Integrated optical isolater Asynchronous parallel communications interface Programmable wavelength control Wavelength stabilized mode with external reference module	RS232 interface for wavelength control directly modulated	Remote programmability RS232 Asynchronous serial interface for module communication Laser bias monitor alarm Module enable/disable input	Integrated wavelength locking Simple to integrate serial interface	Power monitoring Internal optical isolator	

Figure 5.2.27. Tunable laser comparison.

5.2.3.4.7. *Applications*

In an access WDM system, for instance, of 128 wavelengths, there are 128 different lasers designed to produce the exact wavelength for each channel. It is necessary to keep spare lasers to plan for the event when a laser breaks down. To keep in stock 128 spare lasers is too expensive, therefore, tunable lasers are used as spare laser for any of the 128 sources. This is the main application of a tunable laser in respect to savings, as well as decreasing the storage space.

Tunable laser also can be used as continuous-wave optical source with an interferometric wavelength converter as in SONATA ACT project. This subsystem allows the wavelength converter to be agile and presents featuring fast and fine tuning.

In SONATA ACT project, tunable laser is used together with a broadband optical receiver that performs as heterodyning to achieve accurate wavelength control. The output of the tunable laser is coupled to the output of two lasers used as reference sources. The receiver produce a differential frequency of RF between wavelength tunable laser and wavelength reference. A voltage proportional to the RF frequency is produced by a frequency discriminator and compared with a reference voltage and the difference is feedback to the laser driver.

5.2.3.5. *Vertical Cavity Surface Emitting Lasers (VCSELs).*

This is one of the current "hottest technologies" **[VCSE]**. The VCSEL-based transceiver market took hold in 1999 with sales of $247 million, more than three times that of 1998 **[ECAST]**. Fuelled by the recent success of Gigabit Ethernet and Fibre Channel, this explosive growth will continue, supported by a number of all-new applications, reaching $2.8 billion in 2004 and continuing on to $9.6 billion in 2009 as forecast by ECAST (see Figure 5.2.28). Drivers are:

- Rapid advancement of datacom standards
  - Explosive growth of Gigabit Ethernet, followed soon by 10 Gigabit Ethernet.
  - Continued aggressive growth of Fibre Channel, migrating to 2 Gbps Fibre Channel, and on to 10 Gbps.
- Penetration of telecom
  - Low cost solutions for Very Short Reach SONET at 10 Gbps
- New opportunities:
  - Serial and array links within switch fabrics
- Deployment of intra-system array-VCSEL interconnects (VSR).

The most dynamic growth in VCSEL transceiver use is occurring in Gigabit Ethernet and 1–2 Gbps Fibre Channel. These applications will grow by almost

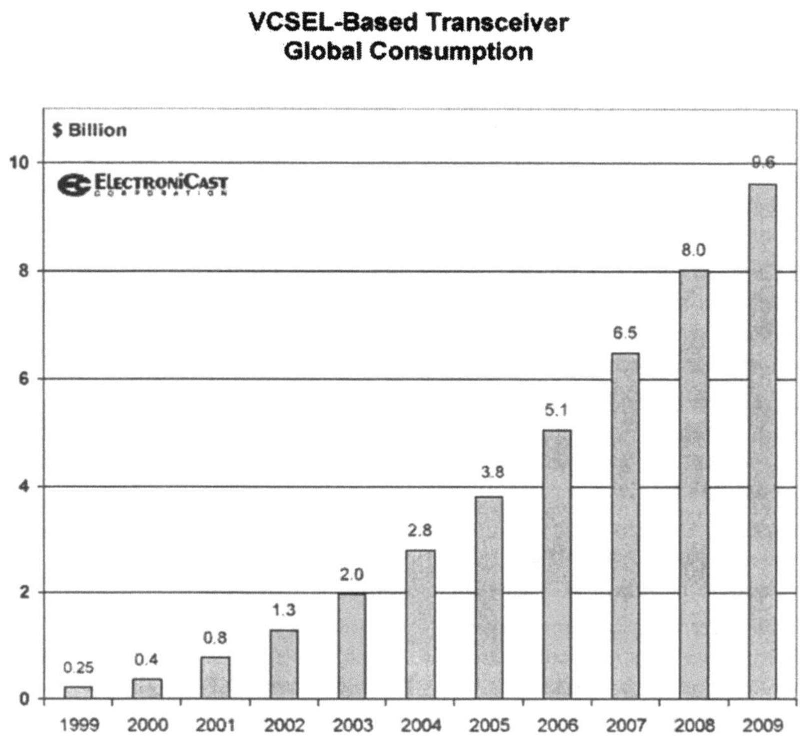

Figure 5.2.28. VCSEL forecasted market (from [ECAST]).

five times by 2004 when 10 gigabit versions will make significant contributions. This high volume production of VCSEL transceivers is fuelling development of multi-wavelength CWDM VCSEL transceivers, long wavelength single-mode VCSELs and 10 gigabit VCSELs opening new opportunities for these standards-based applications. Other technologies where VCSELs will be a very important device are shown in next table.

There are a lot of intentional research papers (see Figure 5.2.29) and companies that have focused their efforts in this device (Agilent, IBM, Methode, Infineon, Blaze, Mitel, Gore, . . .). The following items are other important issues tightly related with VCSELs [OIF]:

- VSR optics: There is a lot of interest (OIF) in defining standards for the Very Short Reach optics, and we have collected many documents concerning this.
- SAN: Storage Area Networks is another "hot technology" which has become a great candidate to concentrate attention and funds.

Title	Source	Company
A comprehensive circuit-level model of vertical-cavity surface-emitting lasers	Journal of Lightwave Tech., Vol. 17, Issue 12, Dec. 1999	Univ. of Illinois at Urbana
A simple rate-equation-based thermal VCSEL model	Journal of Lightwave Tech., Vol. 17, Issue 5, May 1999	Univ. of Illinois at Urbana
An 8-Gb/s optical backplane bus based on microchannel interconnects: design, fabrication, and performance measurements.	Journal of Lightwave Tech., Vol. 18, Issue 11, Nov. 2000	University of Texas, Austin
Design and realization of a 1.55-/spl mu/m patterned vertical cavity surface emitting laser with lattice-mismatched mirror layers	Journal of Lightwave Tech., Vol. 17, Issue 12, Dec. 1999	Univ. of Michigan, Ann Harbor & Corning Inc.
High-speed VCSEL of modulation bandwidth over 7.0 GHz and its application to 100 m PCF datalink	Journal of Lightwave Tech., Vol. 16, Issue 5, May 1998	NEC Corporation
Monolithically integrated low-power phototransceivers for optoelectronic-parallel sensing and processing applications	Journal of Lightwave Tech., Vol. 19, Issue 4, April 2001	Univ. of Michigan, Ann Harbor Univ. of Texas, Austin
Multichannel transmission of a multicore fiber coupled with vertical-cavity surface-emitting lasers	Journal of Lightwave Tech., Vol. 17, Issue 5, May 1999	Rosinski, B. et al.
Noise performance of multimode VCSELs	Journal of Lightwave Tech., Vol. 19, Issue 6, June 2001	TU-Berlin Infineon Tech.
Operation range of VCSEL-interconnect links with "below-threshold"-biasing	Journal of Lightwave Tech., Vol. 18, Issue 4, April 2000	TU-Berlin Univ. of Ulm
Optical interconnection using VCSELs and polymeric waveguide circuits	Journal of Lightwave Tech., Vol. 18, Issue 11, Nov. 2000	NTT Photonics Laboratories
Technology development of a high-density 32-channel 16-Gb/s optical data link for optical interconnection applications for the optoelectronic technology consortium (OETC)	Journal of Lightwave Tech., Vol 13, Issue 6, June 1995	ATT Bell Labs

Figure 5.2.29. Research papers.

Title	Source	Company
A historical perspective of the development of the vertical-cavity surface-emitting laser	IEEE Journal on Sel. Topics in Quantum Elec., Vol. 6, Issue 6, Nov.–Dec. 2000	Towe, E. et al.
Design of low-loss single-mode vertical-cavity surface-emitting lasers	IEEE Journal on Sel. Topics in Quantum Elec., Vol. 5, Issue 3, May–June 1999	Univ. of Southern California, LA
Design of VCSEL's for feedback insensitive data transmission and external cavity active mode-locking	IEEE Journal on Sel. Topics in Quantum Elec., Vol. 1, Issue 2, June 1995	Fiedler, U.; Ebeling, K.J.
GaInNAs: a novel material for long-wavelength semiconductor lasers	IEEE Journal on Sel. Topics in Quantum Elec., Vol. 3, Issue 3, June 1997	Kondow, M. et al.
High performance and novel effects of micromechanical tunable vertical-cavity lasers	IEEE Journal on Sel. Topics in Quantum Elec., Vol. 3, Issue 2, April 1997	SDL Univ. of California at Berkeley
High-performance oxide-confined GaAs VCSELs	IEEE Journal on Sel. Topics in Quantum Elec., Vol. 3, Issue 2, April 1997	Univ. of Ulm
High-performance small vertical-cavity lasers: a comparison of measured improvements in optical and current confinement in devices using tapered apertures	IEEE Journal on Sel. Topics in Quantum Elec., Vol. 5 Issue 3, May–June 1999	Univ. of California at Santa Bárbara
High-power VCSELs: single devices and densely packed 2-D-arrays	IEEE Journal on Sel. Topics in Quantum Elec., Vol. 5 Issue 3, May–June 1999	Univ. of Ulm
Hybrid integration of VCSEL's to CMOS integrated circuits	IEEE Journal on Sel. Topics in Quantum Elec., Vol. 5, Issue 2, March–April 1999	Colorado State Univ.

Figure 5.2.29. *(Continued).*

Title	Source	Company
Long-wavelength vertical-cavity lasers and amplifiers	IEEE Journal on Sel. Topics in Quantum Elec., Vol. 6, Issue 6, Nov.–Dec. 2000	Univ. of California at Santa Bárbara
Modeling of oxide-confined vertical-cavity surface-emitting lasers	IEEE Journal on Sel. Topics in Quantum Elec., Vol. 3, Issue 2, April 1997	Tokio Institute of Technology
Multiple-wavelength vertical-cavity surface-emitting laser arrays	IEEE Journal on Sel. Topics in Quantum Elec., Vol 3, Issue 2, April 1997	Univ. of California at Berkeley
Polarization related properties of vertical-cavity surface-emitting lasers	IEEE Journal on Sel. Topics in Quantum Elec., Vol. 3, Issue 2, April 1997	Texas Tech Univ.
Smart integration and packaging of 2D VCSEL's for high-speed parallel links	IEEE Journal on Sel. Topics in Quantum Elec., Vol. 5, Issue 2, March–April 1999	NEC Opto-Electronics Lab.
Surface-emitting laser-its birth and generation of new optoelectronics field	IEEE Journal on Sel. Topics in Quantum Elec., Vol. 6, Issue 6, Nov.–Dec. 2000	Tokio Institute of Technology
Tunable VCSEL	IEEE Journal on Sel. Topics in Quantum Elec., Vol. 6, Issue 6, Nov.–Dec. 2000	Univ. of California at Berkeley Bandwidth9
Wafer bonding technology and its applications in optoelectronic devices and materials	IEEE Journal on Sel. Topics in Quantum Elec., Vol. 3, Issue 3, June 1997	Cornell University, Ithaca

Figure 5.2.29. *(Continued).*

The driving force for this diode lasers is cost. In order to find out if a laser works well, it is necessary to dice the wafer and mount the chip before it can be powered up to test. If it's good, it will be packaged for sale; if not, it's discarded. For years, researchers have been working on creating a laser that works like a LED, emitting light up out of the wafer, perpendicular to the surface. It took years, but the VCSEL was finally produced. Fig. 5.2.30 shows a comparison between LED and VCSEL.

Instead of coated mirrors on the cavity, it uses epitaxial layers grown on the wafer to create mirrors on the surface with a LED sandwiched in between. These epitaxial layers can even create lenses to focus the light into a tight cone, perfect for coupling to fibres. But the best advantage of the VCSEL is in testing and coupling to the fibre. A completed wafer can be tested in a wafer probe machine, where each device is contacted by microscopic electrical probes and current applied. If it's good, it can be seen on a camera and measured. Thus good devices can be detected before the wafer is diced and bad ones marked automatically for discarding.

The other cost savings with VCSELs comes in coupling to fibers. The tight beam couples easily to fiber without optics, while the divergent beam (due to a non symmetric section) of the standard diode laser requires focusing optics.

Also, the VCSEL does not require a back facet monitor like a diode laser; it can be driven like a LED, but at much higher power levels with a much lower drive current. VCSELs have many advantages when compared with other light sources:

- Simplicity in its fabrication process.
- High power at low current drive.
- Ease of coupling to fibres.
- Good spectral characteristics, similar to single-mode lasers.
- High bandwidth, having BW in the order of ~5 GHZ now.
- Low cost.

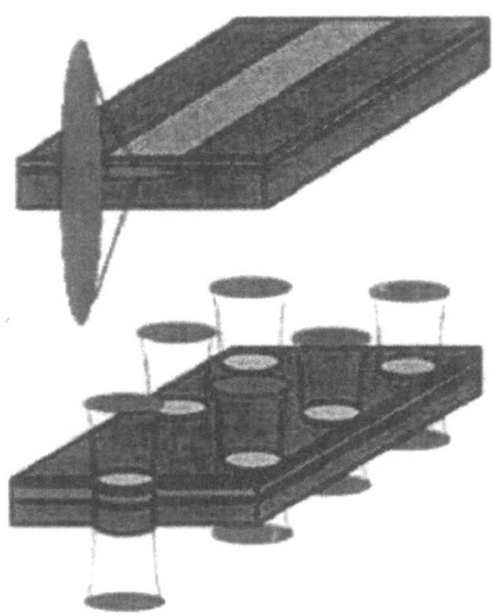

Figure 5.2.30. Comparison between typical edge emitting laser and VCSEL (from **[Chang]**).

They have only an important drawback that precludes their use in all the optical communications systems. They are not yet widely available at the second and third window wavelengths. VCSELs at 850 nm are now readily available, but the first 1310 nm devices are just appearing in the market. Next step, VCSELs at 1550nm, are being developped at labs. Gigabit Ethernet and Fibre Channel already use VCSELs for their low cost version. The 1310 devices will be incorporated in every system appropriate as soon as possible to take advantage of the lower cost (estimated about 25% as much as a F-P laser). 1310 VCSELs could be the key to cost justification of fiber to the home (FTTH). Some 1310nm VCSEL based commercial products have already been announced (Cielo, E20, . . .). Next table shows some interesting fabricants VCSELs:

Company	**Product**	**Web**	**Details**
AMP	ESCON-MM-MTRJ Transceiver	amp.com/fiberoptics	200Mbit/s
Avalon	AP850_308	avalonphotonics.com	VCSEL array
AXT-VCSEL	VCSELs	axt.com/vcsel	
Bandwidth9	VCSELs	bw9.com	
Blaze	Afterburner	www.BlazeNP.com	10Gbit/s 1Gbit/s
Cielo Comm.	VCSELs	cieloinc.com	1300 nm, 2,5Gbit/s
Cognet Microsystems	VCSEL Drivers	www.cognetmicro.com	10Gbit/s
Coretek (Nortel)		coretekinc.com	
E20	GBIC, Paralel-LC, 1 × 9 GbE	e2oinc.com	
Emcore	VCSEL CWDM	emcore.com	10Gbit/s
Gore	nLighteen	gore.com	Tx + Rx, VCSEL, GbE
Honeywell			1550 nm spec
Honeywell	HFT-248x-521	Honeywell.com/ sensing/vcsel	MTRJ SFF 1.25 Gbps OFE
IBM	SFFFamily	ibm.com/chips	
Opt. Comm.Prod	DTR-1250-LC		Transceivers, VCSEL,GbE
Stratos	GBIC	stratoslightwave.com	

Figure 5.2.31. VCSELs manufacturers.

Conclusions
VCSELs are probably the key device that will allow to make FTTH a reality. It performs like a single-mode laser but its price is much lower. E/O converters have been seen as one of the main drawbacks for the proliferation of fiber access networks. VCSELs will probably throw down this myth. The fiber type (multi or single-mode) comes again as an important question. Nowadays VCSELs are only commercial for the 850 nm window but it seems that long wavelength VCSELs are ready to be soon in the market. This implies that arguments against single-mode fiber (if 1300 and 1550 VCSELs are also cheap) will finish. There are a lot of commercial transceivers that use VCSELs. Its penetration into the market is only a question of time. Maybe the key issue to close the circle is the finalization of the FTTH Standard. Anyway, VCSELs are one of the principal ingredients to the optical access success.

5.2.4. Photodetectors

5.2.4.1. *Introduction*

The predominant types of light detector used in communications systems rely on the principle of ionisation in a semiconductor material. There are a number of different kinds of devices but they can all be viewed as variations on a central principle rather than as devices that involve radically different principles. When discussing photodetectors, there are four important parameters:

- *Detector Responsivity:* This is the ratio of output current to input optical power. Hence this is the efficiency of the device.
- *Spectral Response Range:* This is the range of wavelengths over which the device will operate.
- *Response Time:* This is a measure of how quickly the detector can respond to variations in the input light intensity.
- *Noise Characteristics:* The level of noise produced in the device is critical to its operation at low levels of input light.

5.2.4.2. *P-I-N diodes*

The principle involved in a PIN diode is simply the principle of the LED in reverse. That is, light is absorbed at a p-n junction rather than emitted:

- A reverse biased p-n junction passes no current unless something happens to promote electrons from the valence band to the conduction band within the depletion zone. At ordinary room temperatures, there is a small amount of this due to the action of heat, and thus there is a very small current.

- Photons of sufficient energy can be absorbed and cause the promotion of an electron from the valence band to the conduction band (and of course the simultaneous generation of a hole in the valence band).
- The free electron and hole created by the photon absorption now are attracted to their opposite charges at either side of the junction and current flows.

The big problem here is that the depletion zone in a p-n junction is extremely thin. Most light passes through without being absorbed in the junction and instead is absorbed in the doped material at either side of the junction. Ultimately, many of the electron-hole pairs created outside the junction do end up being attracted to the junction and creating current. However, the process is quite slow and p-n junction devices are not fast enough for the current communications applications.

The answer to the problem created by the extreme thinness of a p-n junction is to make it thicker. The junction is extended by the addition of a very lightly doped layer called the intrinsic zone between the p and n doped zones. Thus the device is called a p-i-n diode rather than a p-n diode. This is illustrated in Figure 5.2.32.

The wide intrinsic (i) layer has only a very small amount of dopant and acts as a very wide depletion layer. There are a number of improvements here:

- It increases the chances of an entering photon being absorbed because the volume of absorbent material is significantly increased.
- Because it makes the junction wider, it reduces the capacitance across the junction. The lower the capacitance of the junction the faster the device response.

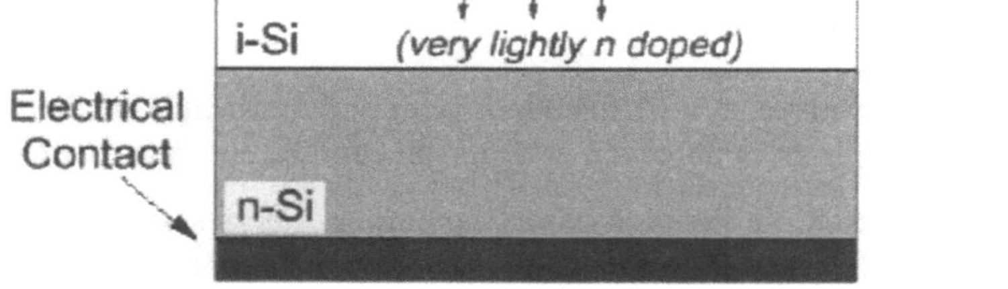

Figure 5.2.32. Typical Silicon P-I-N Diode Schematic.

- There are two ways of current carriage across the junction: Diffusion and Drift. Increasing the width of the depletion layer favors current carriage by the drift process which is faster than the diffusion process.

The result is that the addition of the "i" layer increases the responsivity and decreases the response time of the detector to around a few tens of picoseconds.

The key to the operation of a PIN diode is that the energy of the absorbed photon must be sufficient to promote an electron across the band-gap (otherwise it won't be absorbed). However, the material will absorb photons of any energy higher than its band-gap energy. Thus when discussing PIN diodes, it is common to talk about the "cut-off wavelength". Typically PIN diodes will operate at any wavelength shorter than the cutoff wavelength. This suggests the idea of using a material with low band-gap energy for all PIN diodes regardless of the wavelength.

Unfortunately, the lower the band-gap energy the higher the "dark current" (thermal noise). Indeed but for this characteristic, germanium would be the material of choice for all PIN diodes. It is low in cost and has two useful band-gaps (an indirect band-gap at 0.67 eV and a direct band-gap at 0.81 eV). However, it has a relatively high dark current compared to other materials.

This means that the materials used for PIN diode construction are different depending on the band of wavelengths for which it is to be used. However, this restriction is nowhere near as stringent as it is for lasers and LEDs where the characteristics of the material restrict the device to a very narrow range. The optimal way is to choose a material with a band-gap energy slightly lower than the energy of the longest wavelength you want to detect.

An interesting consequence to note here is that these crystalline semiconductor materials are transparent at wavelengths longer than their cut-off. Thus, if we could "see" with 1500 nm eyes, a crystal of pure silicon (which appears dark grey in visible light) would look like a piece of quartz or diamond.

Typical materials used in the three communication wavelength "windows" are as follows:

500–1000 nm Band. Silicon PIN diodes operate over a range of 500 to 1120 nm as silicon has a band-gap energy of 1.11 eV. Since silicon technology is a very low cost, silicon is the material of choice in this band.

However, silicon is an indirect band-gap material (at the wavelengths we are interested in) and this makes it relatively inefficient. (Silicon PIN diodes are not as sensitive as PIN diodes made from other materials in this band.) This is the same characteristic that prevents the use of silicon for practical lasers.

1300 nm (1250 nm to 1400 nm) Band. In this band, Indium Gallium Arsenide Phosphide (InGaAsP) and Germanium can be used. Germanium has a lower band-gap energy (0.67 eV versus 0.89 eV for InGaAsP) and hence it can theoretically be used at longer wavelengths. However, other effects in Ge limit it to wavelengths below 1400 nm. InGaAsP is significantly more expensive than Ge but it is also significantly more efficient (devices are more sensitive).

1550 nm Band (1500 nm to 1600 nm). The material used here is usually InGaAs (indium gallium arsenide). InGaAs has a bandgap energy of 0.77 eV.

Operation with reverse bias as described above (called "Photoconductive Mode") has a significant problem. At low light levels, the random current produced by ambient heat is a source of noise. At higher light levels however, this is not a problem and the device offers the advantages of much higher speeds (than the alternative mode of operation) and linear response characteristics over a wide range. The higher speed characteristics are the result of lowered capacitance caused by the widening of the depletion layer in the presence of reverse bias.

If the device is operated without an externally applied current, the natural potentials of the p-i-n junction will cause electrons and holes to migrate across the junction allways. This is called "Photovoltaic Mode". Thus you get a small voltage (around 0.15 V) developed across the device. The advantage of this is that there is much less "dark current" caused by ambient heat. Thus in this mode the device is more sensitive, but the output requires immediate amplification because of the low voltage levels produced.

The efficiency of p-i-n diodes at long wavelengths can be improved by the use of heterostructures. For example, using GaAlAs for the p-layer and GaAs for the I and n-layers. Incident light must pass through the p-layer before it enters the i-layer. If it is absorbed by the p-layer, then the energy is lost and does not contribute to the current produced. By using a material with a high band-gap energy for the p-layer (higher than the energy of the incident photons), we can prevent incident long wavelength light from being absorbed. This means that there is less light lost before it reaches the i-layer where we want it to be absorbed.

There are two common measures quoted when the efficiency of PIN photodetectors is discussed.

Quantum Efficiency. This is simply the ratio of the number of electrons collected at the junction over the number of incident photons. In an ideal situation, 1 photon releases 1 electron (and its matching hole, of course). A perfect quantum efficiency is therefore an efficiency of "1".

In real devices, QE is different at each operating wavelength, and so, it should always be quoted in association with a wavelength.

Responsivity. Quantum efficiency does not take into account the energy level of the incident photons. Responsivity is a measure that does take photon energies into account. It is simply the output photocurrent of the device (in amperes) divided by the input optical power (in watts). Thus, responsivity is quoted in amperes per watt. A typical responsivity for a silicon photodiode at a wavelength of 900 nm is 0.44.

Of course, responsivity is very closely linked to quantum efficiency. It is just quantum efficiency adjusted to account for the variation in energy level implied by different wavelengths.

5.2.4.3. *Avalanche photodiodes (APDs)*

APDs amplify the signal during the detection process. They use a similar principle to that of "photo-multiplier" tubes used in nuclear radiation detection. In the photo-multiplier tube:

- A single photon acting on the device releases a single electron.
- This electron is accelerated through an electric field until it strikes a target material.
- This collision with the target causes "impact ionisation" which releases multiple electrons.
- These electrons are then themselves accelerated through the field until they strike another target.
- This releases more electrons and the process is repeated until the electrons finally hit a collector element.

Thus, through several stages, one photon has resulted into a current of many electrons.

APDs are, of course, different from photo-multiplier tubes. Photo-multiplier tubes are vacuum tubes with metallic targets arranged in stages down the length

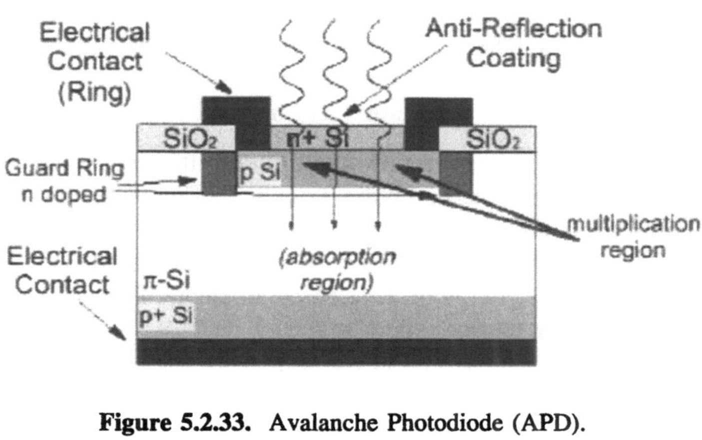

Figure 5.2.33. Avalanche Photodiode (APD).

of the tube. APDs use the same principle, but multiplication takes place within the semiconductor material itself. This process in APDs typically results in an internal amplification of between 10 and 100 times.

In its basic form, an APD is just a p-i-n diode with a very high reverse bias. A reverse bias of 50 volts is typical for these devices compared with regular p-i-n diodes used in the photoconductive mode which is reverse biased to only around 3 volts (or less). In the past, some APDs on the market have required reverse bias of several hundred volts although recently lower voltages have been achieved.

The main structural difference between an APD and a p-i-n diode is that the I zone (which is lightly n-doped in a p-i-n structure) is lightly p-doped and renamed the π layer. It is typically thicker than an i-zone and the device is carefully designed to ensure a uniform electric field across the whole layer. The guard ring shown in the figure serves to prevent unwanted interactions around the edges of the multiplication region.

The device operates as follows:

- Arriving photons generally pass straight through the n + –p junction (because it is very thin) and are absorbed in the π layer. This absorption produces a free electron in the conduction band and a hole in the valence band.
- The electric potential across the π layer is sufficient to attract the electrons towards one contact and the holes towards the other. In the figure, electrons are attracted towards the n + layer at the top of the device because being reverse biased it carries the positive charge. The potential gradient across the layer is not sufficient for the charge carriers to gain enough energy for multiplication to take place.
- Around the junction between the n + and p layers, the electric field is so intense that the charge carriers (in this case electrons only) are strongly accelerated and pick up energy. When these electrons (now moving with a high energy) collide with other atoms in the lattice, they produce new electron-hole pairs. This process is called "impact ionisation".
- The newly released charge carriers (both electrons and holes) are themselves accelerated (in opposite directions) and may collide again.

Both electrons and holes can now contribute to the multiplication process. However, now there is a small problem. Looking at Figure 3.2.4.3, it is obvious that when an electron ionizes an atom, an additional electron-hole pair is produced. The electron moves to the left of the picture and the hole to the right. If the hole now ionises an atom, it releases an electron (and a hole) and the electron moves to the left and starts again.

If holes and electrons have an equal propensity for ionization, we can get an uncontrolled avalanche which will never stop. Thus, devices are built such that one of the charge carriers has a significantly higher propensity for ionization than the other. In silicon, electrons are the dominant carrier. In III–V alloy materials holes are often employed.

The result of the above process is that a single arriving photon can result into the production between 10 and 100 electron-hole pairs.

The important things to note about the device described above are that the multiplication region is very small, and absorption takes place within the π layer rather than near the junction. That is, the absorption and multiplication regions are separated. This is illustrated in Figure 5.2.34. There are two important factors here:

1. The strength of the required electric field is extremely high. (Of the order of 10^6 volts/meter.). In the presence of such a strong field, imperfections in the multiplication region (such as lattice mismatches, impurities and even variations in dopant concentration) can produce small areas of uncontrolled multiplication called "micro plasmas". To control this effect the multiplication region needs to be small.

This is the reason for the guard ring mentioned above. Around the edges of the multiplication region, you tend to get irregularities and imperfections in the material. Without the guard ring, these serve as sites for micro-plasmas.

In addition, to create an electric field of the required strength, we need an applied bias voltage that increases with the thickness of the multiplication region. (Double the thickness of the region – double the required voltage.) High voltages (indeed voltages above about 12 volts) are expensive and difficult to handle in semiconductor devices and so we aim to minimize the applied voltage.

2. For the above reason the junction region is very thin and therefore it can't absorb many incident photons. The π layer is the absorption layer.

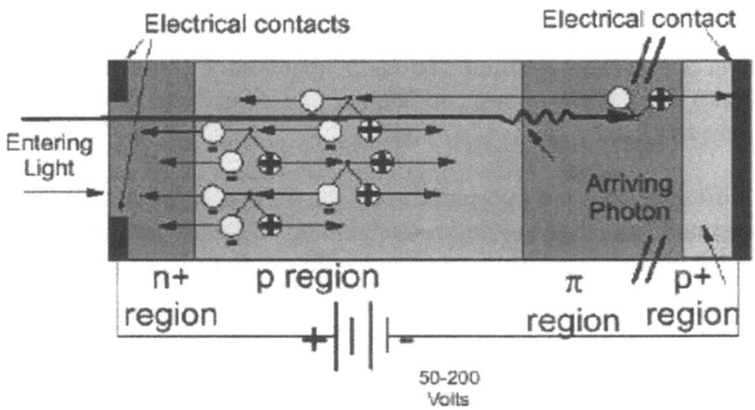

Figure 5.2.34. Avalanche (Amplification) Process.

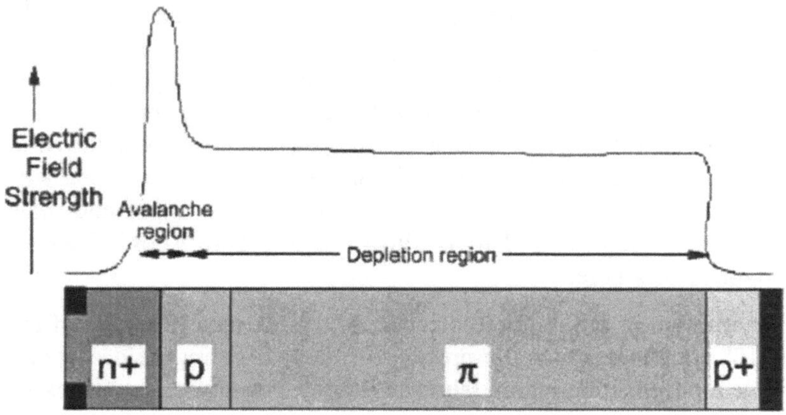

Figure 5.2.35. Electric Field Strengths in an APD.

3. Many APDs are designed so that the depletion layer extends through the entire p-region to the π-region boundary.

As with p-i-n diodes, different materials are usually employed for each of the three important wavelength bands:

800 nm to 1 Micron Band: In this band, silicon is usually employed although germanium will also work reasonably well. Germanium devices however produce higher noise levels than silicon ones.

As mentioned above for p-i-n structures, silicon has a relatively high band-gap energy and is therefore used only for wavelengths shorter than about 1 micron.

In practice, short wavelengths today are only used for very short distance (less than 500 metres) communications. Attenuation over such short distances is generally not great enough to require the sensitivity of an APD (or to justify its cost).

1310 nm Band: This is important as it is the band used by most existing long distance communications systems. Germanium APDs are used extensively, but III–V semiconductor alloys are increasing in usage because of the high noise levels in germanium.

1550 nm Band III–V: APDs are used widely in the 1550 nm band. The most common materials system in use is InGaAs/InP where the majority carrier is holes rather than electrons.

APD Characteristics:

The most important characteristics of APDs are their sensitivity, their operating speed, their gain-bandwidth product and the level of noise.

Sensitivity of APDs: The extreme sensitivity of APDs is the major reason for using them.

Operating Speed: The same factors (such as device capacitance) that limit the speed of p-i-n diodes also influence APDs. However, with APDs there is another factor: "Avalanche Buildup Time". Because (as discussed above) both carriers can create ionisation an avalanche can last quite a long time. This is caused by the to and fro movement of electrons and holes as ionisations occur. Provided the propensity for ionisation in the minority of carrier is relatively low, the avalanche will slow and stop but this takes some time.

Avalanche Buildup Time thus limits the maximum speed of the APD.

Gain-Bandwidth Product: The accepted measure of "goodness" of a photodetector is the gain-bandwidth product. This is usually expressed as a gain figure in Db multiplied by the detector bandwidth (the fastest speed that can be detected) in GHz. A very good current APD might have a gain-bandwidth product of 150 GHz.

Noise: APDs are inherently noisy as the multiplier effect applies to all free electrons including those made free by ambient heat. This is especially a problem in longer wavelength devices where the bandgap energy is low.

In the design of devices, great care is taken to make sure that the electrical potential gradient is sufficient for multiplication (ionisation) to occur but no greater. Greater levels of bias can cause spontaneous ionisation events.

In many long-distances wide-area applications where sensitivity is more important than other factors, the APD has been commonly used. However, as speeds increase and amplified systems are introduced, the noise produced by APDs becomes a limitation. In these systems people today use p-i-n detectors with preamplifiers as you get much higher sensitivity than an APD and much lower noise.

5.4.2.4. *Commercial devices*

Company	Reference	Type	Wavelength Range	-3dB Bandwidth	Data Rate	Sensitivity	Responsivity	Gain	Dark Current	Rise/Fall Time	Temperature Range
Agere	P-172	PIN/APD	1250-1600nm	2GHz		-23/-32dBm	0,89,0A/W				-40 to +85C
Agere	RA2860D	APD	1550nm	10GHz		-26dBm	0,68A/W(M=1)				-5 to +70C
Agere	RA2560A	PIN	1280-1580nm	13GHz			0,8A/W		1nA	<0,5ns	0 to +70C
Agere	P-171	PIN	1100-1600nm	590/150MHz		-33/-38dBm	0,85A/W		1nA	<0,5ns	-40 to +85C
Agere	R2860D	PIN	1280-1580nm	10GHz		-18dBm	0,8A/W		1nA		-5 to +70C
Fujitsu	FRM5J143DS	PIN	1310-1550nm	8GHz		-18dBm	0,85A/W				0 to +70C
Fujitsu	FRM3Z231KT/LT	PIN	1310-1550nm	2,5GHz		-23dBm	0,85A/W				-40 to +85C
JDS Uniphase	ERM577	APD	1550nm	1,8GHz		-32dBm	8,5A/W		10nA		0 to +70C
JDS Uniphase	ERM567	PIN	1550nm	1,8GHz		-24dBm		5V/mW	0,5nA		0 to +70C
JDS Uniphase	EDR515x	PIN	1300nm	200MHz		-38dBm		20V/mW	1nA		-40 to +85C
JDS Uniphase	EDR512x	PIN	1300nm	500MHz		-32dBm		22V/mW	1nA		-40 to +85C
Luminent	Series 4	PIN	1200-1550nm		5Mb/s	-33dBm				12/5ns	
Luminent	Series 16	PIN	1100-1600nm		622Mb/s	-29dBm		2V/mW			-40 to +85C
Luminent	Series 13	PIN	1100-1600nm		160Mb/s	-36dBm		5V/mW			-40 to +85C
Luminent	Series 14	PIN	1200-1550nm		160Mb/s	-41dBm					-40 to +85C
Luminent	Series 10	PIN	1200-1550nm		622Mb/s	-30dBm				<2ns	-40 to +85C
Luminent	Series 11-13	PIN	1200-1550nm		160Mb/s	-32dBm				<2,5ns	-40 to +85C
Nortel	PT-12G	PIN	1300-1575nm	13GHz		-19dBm	0,8A/W			<10ns	0 to +85C
Nortel	AT10G	APD	1300-1575nm	8,1GHz		-26,5dBm				<100ns	0 to +70C
Nortel	PT-10G	PIN	1300-1575nm	11,5GHz		-19dBm	0,8A/W			<10ns	0 to +85C
Nortel	ATA2400-40-U	APD	1300-1575nm	2,4GHz		-34dBm	0,9A/W(M=1)				-20 to +70C
Nortel	ATC2400-40-C	APD	1300-1575nm	2,2GHz		-34dBm	0,9A/W(M=1)				0 to +70C

5.3. AMPLIFIERS

5.3.1. Introduction

In present chapter we will deal with optical amplification issues concerning access networks environment. There exist two main amplification techniques: doped fibre amplifiers and semiconductor optical amplifiers. The main advantages of first ones are the high bandwidth and high gain and saturation power. About semiconductor amplifiers the main advantage is the integrated format and its rapid ON/OFF response making it very attractive in access data networks.

5.3.2. Erbium Doped Fiber Amplifier (EDFA)

5.3.2.1. *Introduction*

Passive Optical Network (PON) system architectures are an excellent solution to fiber access but are limited in range and fan-out by fiber loss, dispersion and optical power. Moreover, optical fiber must be pushed closer to the home in order to facilitate upstream transmissions, increase bandwidth, and improve system reliability. Most solutions propose the use of Erbium Doped Fiber Amplifiers (EDFA). In the past, this was an expensive solution due to the cost of EDFA and its high power consumption in this cost sensitive part of the network. Fortunately, the cost/mW drops as the saturation output power of high-power EDFAs increases. Actually, techniques combined with EDFA are essential for realizing high capacity lightwave transmission, and different configurations have been deployed to built flexible optical networks.

EDFAs can offer 40 nm typical optical bandwidth that corresponds to 5000 GHz electrical bandwidth. WDM is considered as a suitable technology to make use of this bandwidth and to increase transmission capacity of an optical fiber system, but does not hardly exploit these capabilities, even WDM systems usually do not occupy more than 50 GHz bandwidth.

5.3.2.2. *EDFA operation*

Optical amplifiers amplify incident light through stimulated emission, the same mechanism used by lasers. Its primary parameter is the optical gain reached when the amplifier is pumped optically to achieve population inversion. The gain in general depends not only on the frequency or wavelength of the incident signal, but also on the local beam intensity at any point inside the amplifier.

Block diagram of a single-stage co-directionally pumped EDFA is shown in Figure 5.3.1. The pump laser (typically either 980 nm or 1480 nm) is coupled

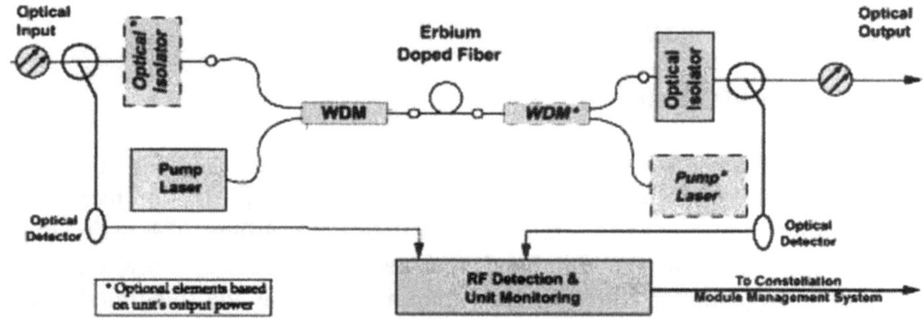

Figure 5.3.1. Erbium-doped fiber amplifier block diagram.

to the erbium-doped fiber with a wavelength division multiplexer. The erbium-doped fiber is highly absorptive at the pump wavelengths, and is a good emitter of light in the 1550 nm. region.

In Figure 5.3.2. the amplification process can be seen. The energy from the pump laser boosts the erbium ions in the fiber from the ground state to a higher, metastable energy level. Once in the metastable state, the ions are stimulated by the signal propagating along the erbium-doped fiber, causing them to decay back to the ground state. It results in the emission of a photon in phase with the signal, thus amplifying the signal by stimulated emission.

5.3.2.3. *EDFA parameters*

Gain

The purpose of the EDFA is to provide gain (G), which is defined as the ratio of the output signal power (P_{out}) to the input signal power (P_{in}). The input and output power levels measured are actually the sum of the signal power, and the small amount of spontaneous emission power within the optical spectrum analyzer's resolution bandwidth at the signal wavelength. This additional measured power usually has a negligible impact on the gain calculation,

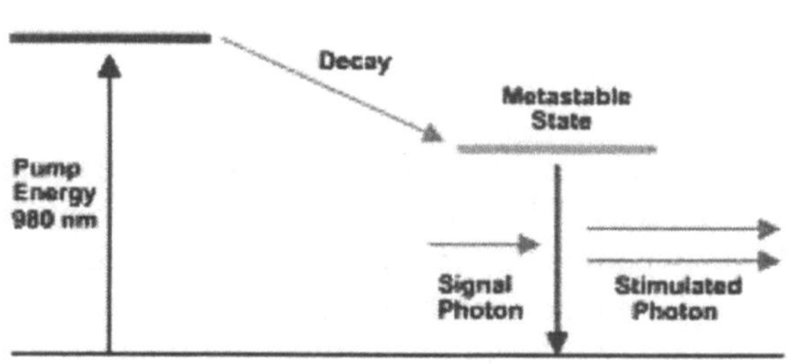

Figure 5.3.2. EDFA optical amplification.

but it can be a factor when high spontaneous emission levels are present. It is also mentioned by Hewlett Packard Company in **[Hentschel 95]**. The gain bandwidth refers to the range of wavelengths over which the amplifier is effective. Figure 5.3.4 shows a typical spectrum of gain in an EDFA.

Saturation power

The saturation power (P_s) is typically defined as the output power at which there is a 3-dB reduction in the amplification factor G, which is the same as P_{out}/P_{in} ratio (the small-signal gain). When $P_{in} << P_s$ the amplifier is in the unsaturated region and $G = G_o$ (see Figure 5.3.3). This is referred to as the small-signal gain, since the input power should be small so that the gain remains unsaturated during amplification. When P_{in} is increased enough, the amplifier reaches the gain saturation point. Gain saturation is the value of output power at which the output power no longer increases with an increase in the input power because the carriers in the amplifier are unable to output any additional light energy. Therefore, when P_{in} is comparable to P_s, amplification factor G is also expected to decrease. A Typical saturation power for EDFA is about 10 to 20 dBm.

Amplified spontaneous emission (ASE)

The EDFA also produces amplified spontaneous emission which adds to the spontaneous emission produced by the source. Because the output spectrum contains spontaneous emission from both the source and the EDFA under test, the EDFA ASE cannot be determined directly from the output spectrum measurement. The calculation of EDFA noise figure requires that the portion of the output ASE level that is generated by the EDFA is known. This is

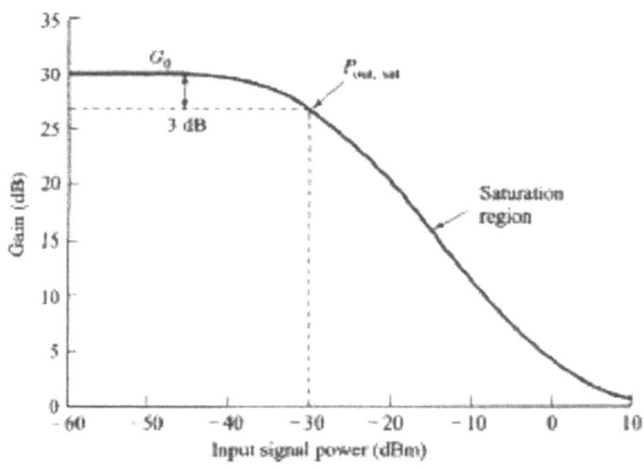

Figure 5.3.3. EDFA amplification factor G.

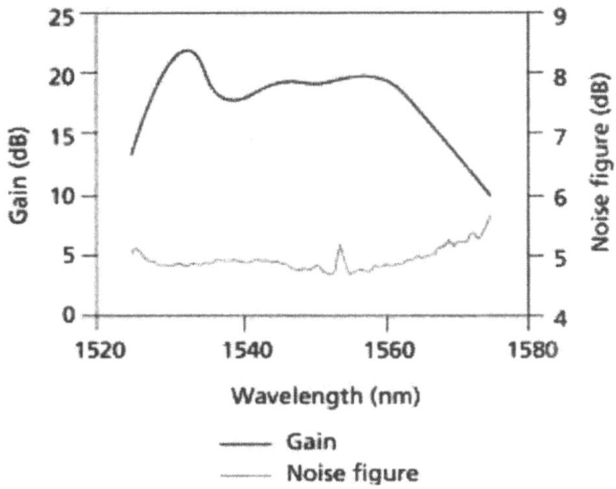

Figure 5.3.4. EDFA gain and noise figure spectrum.

calculated as the difference between the output spontaneous emission power and the equivalent source spontaneous emission power at the amplifier output. A method for remote fault localization in a link with concatenated EDFAs by simple monitoring of the ASE noise accumulated in the receiver has been developed in **[Prat 00]** in the Signal Theory and Communications Department at the Politechnique University of Catalonia, Spain.

Noise figure
The noise figure of an optical amplifier is particularly important for analog systems. The signal-to-noise ratios of analog systems are often so demanding that noise figures of less than 5 dB must be measured reliably. Another important purpose of the noise figure is to characterize the amount of ASE produced by the amplifier. Figure 5.3.4 shows the spectrum of the noise figure in an EDFA.

5.3.2.4. *EDFA specifications*

The EDFA makes use of rare-earth ions as a gain medium. These ions are doped inside the fiber core during the manufacturing process and pumped optically to provide the gain. The amplifier characteristics such as the operating wavelength and the bandwidth are determined by the dopants rather than by the silica fiber which plays the role of a host medium. Many different rare-earth ions, such as erbium, holomium, neodymium, samarium, thulium, and ytterbium, can be used to realize fiber amplifiers operating at different wavelengths covering visible to infrared region. EDFAs have attracted the most

Parameters	Unit	Typical Value
Wavelength range	nm	1528-1562
Nominal saturated output power range	dBm	From 14 to 23
Nominal gain range	dB	From 13 to 23
Nominal gain flatness	dB	<1.0
Nominal noise figure	dB	<6.0
PMD	ps	<0.5
PDL	dB	<0.3
Input / output tap responsivity	µA/mW	From 10 to 80
Input / output return loss	dB	>40.0
Operating temperature range	°C	-5 to 70

Figure 5.3.5. Typical specifications for C-Band EDFA.

attention among them simply because they operate near 1.55 µm, the wavelength region in which the fiber loss is minimum. Typical values for an EDFA in C-Band are shown in Figure 5.3.5.

5.3.2.5. *Research*

Bi-directional EDFA

Most of bi-directional optical amplifiers applicable to WDM employ two unidirectional optical amplifiers together with optical circulators or WDM filters to separate the signals propagating in the opposite directions. However, those devices tend to be costly because EDFAs, circulators, WDM filters, and other accessories are expensive. This hinders the bi-directional WDM transmission systems to penetrate deep into the metro and access networks whose traffic is rapidly growing due to the Internet. Figure 5.3.6 represents the design developed by Corning S.A., France, of a bi-directional multiwavelength two wavelength optical branching amplifier (TWOBA). This is used to compensate the propagation and the splitting losses along the link for bi-directional digital signals and downstream analogue CATV signal **[Jacob 98]**.

The TWOBA comprises of two parallel optical amplifier sections. The first erbium-doped amplifying section, associated with the blue band of the gain spectrum, is made of two coils, namely EDFA#1 and EDFA#2, and a gain flattening filter inserted between the two coils. This bi-directional section (no optical isolator) is used to amplify four upstream and four downstream digital channels simultaneously. The second amplifying section, associated with the red band, is unidirectional and is used to amplify the analogue CATV channel. Two 120-mW pump laser diodes 980 nm are used. Two medium band splitters (MB-WDM) located at the input and the output of the EDFA are used to separate or combine analogue CATV and digital interactive signals and to improve NF. Finally, a 1×4 splitter is added at the output of the

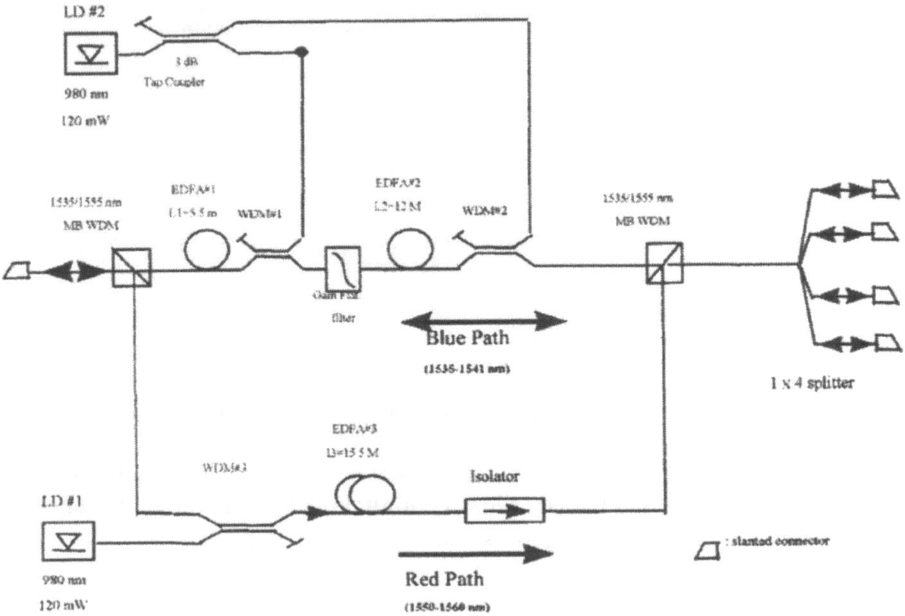

Figure 5.3.6. Configuration of TWOBA (bi-directional multiwavelength two wavelength optical branching amplifier).

TWOBA to allow its connection with four different optical network units (ONUs).

Recently, a low-cost bi-directional all-optical amplifier for bi-directional transmission into metro and access networks has been deployed by the Korea Telecom Access Networks Laboratory **[Seung-Tak 01]**. Figure 5.3.7 shows the configuration of this amplifier. It employs only one unidirectional EDFA to amplify signals of both directions and a fiber Mach-Zehnder (MZI) interferometer that acts like a four-port symmetric wavelength interleave to combine and then separate the signals.

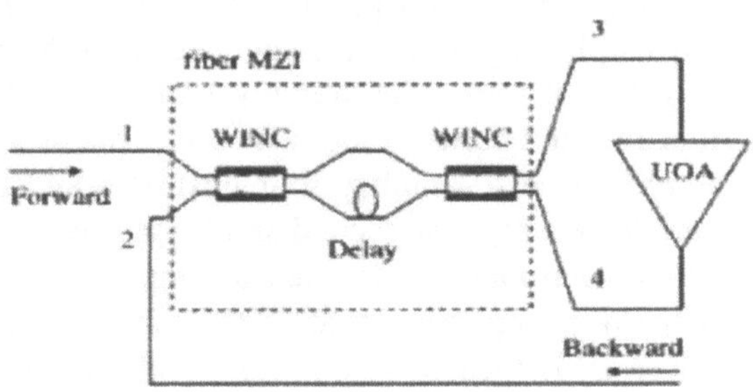

Figure 5.3.7. Configuration of a bi-directional amplifier.

Forward and backward signals enter port 1 and port 2 of the perfect (infinite isolation) fiber MZI, respectively. The combined signals appear on port 3 to be fed to the amplifier. The amplified signals then enter port 4 to be separated. The forward and backward wavelengths come out of port 2 and port 1, respectively. These signals propagate along the fiber experiencing backscattering and reflection at the optical discontinuity. Backscattered or reflected lights of both the forward and backward signals are collected in port 4, but the isolator inside the amplifier blocks them. No returned lights will enter port 3 of a perfect (infinite isolation) fiber MZI.

Channels are separated by 1.7 nm, and channel spacing between adjacent forward and backward signals is 0.85 nm. Insertion loss and isolation of the fiber MZI were estimated to 2 and 12.6 dB, respectively. The experimental setup was constructed as shown in Figure 5.3.8.

Figure 5.3.9 shows the optical spectra observed at various points. The combined bi-directional signals at the input and the output of the UOA in the BOA are shown in Figure 5.3.9 (a) and (b), respectively. The optical spectra observed on dropping ports of the circulators are shown in Figure 5.3.9(c) and (d). The unwanted signal, which should have directed to the other end, appeared on the spectra. This is the result of incomplete isolation (12.6 dB) at the fiber MZI. Figure 5.3.9(e) shows the optical spectra after the bandpass filter whose 3 dB bandwidth is 0.6 nm.

Optimum optical amplifier location

Spectrum slicing is an attractive scheme for implementing wavelength division multiplexing (WDM) passive optical networks (PONs) in the customer access environment. Recent demonstrations of spectrum-sliced customer access

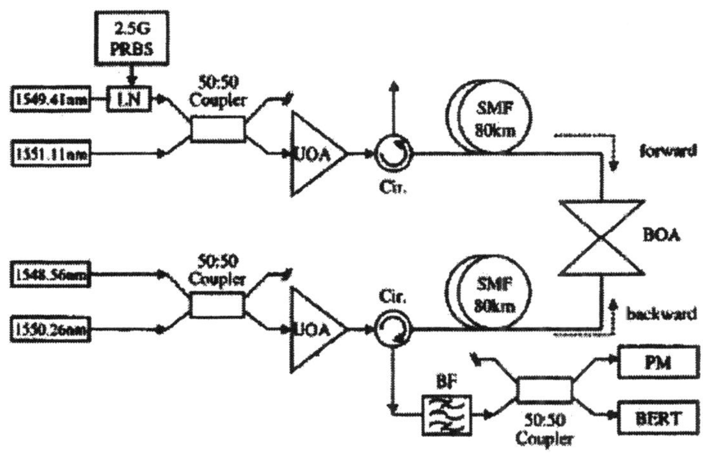

Figure 5.3.8. Experimental setup.

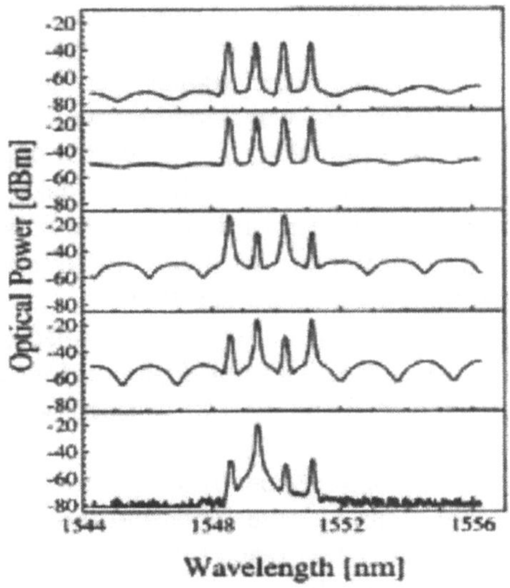

Figure 5.3.9. Optical spectra observed.

networks have employed one or more optical amplifiers at the central office to improve the power budget for downstream transmission. Optical amplification before the demultiplexer at the remote node, as illustrated in Figure 5.3.10, provides cost sharing between all users **[Cahill 98]**. A similar approach can also be used to improve the upstream budget. However, the access environment is very sensitive to infrastructure and maintenance costs, and these would vary depending on the location of amplifiers. It is, therefore, important to characterize for a variety of amplifier locations the obtainable improvement in network performance.

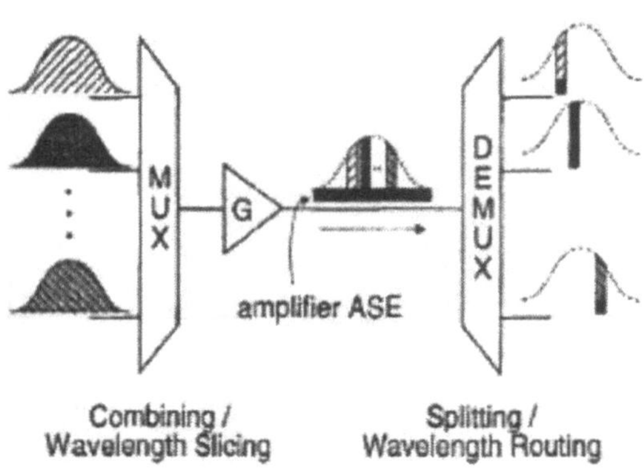

Figure 5.3.10. Optically-amplified spectrum-sliced WDM link.

When the amplifier is located close to the source, i.e., as a "poweramplifier", gain saturation reduces the loss budget from its optimal value. Moving the amplifier further from the source reduces the amplifier input power and the gain saturation. However, as the amplifier input power decreases, the noise generated by ASE in the amplifier (denoted as the solid black fill in Figure 5.3.10) degrades the SNR and reduces the loss budget.

Consequently, there is an amplifier location where the combined effects of gain saturation and amplifier ASE noise are minimized. As shown in Figure 5.3.11, adjusting the amplifier input power through ATTN1 simulates amplifier location. ATTN2 is then varied to achieve a BER of 10^{-9}, and the loss budget is then calculated as the sum of these attenuations (dBs).

There is a wide range of a amplifier input power values where the improvement in loss budget is close to optimum. However, the optimum location for an amplifier in either direction, is likely to be nearer the remote node than the central office, see Figure 5.3.12.

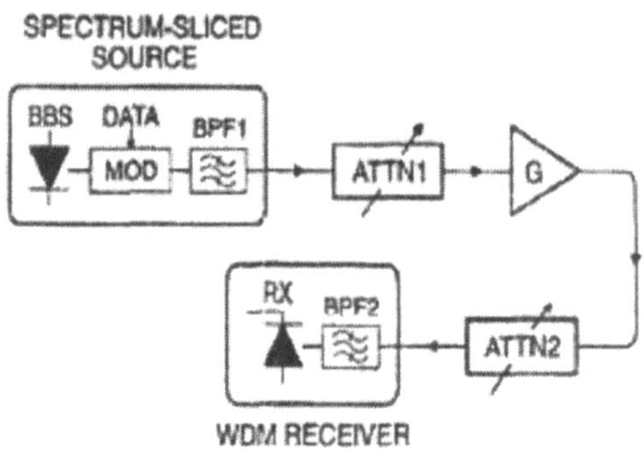

Figure 5.3.11. Single-channel spectrum slicing link.

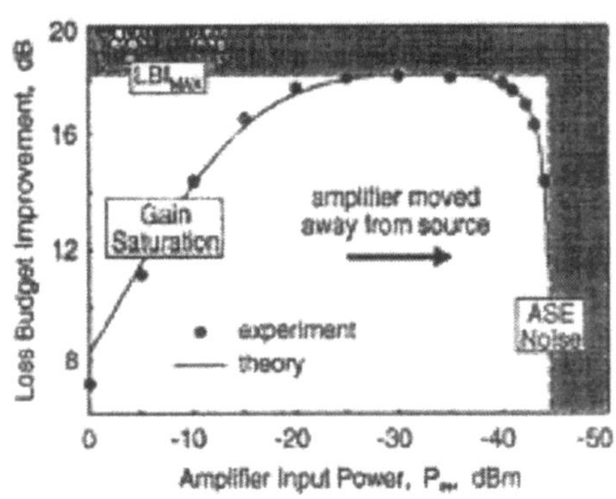

Figure 5.3.12. Loss budget improvement vs. amplifier input power.

A FTTH system based on EDFA
Fiber to the home (FTTH) access systems offer excellent upgradability and the advantages of a passive outside plant, but they have been relatively unsuccessful in the marketplace because of their large deployment cost. A new FTTH access PON that differs from the FSAN system has been proposed and demonstrated by Bell Laboratories, Lucent Technologies, USA **[Wood 99]**. In this system, high-speed data service, such as Internet access, can be added in small increments as customer demand grows and can deliver broadcast analog and digital video.

Despite the fact that this system requires each home receives a relatively large optical power; it can be economical because of the rapidly falling cost of EDFAs on a cost/mW basis. Figure 4.3.11 shows a schematic of the system. The modulator output is amplified by a first EDFA, split 4 ways, and ten transmitted through 13 km of standard single-mode fiber. A second EDFA, at a hub location, reamplifies the downstream signal. The EDFA's have noise figures of ~6 dB. The output of the second EDFA is split 64 ways, and ten are sent through drop fibers (which ranged from 1 m to 1 km in length) to optical network units (ONU's).

Using EDFAs, a reasonable optical power can be delivered to the ONUs, even with the large splitting ratios shown in Figure 5.3.13.

In addition, as fiber is pushed closer to the home, more optical power is

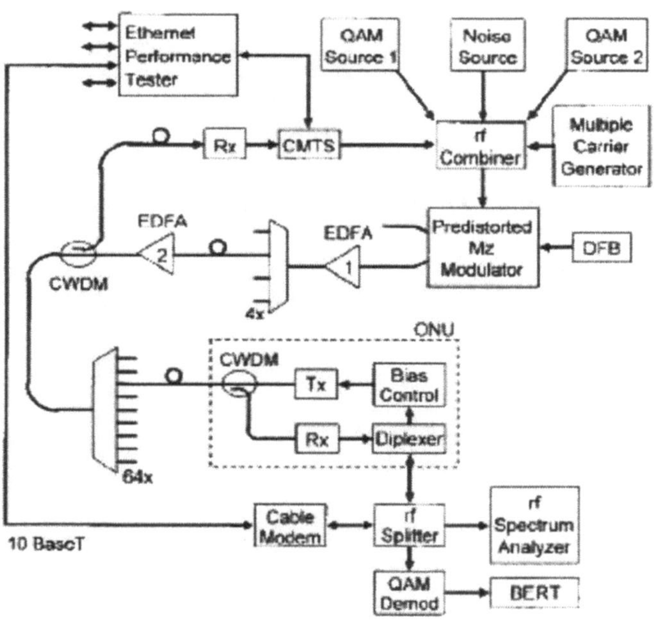

Figure 5.3.13. Experimental arrangement.

required. For FTTH roughly 2 mW of amplifier output power must be allocated for each home passed. Fortunately, the cost/mW drops as the saturation output power of high-power EDFA increases.

5.3.3. Semiconductor Optical Amplifiers (SOAs)

5.3.3.1. *Introduction*

After some years where EDFAs have dominated optical amplification, Semiconductor OAs have a renewed interest thanks to their diverse functionality. Although they cannot compete with EDFAs in long haul amplification, they may provide a high number of interesting optical applications for future access:

- optical switching;
- wavelength conversion;
- optical modulation;
- detectors;
- amplification at 1300 nm;
- all in conjunction with its capacity for integration.

The SOA is based on the same principle as the semiconductor laser, but where the facet reflectivities that provide the optical feedback for oscillation have been eliminated. An input fiber is coupled to one waveguide facet and the output fiber to the other. In between, the light is progressively amplified by the active region through stimulated emission with current pumping **[O'Mahony 88]**. As amplifier, the performances are limited because of the residual reflectivity, the polarization dependent gain and the coupling loss. Typical gain ranges from 10 to 15 dB in a spectral window of about 40 nm.

Up to today, SOAs have been rarely included in optical access networks, although they have promising potentialities because of their integrability, amplification also at 1300 nm and the more advanced mentioned functions.

5.3.3.2. *SOAs in access*

Interesting works have employed SOAs as modulators/amplifiers in a metro/access network experiments. The architecture proposed in **[Iannone 00]** is a WDM ring, using individual wavelengths to provision services to a geographically diverse set of users. Each wavelength forms a virtual ring and operates independently of the other rings.

Figure 5.3.14 shows a schematic of the basic layout. It hubbed ring with a Network Node (NN) connected "feeder" fiber to several Access Nodes (ANs) which turn serve subscribers at End Stations (ES) over "distribution" fibers.

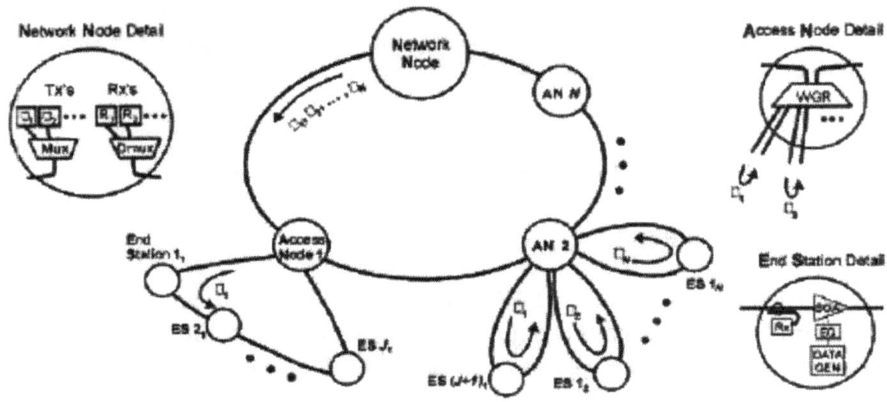

Figure 5.3.14. Network architecture.

The distribution scheme is based on an earlier "RITE-Net" WDM star architecture for passive optical networks (PONs). The NN contains a WDM transmitter (represented as multiple sources and a multiplexer) to create data to be sent downstream for each virtual ring, and a corresponding WDM receiver to detect each of the upstream signals. They implement the AN by using a waveguide grating router (WGR). Light from the feeder ring (left top fiber in AN inset) enters a waveguide grating router input port and is demultiplexed according to its wavelength.

The ES consist of receivers (Rx) for downstream information and semiconductor optical amplifiers (SOAs) which amplify and modulate light to create

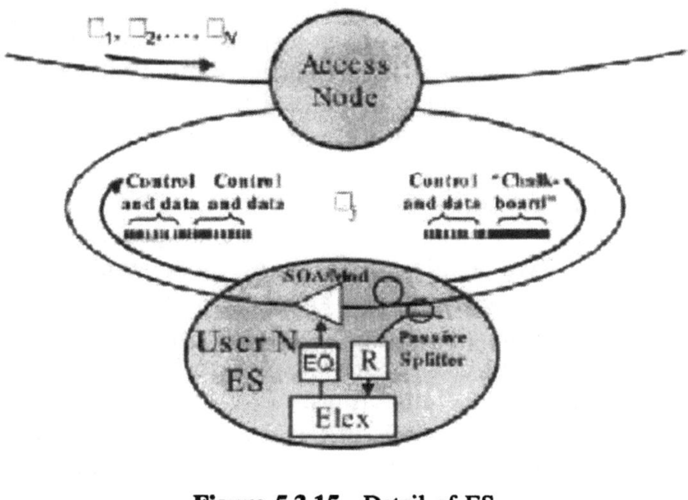

Figure 5.3.15. Detail of ES.

upstream signals. When the light entering the SOA is bearing data for this or another ES, the SOA acts simply as a single channel optical amplifier. However, when the ES must modulate an optical chalkboard, its data is pre-equalized to compensate for the SOA's frequency response, and applied to the SOA.

Modulated light enters the ES for user, where portion is split off to a receiver, which recovers the downstream information, and the remainder is passed to a SOA/modulator. All wavelengths are generated at the NN. Like a broadband optical modulator, the SOA is essentially indifferent to the wavelength of the light used: the stimulated emission will be at the same wavelength as the input light.

Three experimental configurations of a 120-km three-node ring demonstrate the flexibility of the architecture and the feasibility of using SOAs as remote modulator/amplifiers.

The first, shown schematically in Figure 5.3.16(a), is a single-wavelength virtual ring shared by two users with one user per AN serving area; the second, shown in Figure (b) demonstrates an increase in the number of users per virtual ring from two to four; the third configuration (Figure c) shows the same four users sharing two virtual rings (two users per virtual ring), thus permitting higher per-user data rates than in the previous single virtual ring configuration.

BER measurements were performed in continuous mode to test the SOAs response to high-speed modulation and confirmed cascaded operation in these systems. The upstream power penalties are negligible for three configurations of a 120-km ring: a single wavelength ring serving 2 and 4 users, respectively, and a two-wavelength, two-user-per-wavelength configuration.

The use of wavelength channels allows flexibility with regards to user demand for bandwidth, and helps integrate different protocols. To realize such networks, it is essential to increase the number of wavelength channels by employing dense wavelength-division multiplexing (DWDM) technologies. However, DWDM networks require precise wavelength control at every node that makes the cost per channel very high. A promising way to solve the problem is to use a centralized multi-wavelength light source (CMLS)

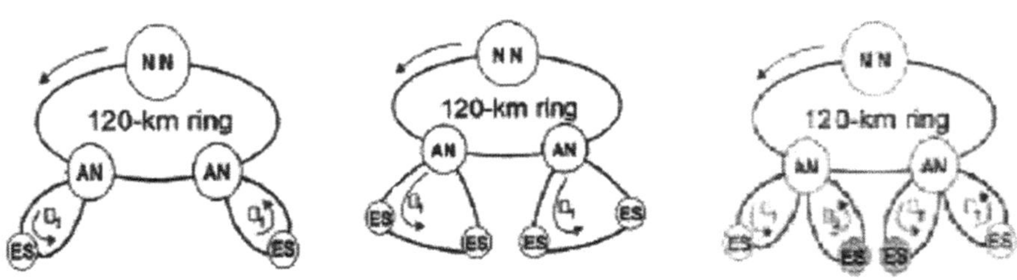

Figure 5.3.16.

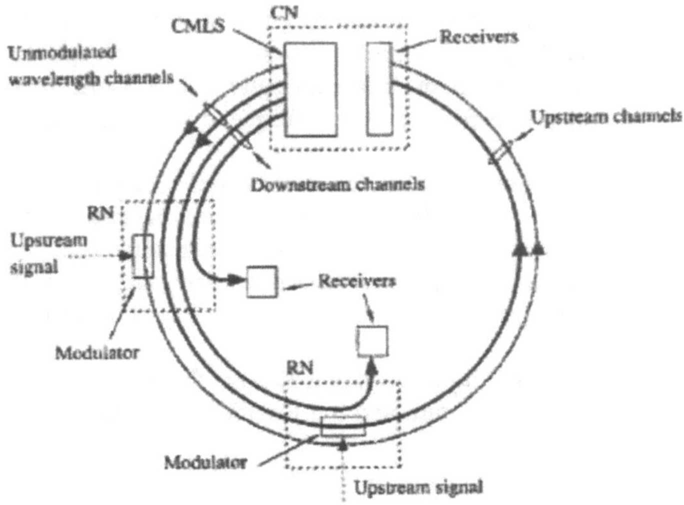

Figure 5.3.17. WDM access ring using CMLS.

[Takesue 00]. Figure 5.3.17 shows an example of a wavelength-division multiplexing (WDM) access ring with a CMLS.

A simple node configuration is proposed which is called wavelength channel modulator (WCM) for an access ring network. The WCM features a potentially inexpensive configuration that uses FBGs and a SOA with low crosstalk and stable operation. They calculated the intrachannel crosstalk arising from WCMs in a ring network and show the good cascadability of the WCM. They then constructed a WCM test model to confirm its low-crosstalk characteristics experimentally.

Figure 5.3.18 shows the basic configuration of the proposed WCM which consists of two circulators, two FBGs with the same reflection wavelength,

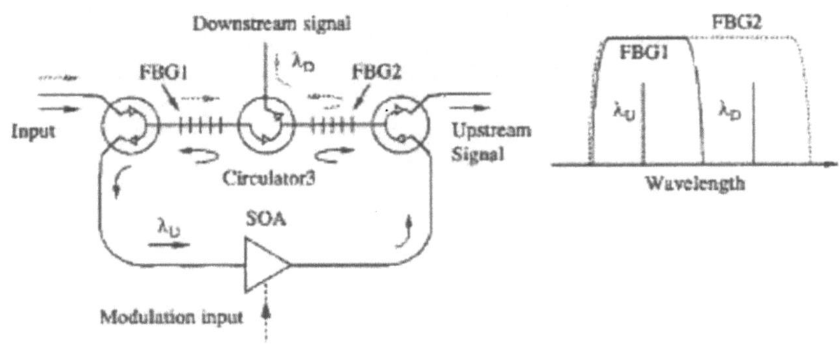

Figure 5.3.18. WCM configuration.

an optical isolator and a SOA. A WDM signal, which has an unmodulated channel with a wavelength, is launched into the input port of circulator 1. The WDM signal passes through circulator 1 and is input into FBG1 where the channel is reflected. Other channels pass through FBG1, the isolator and FBG2. The channel reflected by FBG1 is returned to circulator 1, and led to the SOA. Channel is intensity-modulated by the SOA with the upstream signal and led to FBG2 after passing through circulator 2. The channel is then reflected by FBG2 and consequently multiplexed with other channels. The WDM signal finally passes through circulator 2 and is output.

FBGs are chosen because they are relatively inexpensive and allow easy fabrication. Moreover, even when there are many nodes in a ring network, the use of FBGs will not lead to a narrowing of the optical bandwidth of each channel. By contrast, when a wavelength multiplexer such as a WGR is used in each node, the optical bandwidth for each channel becomes narrower as the number of nodes increases. Compared with conventional waveguide modulators, a SOA intensity modulator has several advantages such as low polarization sensitivity, compactness and potential cost effectiveness.

The ring network uses the C-band and 50-GHz channel spacing. Then, approximately 80 wavelength channels are available and at most 40 RNs can be integrated into the ring using different wavelengths for up- and downstream transmission. Such a network requires in-line optical amplifiers to compensate for node losses.

A test model is constructed by the proposed WCM and confirmed its low-crosstalk performance using the setup shown in Figure 5.3.19. The peak reflection wavelength and the 3-dB bandwidth of both FBGs were 1551.0 and 3.0 nm, respectively. The 3-dB modulation bandwidth of the SOA used in the experiment was approximately 600 MHz and the peak gain was set at

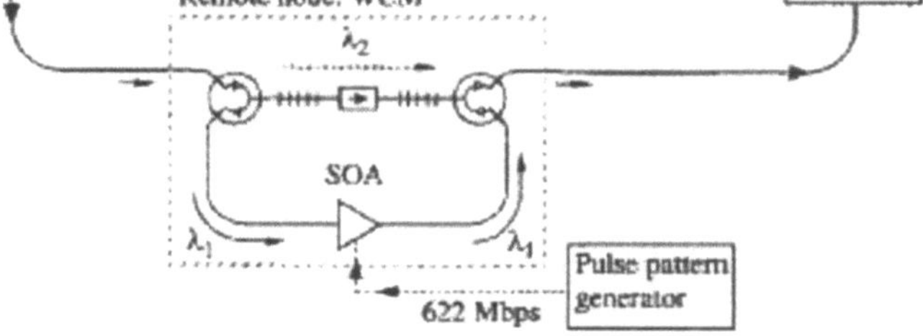

Figure 5.3.19. Experimental setup.

7 dB. A continuous lightwave from a tunable laser, set at 1551.0 nm, was polarization-adjusted to obtain the maximum SOA gain and input into the WCM. The lightwave was modulated at 622.08 Mbps by the WCM in non-return-to-zero format with a pseudorandom bit sequence, and output as an upstream signal.

In another work **[Buldawoo 97]**, a reflective SOA is experimentally built and tested to act as both receiver and modulator for FTTH subscribers. The active layer is InGaAs Multi-Quantum-Wave on an InP substrate, with a 6° tilt to minimize parasitic reinjection of light. 3 dB fiber-to-fiber gain is obtained with a polarization sensitivity lower than 1 dB. The optical bandwidth is 40 nm.

A different type of SOA has recently appeared in industry: the Linear Optical Amplifier (LOA). It basically consists on a SOA integrated in a VCSEL, sharing the same active region. The laser acts as an optical pump source of very high intensity and efficiency. This results in a very stable gain in front of varying optical signals of relative power. A commercial LOA has constant 17 dB gain and output power of 10 dBm.

5.4. PASSIVE COMPONENTS

5.4.1. Introduction

In this last section of chapter five, we present a state-of-art of optical passive components. Tere is a lot of passive components but probabliy the most important are the following: couplers, multiplexers and connectors, being the last ones the most determining in access networks. The study has been divided into research & development devices and commercial devices.

5.4.2. Optical couplers

5.4.2.1. *Introduction*

For access, optical couplers are mainly used in Passive Optical Network (PON) to split or combine the optical signals. These are reciprocal and symmetric components intrinsically, so they can be seen as both combiners and splitters. The number of used input/output ports and direction of the light depends on the application.

Although the most standard coupler has 2 × 2 ports, in the PON the communication is point-to-multipoint: all users (N) communicate with the Head-End through the coupler, which has a split ratio of 1:N. In typical PONs, N ranges from 4 to 32, normally being raised-by-the-power-of-2-

numbers (4, 8, 16, 32). It is limited by the power loss, the bandwidth division and the cost.

In this section, the different couplers technologies that are used in access networks are explained, highlighting the main features that are of interest in access, with the practical parameters values that have been obtained in the R&D laboratories.

5.4.2.2. *Coupler types*

Competing technologies to produce these couplers are:

1. Combinations of the basic 1 × 2 (or 2 × 2) Fused Biconical Taper (FBT) coupler to form the 1:N coupler.
2. 1:4 single fused fibers.
3. Planar Lightwave Circuit (PLC) waveguides with ribbon fiber alignment.

These and other techniques are also investigated at the laboratory level to reduce fabrication costs and/or increase performances: excess loss, split ratio uniformity, bandwidth and temperature range. Other parameters like Polarization Dependent Loss, return loss and directivity are less important here.

5.4.2.3. *Research*

Regarding the first commercial category, the research focuses on increasing the reliability and simplifying the packaging. In this direction, the work of Hitachi **[Sasaki 00]** describes the new technique of laser irradiation using a CO_2 laser as a heat source in the fusion of the elongated fibers. This also solves the contamination problems, and speeds the process. An excess loss of 0.79 dB is obtained in a tapered region of only 6.6 mm. Figure 5.4.1 shows the scheme of the fused coupler and a photograph detail of the elongated fibers.

In this type of couplers, there is a trade-off between the excess loss and the taper angle and the coupling region length. An accurate measurement and control of these parameters are basic for a good yield fabrication. This was

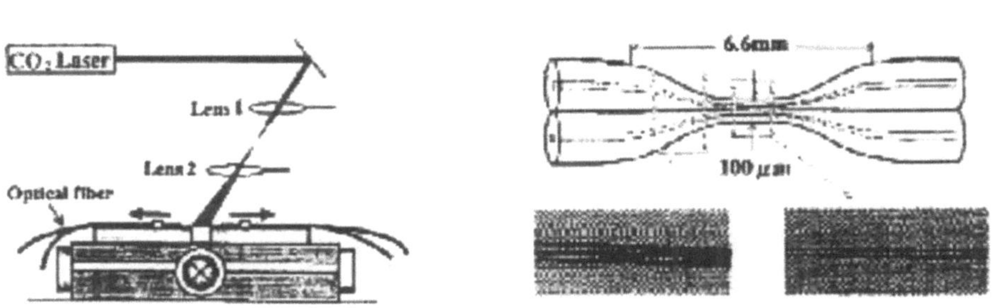

Figure 5.4.1. Fabricated fiber coupler scheme.

performed using metallographic analysis. With this technique, very low losses of 0.13 dB have been obtained, and the fabrication parameters were derived for different designs **[Hsieh 99]**.

An alternative technique to obtain the tapered coupler is by fiber lateral polishing instead of fusion, but normally present worse performances and cost.

A relevant published paper corresponding to the second category, 1:4 single fused fibers, is the work of E-Tek Dynamics **[Pan 98]**. It describes this technique with the theoretical model, optimization and fabrication results. The 1:4 fused coupler is made monolithically of 5 identical fibers colinearly arranged, simultaneously heated and pulled to the form biconical taper (Figure 5.4.2). The input fiber is the central one (3). The experimental obtained excess loss is 0.5 dB maximum in a wide band from 1200 to 1600 nm. The uniformity is better than 1 dB (Figure 5.4.3).

The third category method is based on integrated optics, also called Planar Lightwave Circuits (PLC). The PLC allows the implementation of advanced functionality in the optical domain in a simple way. Coupling, multiplexing and switching are examples.

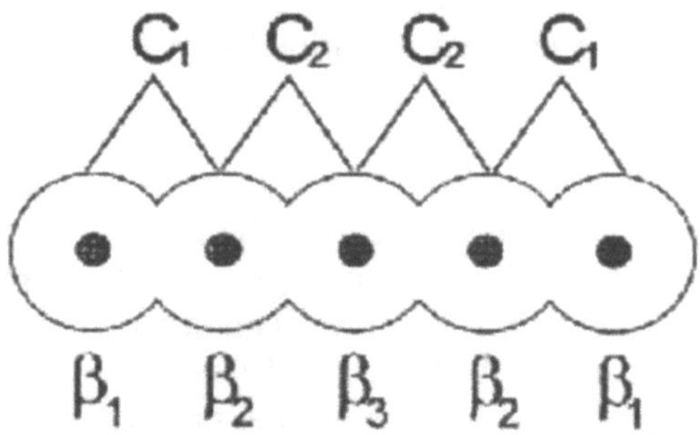

Figure 5.4.2. 1 × 4 fiber fusion coupler scheme after pulling process.

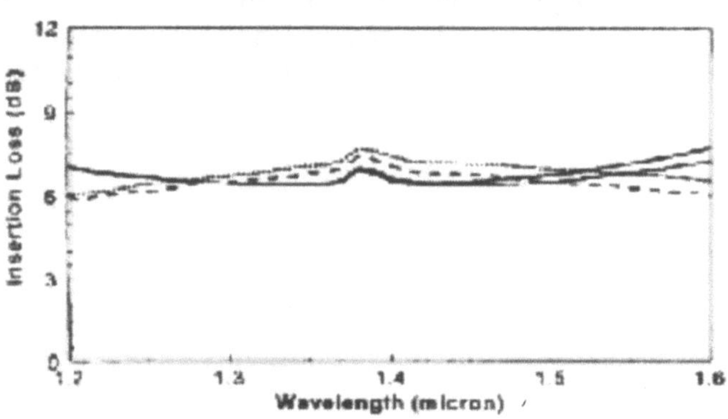

Figure 5.4.3. Measured insertion loss for all output ports, from 1.2 to 1.6 μm.

The critical point is the fiber alignment to the planar waveguides. Normally, the multiple waveguides are aligned with ribbon fiber. This is conventionally a manual and a very accurate process with high cost. Automatic processes for passive alignment are developed with different techniques for the PLCs:

Integrated mode shape adapters or waveguide tapers increase the coupling efficiency with the fibers, increase the alignment tolerances and avoid discrete optical elements as shown in **[Hunziker 98]** by the Swiss ETH. Figure 5.4.4 shows a taper fabrication using the etch process and selective area regrowth to obtain a down-tapered waveguide core, an up-tapered upper waveguide cladding, and a rib waveguide. The coupling loss is reduced by a factor of three approximately due to a better adaptation to the gaussian beam of the optical fiber. In Figure 5.4.5, this improvement with the larger mode and a better alignment tolerance can be observed.

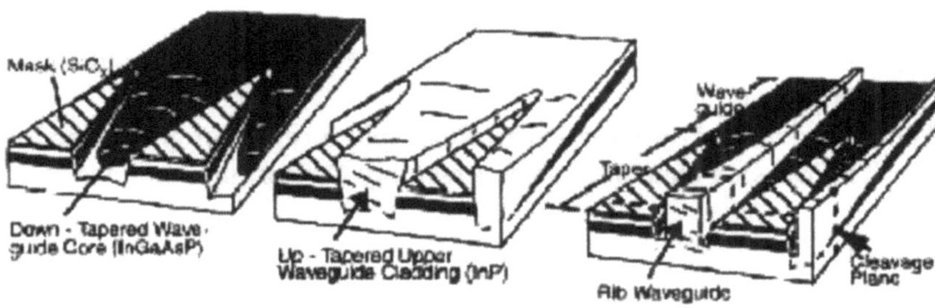

Figure 5.4.4. Fabrication sequence of optical waveguide mode shape transformers.

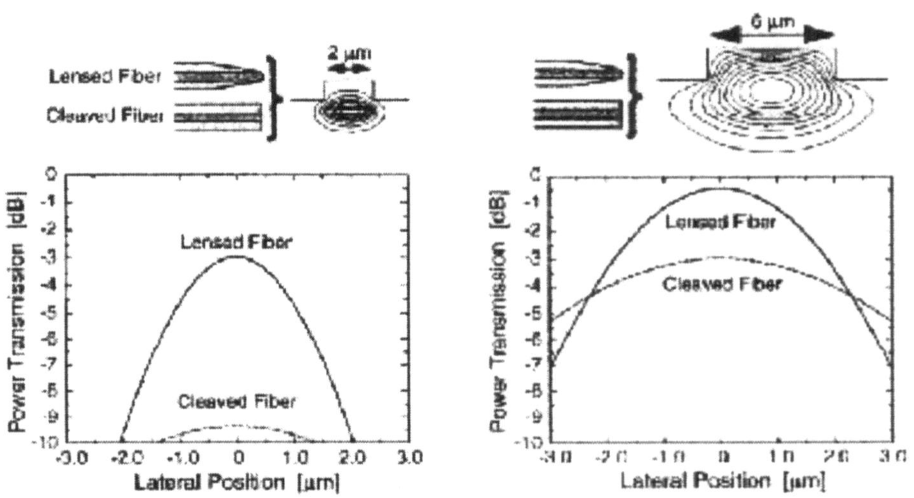

Figure 5.4.5. Calculated coupling efficiencies as a function of the lateral alignment between cleaved/lensed fibers and waveguide modes.

Also, the use of a mounting board allowing passive self-alignment features like integrated V-grooves and alignment trenches reduces the number of alignment steps and the need for precision tools.

Furukawa **[Iwashita 97]** has obtained a 1 × 8 star coupler in an integrated PLC waveguide device compatible with a MT connector and ribbon fiber (scheme in Figure 5.4.6). It was fabricated on a silicon wafer, without active alignment in a very compact package for mass production. It accomplishes repeatability and environmental tests. The mean excess loss is 0.54 dB with a change of 0.12 dB and return loss of 40 dB.

Another design, a 1:16 splitter from NTT **[Yoshida 92]**, exhibited insertion loss of 14.2 dB ± 0.6 dB uniformity among the branching paths. The 1:16 splitter chip was 41 × 6 mm.

The PLC technology can be used also to improve some parameters like coupling ratio flattening. A Mach-Zenhder structure with a path difference shorter than 1um can operate as a wavelength-insensitive coupler. In **[Jinguji 90]**, the 20% coupling ratio only varies by 2% in the wavelength range of 1.3 to 1.65 μm.

Another advance is the low-cost robotic platform development under the NIST ATP Precision Optoelectronics Assembly Consortium, that has obtained

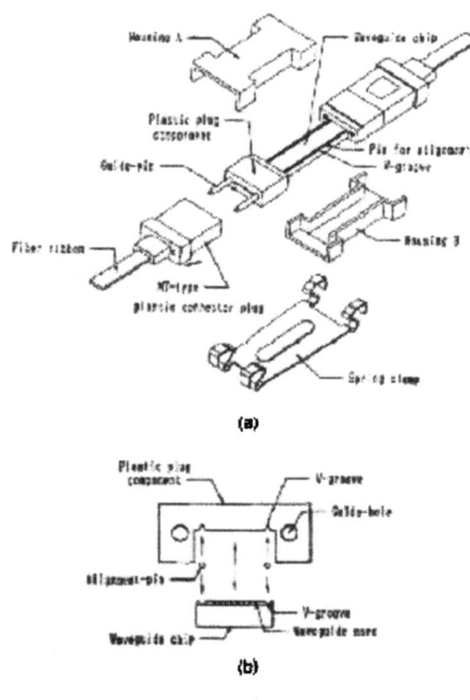

Figure 5.4.6. Mechanical scheme of the 1:8 PLC-MT splitter device with ribbon fiber.

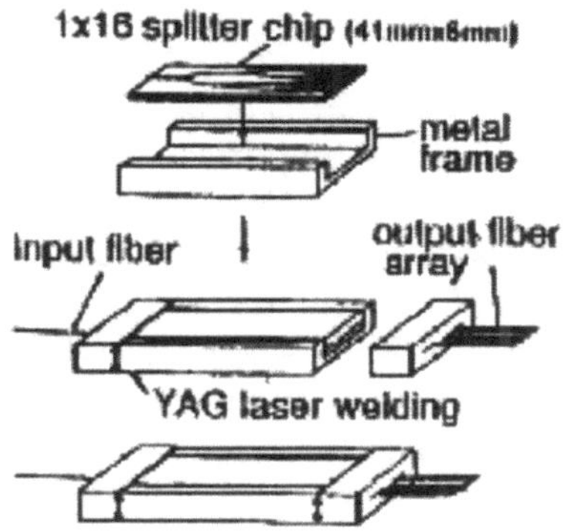

Figure 5.4.7. 1:16 PLC splitter.

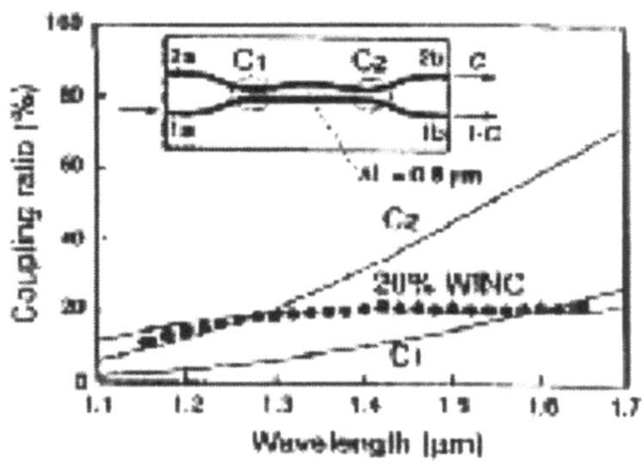

Figure 5.4.8. Wavelength insensitive coupler.

reproducible packaging results within ±1 μm **[Witham 00]** at optical alignment and bonding.

A different demonstrated interesting technique for access employs a diffractive optical element to equally couple the light from an input fiber to the output array fibers. The device, shown in Figure 5.4.9, of E-Tek Dynamics **[Pan 99]** uses this element between two 4f-aspherical lenses and short pieces

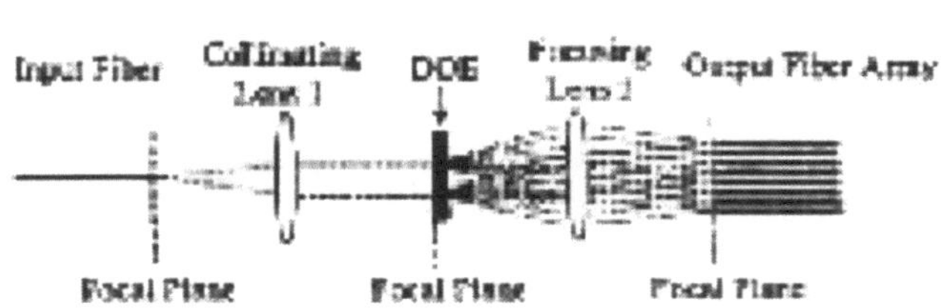

Figure 5.4.9. Difractive 1:8 coupler.

of graded index fibers to mount a cost-effective 1:8 coupler with excess loss lower than 2.4 dB, a broad good uniformity of ±0.25 dB from 1510 and 1570 nm (graph of Figure 5.4.10), and PDL of only 0.05 dB.

For larger port-counts, optical amplification becomes necessary to compensate for the power division and excess losses. A novel configuration **[Gan 00]** to obtain a coupler with very large port count has been proposed. It combines a number (m) of N × N couplers, and 2 optical amplifiers and several filters to obtain a structure of [m(N – 2)] × [m(N – 2)] star coupler for the WDM applications. The structure (Figure 5.4.11) shows a highly scalable port increase, but is too complex for access.

Another research approach was a 1:64 coupler made of 1:8 couplers and EDFA's in a reflective star configuration **[Zeng 94]**, as shown in Figure 5.4.12. The measured variations of the insertion loss of the ports are less than 3 dB. The Noise Figure is about 12 dB in the lossless coupler.

5.4.2.4. *Commercial*

A large number of manufacturers produce 1:N couplers using these technologies. An example of the best performances at the commercial level is: excess loss of 0.3 dB, ±2% ratio tolerance, temperature range from –20° to +70° and 40 nm bandwidth for a 1:4 splitter. For a 1:16 splitter, we may have 2 dB of insertion loss and 1.8 dB of uniformity. Some perform dual window operation.

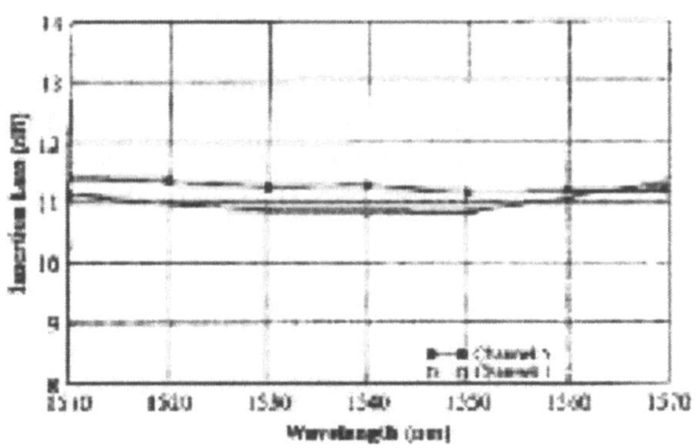

Figure 5.4.10. Measured insertion loss with the wavelength, for the best and worst channels.

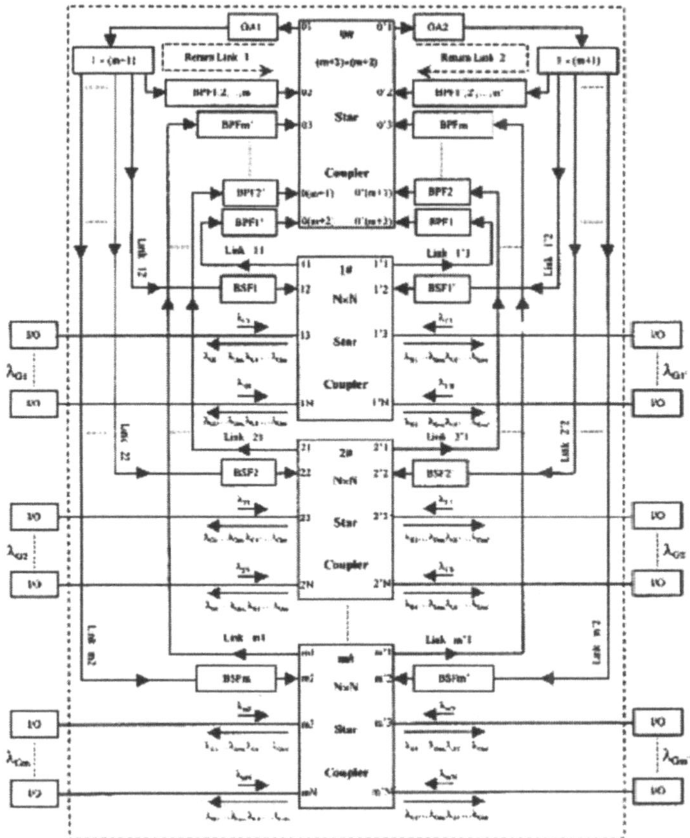

Figure 5.4.11. Highly scalable coupler architecture.

5.4.3. Optical multiplexers

5.4.3.1. *Introduction*

Most advanced PON systems use the WDM technology in order to offer a "Lambda-to-the-user" service. With the use of a remote optical multiplexer, instead of an optical coupler, the optical wavelengths are route to each user, and a high reduction of the power loss is obtained, as well as independence and privacy between users, that can make use of a full optical wavelength channel. Optical multiplexers are also bidirectional devices that both split or couple the different wavelengths, depending on the light direction.

The main parameters are the insertion loss, isolation (crosstalk from other channels), channel spacing, and channel bandwidth.

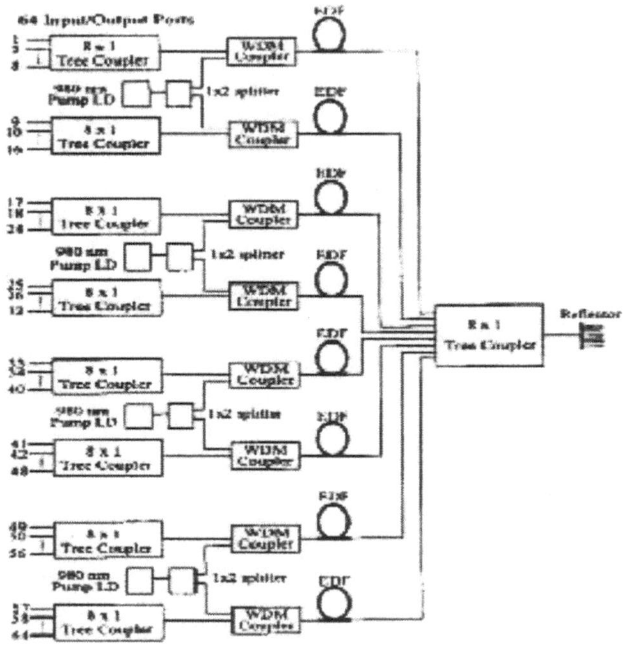

Figure 5.4.12. Reflective and amplified 64-port coupler.

Three different levels of wavelength density or spacing may be used:

- Wide-WDM: to separate/couple different bands, for example the 1300 and 1550 nm bands. This is used in bidirectional PONs using only one fiber per user.
- Coarse-WDM: the new standard for low cost wavelength multiplexing where temperature control of laser can be avoided. The channel spacing is standardized to 20 nm, for the second and third window, and even in between.
- Dense-WDM: the ITU-T grid for transport WDM systems is used. Channel spacing may be as narrow as 100 GHz (and multiples) in the fiber third transmission window.

They involve different level of technology and complexity, while great advances have recently been made.

5.4.3.2. *Research*

For simple two-window multiplexing or low-channel count, conventional techniques like multi-layer thin-film coated glass are used usually with an extra effort for integration in OEIC and small transceiver packages **[Tanaka 99]**.

Low channel spacing in available devices is 200 GHz and the port-count is up to 8.

Also, all-fiber devices may be performed by combining the fused biconical taper couplers technique with the two-beam holographic method for fiber-grating (Figure 4.6.13). Low cost add-drop multiplexers were demonstrated with a measured isolation of 18 dB **[Sasaki 00]**.

Another interesting technique is the fiber-based lattice component formed by the integrated cascade of 3 fused couplers, like the demonstrated by Corning in **[Nolan 98]**. The device is fabricated by inserting two fibers of different propagation constants into a borom-doped silica tube that is collapsed and tapered in three locations. It is 12-cm long, and acts as a band-splitter (Figure 5.4.14).

A newer technique is based on a tilted Bragg grating written of a two-mode fiber and a mode-selective coupler. This is an all-fiber design that performs both add&drop multiplexing functions with low loss and low cost for a limited number of WDM channels **[Park 01]**. Because of a tilt in the grating, the reflected light changes from the LP11 to the LP01 mode towards the drop port in the configuration shown in Figure 5.4.15. The dropped spectrum 3dB-width was 0.5 nm, with 1.5 dB loss. The coupling efficiency was 90% and the extinction ratio 20 dB.

The integrated planar lightwave circuit (PLC) technology can be also used here especially in combination with other functions. A Mach-Zenhder interferometer with a differential branch delay of several to tens of micrometers can be applied to two-channel wavelength division multiplexers. In a specific 1.3–1.55 μm design, NTT **[Yoshida 92]** got 1–4 dB insertion loss with 20 dB of crosstalk in combination with a Y-branch for a monitor photodiode (Fig. 5.4.16).

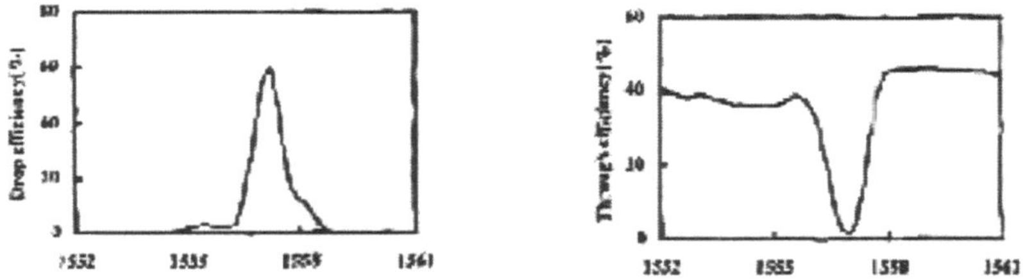

Figure 5.4.13. Scheme and function transfers of the grating coupler.

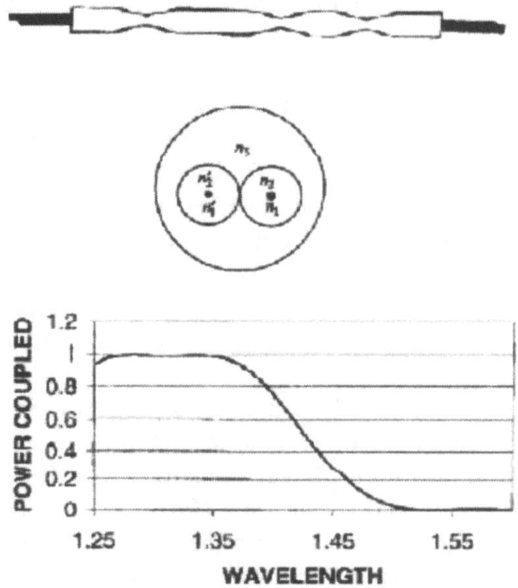

Figure 5.4.14. Lattice fiber band-splitter.

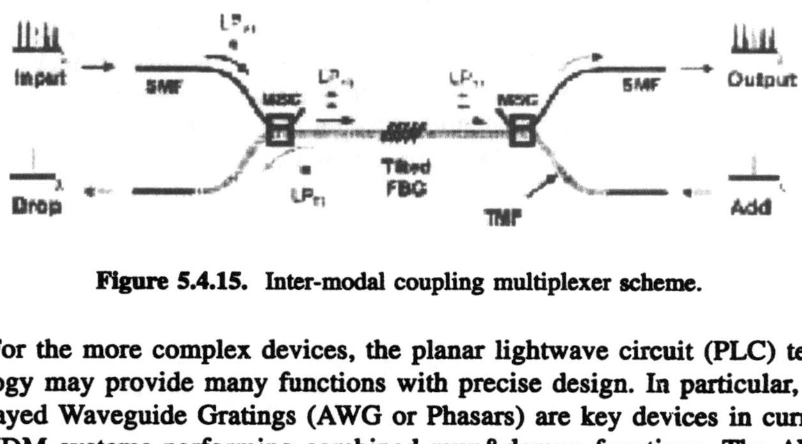

Figure 5.4.15. Inter-modal coupling multiplexer scheme.

For the more complex devices, the planar lightwave circuit (PLC) technology may provide many functions with precise design. In particular, the Arrayed Waveguide Gratings (AWG or Phasars) are key devices in current DWDM systems performing combined mux&demux functions. The AWG consist of input/output waveguides, 2 slab waveguides, and the arrayed waveguides in between. Different channel number and spacing may be obtained by modifying the optical circuit parameters **[Uetsuka 99, Boos 99]**. Spacings as narrow as 8.5 GHz have been experimentally obtained. A typical configuration is a 1 × 32 or 32 × 32 channel AWG, 100 GHz spacing with insertion loss around 7 dB and crosstalk better than 25 dB (Figure 5.4.17). In the Corning process, a core layer of doped silica is deposited by flame hydrolysis and consolidated on a silica substrate (SiO2/SiO2 chip). The pattern of the wave-

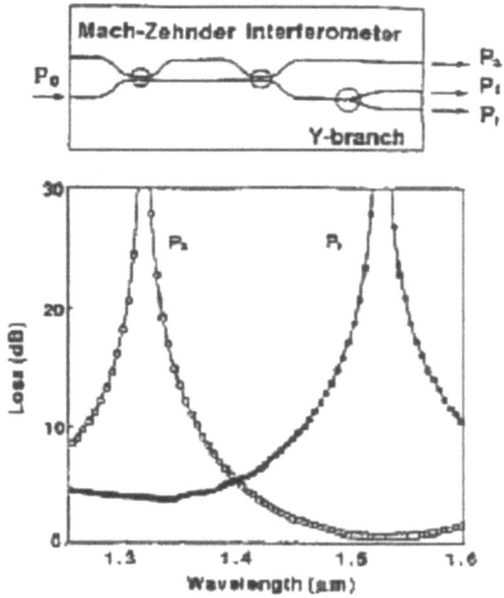

Figure 5.4.16. Band multiplexer and Y-branch, scheme and response.

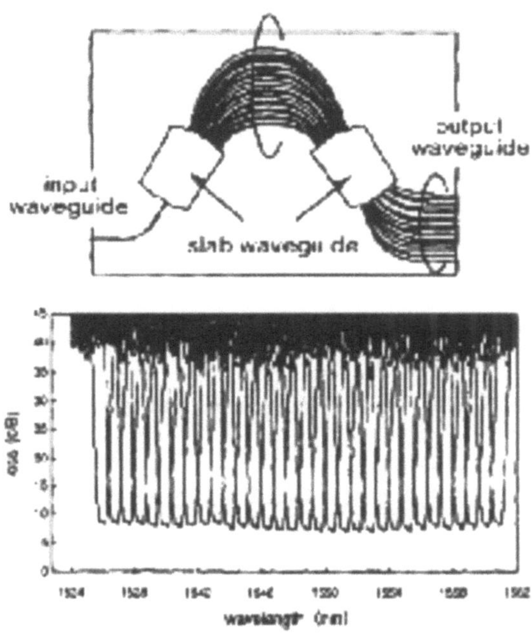

Figure 5.4.17. 1 × 16 AWG multiplexer scheme and spectral response.

guiding structure is transferred to the core layer by photolithography and reactive ion etching techniques. Also, flat top or gaussian responses can be obtained for a better cascadability. The Polarization Dependent Loss is usually better than 0.5 dB. In other design **[Nadler 99]**, stress release grooves have been used to reduce polarization sensitivity of the wavelength from 0.66 to 0.05 nm; the crosstalk was 40 dB and the insertion loss of only 5 dB for a 4 × 4 AWG as shown in Figure 5.4.18.

On the other hand, bulk optics devices can also operate with good response characteristics like the bulk-grating design, but can be normally large and vibration sensitive. An application example of a 20-channel 1 nm-spacing diffraction grating multiplexer is found in **[Hill 90]** using multiple-lens. The length is 190 mm. A broadband WDM PON is demonstrated using these devices.

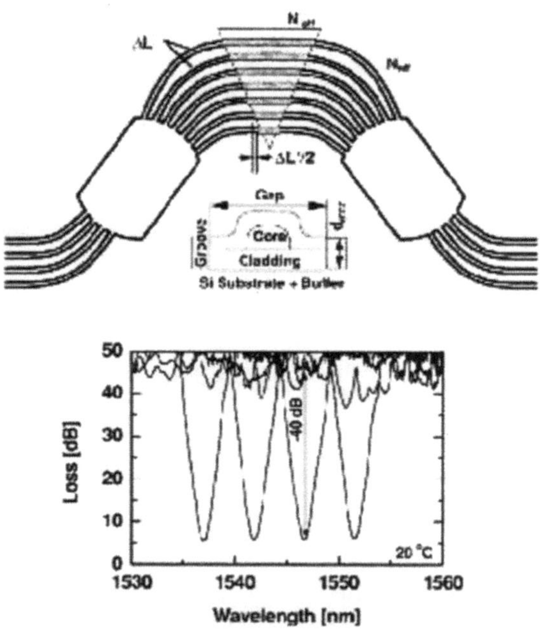

Figure 5.4.18. 4 × 4 AWG scheme and spectral response.

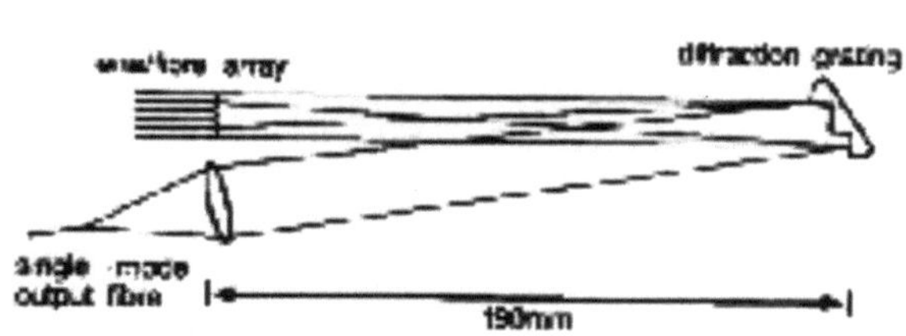

Figure 5.4.19. Scheme of a 20-channel 1 nm-spacing diffraction grating multiplexer.

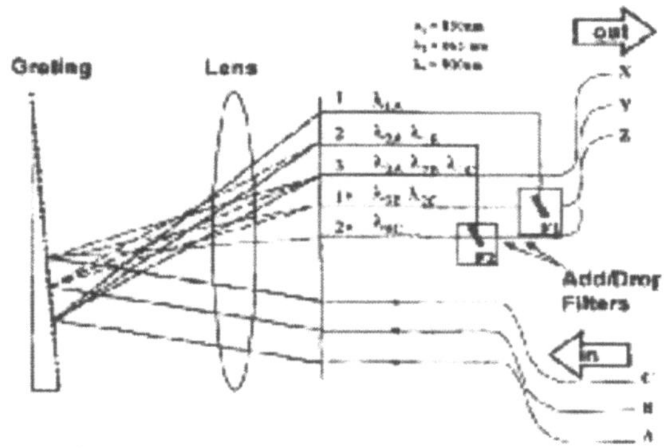

Figure 5.4.20. Scheme of a multimode fiber 3 × 3 diffraction grating multiplexer/router.

Also, a bulk grating for coarse multiplexing/routing for multimode fiber at the first window has been successfully demonstrated in **[Patel 99]**, with 3 × 3 ports at 37 nm spacing (Figure 5.4.20).

5.4.4. Optical filters

The technology used for the optical filters is similar to the one used for the multiplexers just detailed above. In fact, a filter can be seen as the simple case of a multiplexer with only one input and output. However, there is an added functionality that is often pursued in an optical filter: the tuning capability.

The most common tunable filter is the Fabry-Perot filter. It consists of an optical cavity (often fiber itself) with highly reflective facets. It exhibits a very selective band-pass transfer function, although with periodic behavior. The peaks are much narrower than the valleys if the reflective facets are of high quality. Typically, a channel can be tuned and filtered out from a WDM of about 40 channels. The tunability is normally obtained by mechanical elongation of the cavity using the piezo-electric effect. Tunable filters can be used in access for WDM PONs since Fabry Perot filters are quite compact. A subscriber can select any WDM channel only by tuning its filter peak wavelength.

Another interesting type of filter is the Fiber Bragg Grating. It is easy to implement via UV-irradiation and shows good thermal stability, but it is not (easily) tunable.

Other kind of filtering is by absorbing materials. Although they are only used for wide WDM (for example, to separate a second window signal from a third one), they are easily integrated in opto-electronic ICs.

Another kind of filter is the Acousto-optic tunable filter. The operation is based on the interaction between the lightwave and a surface acoustic wave (SAW). It presents a very wide tuning range, by changing the applied acoustic frequency, but suffers from strong side-lobes.

5.4.5. Connectors

5.4.5.1. *Introduction*

In the past, the most commonly used fiber optic connectors in networking applications are the SC-type[3] and ST[4]-style connectors. While they perform well for many applications, both types of connectors have limitations with respect to cost-effective networking (they are difficult to integrate and this is a major drawback in access networks where a lot of connectors have to be placed).

Two main purposes are suited by the Small Form factor connectors (SFF):

- High densities, crucial at the network headend where space savings is very important.
- Easy to install and use. As every customer has to be equipped with a connector, it is an important economical issue, the availability of connectors which could be installed without expertise workforce or very expensive instrumentation.

Unlike ST or SC connectors, SFF connectors use a latching mechanism similar to that used in the copper RJ-45 making their use easy and intuitive. All the SFF connectors can be quickly terminated and are, as its name suggests, small size. All of them are standardized by TIA/EIA-568. After two years of intense debate and development, the Telecommunication Industry Association's TR 42 Committee has completed revisions to the section of ANSI/TIA/EIA 568 that covers optical fiber technology. During the revision process, the standard was divided into three sections. The 568-B.3 document was completed in March 2000 and 568-B.1 is being released in advance (containing general requirements) and 568-B.2 (on balanced copper cabling). The document is available in the web site, www.tiaonline.org. The revised standard is important for the design of premises networks on several levels. Annex A of the pending TIA 568-B.1 will elevate the status of centralized optical fiber cabling from a technical service bulletin (TSB75) to a normative annex. TIA

---

[3] SC: Square, keyed connector with push-pull mating 2.5 mm ferrule and molded housing for protection.

[4] ST: Cylindrical with twist lock coupling, 2.5 mm keyed ferrule. (For both short distance applications and long line systems).

568-B.3 addresses the use of such critical technologies as 50/125 micron multimode fiber and small-form-factor connectors. According to Tony Beam, who chairs the Fiber Optic Task Group and served as Chair of TIA TR 42.8 during this process, "the new standard takes into consideration the need for companies to support higher bandwidths all the way to the workstation and gives users the protection of standards-based solutions."

TIA 568-B.3 includes multi-mode 50/125 μm optical fiber in the standard, which is useful for the network designers because it offers great flexibility. Although there is not a recommended media in 568-A, 50 μm multimode fiber is a clear possibility. In fact, its birth was previous to that of widely used 62.5 μm fiber. The standard includes all specifications for fiber performance including physical, mechanical, optical and environmental. The minimum bandwidth for 62.5/125 μm optical fiber is 500/500 MHz·km at 850/1300 nm. The minimum band-width for 62.5/125 μm optical fiber remains at 160/500 MHz·km. The standards for cabled fiber performance for the two fiber types are the same. In spite of this and as was stated in section FIBRA, lately, much effort in improving the performance of multi-mode optical fiber has been made.

Another significant part of the standard is related to the connectors. Rather tha specifying a concrete connector design (to replace or supplement the SC-duplex connector), the standard defines connector choice through performance requirements (optical, mechanical and environmental) that an optical fiber connector used in the premises that the environment must meet. Thus, TIA 568-B.3 allows the new small-form-factor (SFF) connectors as long as two main requirements are met. First, a Fiber Optic Connector Intermateability Standard (FOCIS) document (TIA/EIA-604-XX) is required to ensure that products made by different manufacturers will be made with good results. Second, the connectors must meet the mechanical and environmental performance requirements set forth in Annex A of TIA 568-B.3. In this way, the complete performance requirements for premises optical fiber connectors will be established for the first time

The basic minimum requirements for an optical connector are maximum loss of 0.75 dB for multimode or single-mode fibers, and a minimum return loss of 20 dB for multimode and 26 dB for single-mode fiber. SC duplex and ST-compatible connectors are still allowed under these same constraints. More information on the technical advantages and affordability that optical transmission brings to local area networks and fiber-to-the-desk applications, can be found in the Fiber Optical LAN Section (FOLS) of the TIA. The FOLS is a consortium of leading fiber optic cable, component and electronics manufacturers. Member companies include 3M, AMP, Allied Telesyn, Berk-Tek, Belden Wire & Cable, CommScope Inc., Corning, LANCAST, Lucent Technologies, Ortronics, Panduit, Siecor, The Siemon Company, SpecTran, Sumitomo Electric Lightwave and Transition Networks.

The TIA/EIA-568-B.3 specification on optical fiber cabling consists of one recognized cable type for horizontal subsystems and two cable types for backbone subsystems:

- Horizontal: 50/125 μm or 62.5/125 μm multimode (two fibers per outlet).
- Backbone: Backbone – 9/125 μm or 62.5/125 μm singlemode or multimode.

All optical fiber components and installation practices shall meet applicable building and safety codes. Some features are:

Optical Fiber Patch Cords:
They shall be a two-fiber (duplex) indoor cable of the same type as the cables to which they connect. They shall be configured so that "A" connects to "B" and "B" connects to "A".

Figure 5.4.21.

Connecting Hardware:
Connectors shall be protected from physical damage and moisture. Optical fiber cable connecting hardware should incorporate high-density termination to conserve space and provide for ease of optical fiber cable and patch cord management upon installation.

Optical fiber cable connecting hardware should be designed to provide flexibility for mounting on walls, in racks, or on other types of distribution frames and standard mounting hardware.

Optical Fiber Cabling Installation:
It is recommended that a minimum of 1 m (3.28 ft.) of two-fiber cable (or two buffered fibers) be accessible for termination purposes. Testing is recommended to assure correct polarity and acceptable link performance. Clause II of 568 B.1 provides recommended optical fiber link performance testing criteria.

Optical Fiber Connections:
Connector designs shall meet the requirements of the corresponding TIA FOCIS documents. Telecommunications outlet/connector boxes shall be securely mounted at planned locations. The telecommunications outlet/connector box shall have:

- Cable management means to assure a minimum bend radius of 25 mm (1.00 in.) and should have slack storage capability.

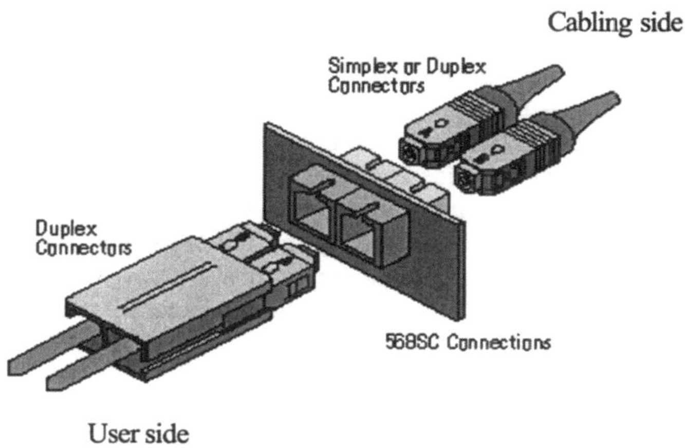

Figure 5.4.22.

- Provisions for terminating and housing a minimum of two optical fibers. Identification of fiber types.
- Multimode connectors or a visible portion of it and adapters shall be identified by the color beige.
- Single mode connectors or a visible portion of it and adapters shall be identified by the color blue.
- The two positions in a duplex connector are referred to as "position A" and "position B".

The next subsections will deal with the most common type of connectors created for the access (and short reach in general) environment: the SFF connectors. After seeing the main characteristics of SFF connectors, some issues concerning connection termination will be given in section 4.6.3.3. Some interesting properties of fusion and mechanical splices (section 4.6.4) will also be treated and, finally, the main conclusions will be highlighted.

5.4.5.2. *Small Form Factor (SFF) Connectors*

Qualified SFF duplex and multi-fiber connector designs may be used in the main cross-connect, intermediate cross-connect, horizontal cross-connect, and consolidation points.

The TIA Fiber Optic Connect Intermateability Standard (FOCIS) describes all the SFF designs which have to satisfy the requirements specified in Annex A of the proposed 568-B.3 standard.

Some advantages of SFF connectors include compact size, modular compatibility with the eight position modular copper interface, and adaptability to high-density network electronics.

According to a study realized by ElectroniCast **[ECAST]**, the global consumption value of fiber optic connectors used in LAN/premises private data networks will increase from $242 million in 2000 to over $2.1 billion in 2010.

It is clear that, if optical connectors have to proliferate, the old standard SC connector is a poor choice and needs a substitute, since it has had poor market penetration when compared to the less expensive ST. Six "small form factor" connectors were considered by the EIA/TIA, but a combination of politics and competition, not technology, probably drove the decision not to be taken for the adoption of a new fiber optic connector for the "568" standard.

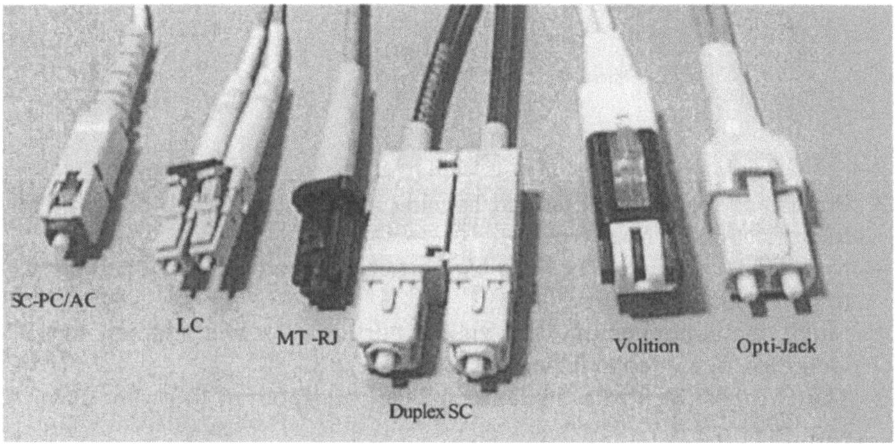

Figure 5.4.23.

All of the above 6 connectors offer a much smaller size, about the same as the modular 8 pin connector used on Cat 5 copper cable. All are oriented toward fast termination for field installation, although all require special tool sets. All of them are so small that they require special transceivers to be designed especially for that connector.

5.4.5.2.1. *Types of connectors SF*F

Following, some of the most used connectors with their principal characteristics will be presented:

MT-RJ

The MT-RJ connector almost became the new standard, with 59% of the committee vote, just 8% less than of the 2/3 majority needed for approval. AMP and Siecor promote this connector and several other vendors are tooling up to offer it.

The MT-RJ is a very small duplex connector with a molded body with two fibers spaced 0.750 mm apart. While it could be used for epoxy-polish termination in the field, the current designs use "cleave and leave" technology with a prepolished fiber stub.

Success at the market is due to the MT-RJ design wich is in the public domain, and a number of vendors have already pledged to support it by offering connectors and transceivers.

3M Volition
By far the most radical design, the 3M Volition does not use an alignment ferrule but "v-grooves" which still require polishing the bare fiber in a special fixture. It is comprised of a unique jack and plug design. Field termination has been limited to jacks, but will soon be expanded to patch cords.

Even a single mode Volition will be available soon. Its simple design, lack of expensive parts and easy termination makes the 3M Volition connector appear to be the least expensive design.

Lucent LC
The Lucent LC connector is the most "normal" design with a ceramic ferrule and plastic body. In fact, it looks like a mini-SC. Performance of the LC is probably the best of the bunch, packing density is as good as the others, and the termination process is totally standard.

A special cable, likewise shrunk to mini-size is available for the LC, but it is compatible with normal 3 mm jackets and 900 micron buffered fiber.

Panduit Opti-Jack
The oldest design, Panduit has been selling the Opti-Jack for quite a while now. It is a duplex design based on the ST/SC 2.5 mm ferrule, with a plug/jack configuration, although a plug/plug adapter is now available. Active device mounts, to develop transceivers are in the works, as is a POF (plastic optical fiber) version. The Opti-Jack is easily field terminated with common fiber optic tools, and it is designed around regular zipcord cable. Of all these connectors, it is the one that looks the least threatening to the average installer.

Siecor/IBM SC-DC or SC-QC
This one is easy to describe – just imagine a SC connector with two or four fibers in a line across the end of a standard 2.5 mm ceramic ferrule. Terminate both at once (although the Siecor version is a "cleave and leave" type) and you are in business. It probably needs a special cable (like a 2 fiber distribution or ribbon cable) but it is a neat idea.

The IBM connection means "Big Blue" has something planned for this connector, so it will be a factor in the marketplace too.

MU
There is yet another contender – the MU. Created by NTT in Japan, it is just now showing up in the US. It really is a shrunken SC, with a ferrule compatible to the LC. With Japan, Inc. behind it, it has got to be a factor too!

To date, there have been almost 100 fiber optic connector designs in the last 20 years. Only about 5 or 6 have had significant market acceptance. Of these 6 new designs, which will make it? **[SFF2]**.

By not choosing a new EIA/TIA 568 standard, the 568 committee has created a situation that in five years will leave both vendors and end users unhappy, because some of these connectors will no longer be supported by the vendors. End users will be left with products that will no longer be possible to maintain. On the other hand, distributors have to offer whatever the customer wants, but they also have to worry about long term

The next table shows some information found in the web about connectors:

Company	**Products**	**URL**	**Information**
3M	Volition system	http://www.3m.com/volition/index.html	volition VF-45 connector
Alcoa Fujikura	MT-RJ connectors, MU, LC Mechanical splices	http://www.aflfiber.com/	
AMP	AMP ESCON Multimode MT-RJ	www.amp.com/fiberoptics	200 Mbit/s multi mode 1300 nm
Go4Fiber	MT-RJ connectors, MU	www.go4fiber.com	Fujikura fusion splicer . . .
Krone	Cables, connectors	www.kroneamericas.com	
Molex	MT-RJ	www.molexpn.com	
Optokon	Connectors, adapters	http://www.optokon.cz/default_en.htm	MT-RJ, MU . . .

5.4.5.3. *Installing Connectors*

All of the common types of connectors are fairly simple to install, (at least at the laboratories), although you can expect a 10-percent loss until installers have a few days worth of experience. After that, figure on losses of 2 to 5 percent, depending on the cleanliness of the area in which the connections are made.

Before the installation of connectors onto a fiber cable, a breakout kit may have to be installed. This procedure will not be necessary on breakout cables having 2-mm buffered fibers, but will be required on 250-, 500-, and 900-micron tight-buffer cables. The breakout kit consists of a buffer tubing (usually 2 mm) over a 900-micron inner tubing. The bare fibers are inserted into these buffer tubes to provide handling protection and strength when mounted onto connectors.

Installing a fiber connector onto a pigtail or unbuffered fiber can be done in several ways. The three most common are epoxy glue with oven-cure, then polish; Hot Melt pre-glue, then polish; and cleave and crimp, unpolish. The epoxy-glue method is the oldest and is still widely used today. This process involves filling the connector with a mixed two-part epoxy, then insetting the prepared and cleaned fiber into the connector. After curing the epoxy in an oven for the specified period of time (usually 5 to 20 minutes) the fiber is scribed and cleaved nearly flush with the end of the connector. Finally, it is polished with a succession of finer and finer lapping papers (typically ranging from 3-micron grit down to 0.3-micron grit).

With the Hot Melt method (a trademark of 3M Co.), the connector comes preloaded with glue and must be placed into an oven to soften the glue. Clean, prepared fiber is then inserted into the connector, then left to cool. After cooling, the fiber is scribed and polished in the same process as was used in the epoxy method.

Cleave and crimp connectors do not require a polish procedure since these connectors already have a polished ferrule tip. Thus, installation simply involves inserting a properly cleaved fiber to butt against the connector's internal fiber "stub." The fiber connector is then crimped to hold the fiber in place. Each mounting method has advantages and disadvantages, varying from ease of installation to cost per connector to performance qualities.

5.4.5.4. *Optical Splices*

In contrast to optical connectors, optical splices are used for fixed (more-or-less permanent) joints. Generally speaking, splices offer a lower return loss, lower attenuation, and greater physical strength than connectors. Also, because only one splice is required where, there might be two connectors. Splices are usually less expensive and less labor-intensive than connectors. In addition, they offer a better hermetic seal than connectors which constitute a smaller joint for inclusion into splice closures, and allow either individual or mass splicing.

The two basic types of splices are fusion and mechanical. Fusion splices are made by using an electric arc to ionize the space between prepared fibers to eliminate air and to heat the fibers to 2000 °F. The heated fibers take a semi-liquid form and meld them together, producing a single fiber (rather than two joined fibers) when done properly. The splice is then covered with a plastic sleeve or other protective device.

The completed splice is placed in an appropriate splicing tray, which is then fit into a splice organizer and, in turn, into a splice closure. One downside to fusion splicing is that it generally must be performed in a dust- and contaminant-free environment, such as a splicing van or trailer. In addition, fusion splicing is prohibited in manholes because the gases frequently found in such locations could create a deadly explosion when exposed to the electric arc generated during the splicing process.

Mechanical splicing is considered a quick and easy method for repairing joints, though it is also used in new construction, especially in small projects. In this method, cleaned and prepared fibers are glued, crimped, or faced together. The splice is then covered with some type of index matching gel or liquid and placed in a V-groove or tube-type device to align the fibers. Finally, the groove or tube is either crimped or snapped to hold the fibers in place. As with fusion splicing, the completed splice is placed in an appropriate splicing tray, then splice organizer, then splice closure. Unlike fusion splicing, mechanical splicing does not require a controlled environment other than a reasonable level of dust control.

Mechanical splices are not without drawbacks, however. For instance, the index matching gel or liquid is subject to contamination and aging, as is any adhesive glue used. Moreover, good performance of mechanical splices when connecting single-mode fibers is not guaranteed (mainly due to core eccentricity tolerances).

These factors, along with the following, usually determine which splicing method to use.

- Type of fiber: Most single-mode fiber is fusion-spliced. Multi-mode fiber, with its complicated core structure, does not always fusion-splice easily, so mechanical splices are generally used.
- Attenuation, including return loss: New fusion splicers provide incredibly low loss when used properly, while a mechanical splice may allow as much loss as a properly terminated connector.
- Physical durability: The "welding" process used in the fusion splice results in higher strength and greater durability than a mechanical splice. In addition, with fusion splices, the mechanical tensile strength of the fiber gets near to that of the original fiber.

- Ease of installation: Today's fusion splicers are at least partially automated if not fully automated in a one-button process. Mechanical splicing devices vary, but are usually easier to use than connector kits.
- Cost per splice: While mechanical splices typically cost only $7 to $12 a piece, a fusion splice can run anywhere from $3,000 to $30,000. Once this machine is purchased, however, the cost per splice is negligible.

SUMMARY

Nowadays, technologies and techniques in optical fiber cables are changing and improving really fast. As exchange of information between users is also growing vertiginously, it is widely accepted that fiber in the last mile is the definite solution for the access networks (the only issue to be clarified is when it will be a reality). This section has explained some of the most important types of fibers, cables, and techniques used to bring fiber to the last mile. Assuming the well known advantages of optical fiber against other physical mediums, there are some issues which have to be yet defined. Maybe the most important question at this moment is what kind of fiber (multi or single-mode) is the appropriate to implement FTTH. Every one of them has its advantages and drawbacks (easier connectivity and availability of cheap devices for the multi-mode in front of assured future proof solution for the single-mode one).

A third and not very probable solution could be the use of plastic optical fiber (POF). Its main advantage is again its easy handling, but current POF designs are not adequated for high bandwidth applications. In spite of this, some companies are developing new POF types so it can not be discarded as a solution. An important issue in defining the required characteristics of optical access networks is what would be the maximum distance to cover. It seems clear that, in the case of GEPON and supposing a 10 km limit, multimode (silica or plastic) fibers should be precluded.

With respect to the cabling systems, ribbon cables are potentially good candidates to the implementation of optical access networks. Their main advantages are:

- Possibility to splice multiple fibers simultaneously
- Easier handling which allows time and labor savings

Finally, concerning the cable placement and in addition to the traditional methods (blowing fiber/cable or pulling), a new method has to be highlighted which consists in utilizing the sewer conduits to arrive to every home. This technique, carried out by companies like Citynet Telecommunications **[CITYNET02]** or CA-BOTICS Fiber Systems **[CAB02]**, Alcatel and others, has had a great success in the US and is being introduced in Europe where its architectural characteristics (old buildings and narrow streets) make it still more attractive.

Lately, a lot of different products that integrate in the same substrate the E/O and the O/E converters have appeared. This is another factor that helps to understand how the optical access is in the pre-explosion era.

Automation improvements in the manufacturing processes are an important issue to consider in this field. EO/OE converters have been until now hand made. Costs are highly correlated to time spent in making the devices. Main steps in the optoelectronic area must improve time (automation) and space (integration) savings. Current transceivers, mainly thought for LAN and Access environments are emerging as good performing and cheap devices.

Several companies are launching a lot of similar transceivers oriented to the GbE market using VCSELs and SFF connectors and as the main characteristic, cheap and integrated.

Some issues not completely solved yet are: cross-talk between transmitter and receiver, design of new structures that allow easier integration of the different parts, automated processes, . . .

Single-wavelength source is the basic component of an optical transmission system. There are several types of sources and each of these has a more specific application in the optical telecommunication field. These applications depend on the source parameters such as out power, linewidth, bandwidth and wavelength stability.

The simplest source is the light emitting diode (LED) which is used for short distances due that principally the most of the light power generated by LED is not coupled to the fiber. Moreover, the light spectrum is too broad for some applications. A more specialized source is the laser diode which can generate greater power and better wavelength stability. Fabry-Perot and distributed feedback Bragg (DFB) lasers are the most principal laser diodes in the optical telecommunication market.

Sources as Fabry-Perot laser generates a spectrum with several wavelengths which are very near among them. This kind of source is not used for long distance due to the dispersion, but is useful in systems where a single-wavelength is required. Thus, FP laser is not suitable for WDM systems, however, a multi-wavelength source with several wavelengths uniformly separate in frequency can be obtain from its spectrum. In contrast, DFB laser generates a narrow linewidth by including a periodic grating structure into the cavity laser. This feature allows that DFB laser is used in most of the optical systems.

DFB laser has been the most important single-wavelength source. To improve the efficiency and performance of DFB lasers, modules that integrate the laser cavity with a modulator have been developed. DFB laser can have the grating structure outside the active region. This structure is known as distributed Bragg reflector (DBR) laser which is also used to build tunable lasers.

In the last years, various technologies have been developed in tunable sources. Among them, distributed Bragg reflector (DBR) tunable laser of three sections, tunable laser of four sections as grating coupler sample reflector (GCSR) and super structure grating DBR (SSG-DBR), vertical cavity surface emitting laser (VCSELs) and tunable laser with MEMS technology.

The use of the optical amplifiers in optical transport networks and optical access networks is an issue of vital importance because they can save costs in the use of components between node and user, eliminate the optical-electrical-optical conversion because signal regeneration is not necessary, and allow arriving at a greater number of users.

The EDFAs is considered one of the most important components in optical networks for increasing the capacity of optical links due to the saturation power output increase every time while the cost/mW decrease. However, EDFA introduces amplified spontaneous emission (ASE) noise that increases when the amplify gain is greater which can degrade the signal quality.

In FTTH, each user requires of a bi-directional communication, for that reason, it is important to reduce the amplification costs as much as it is possible without degradation of signal. Taking into account this aspect, various bi-directional optical amplifiers have been proposed. Some amplifiers employ two unidirectional optical amplifiers together with optical circulators or WDM filters to separate the signals that make them expensive. However, low-cost bi-directional all-optical repeaters have been developed in the Korea Telecom Access Networks Laboratory. This amplifier utilizes only one unidirectional and one Mach-Zehnder interferometer to combine and separate the signal.

Another important subject is the optimum amplifier placement in an access network in order to maximize the power budget. Different methods have been created to discover this place where the power budget is maximum and the degradation of signal is minimum. This requirements can be achieved by using of variable attenuators to simulate the loss of an optical link. Also, it is known that some algorithms have been developed for this purpose.

Semiconductor optical amplifiers may provide a number of new functionality in advanced optical access networks, apart from the amplification itself. The most interesting has consisted of its fast modulation capability: all the light sources are located in a central node, and the user nodes only modulate this passing light using a local SOA. This avoids the temperature effects on the laser wavelengths of the conventional remote transceivers.

Constituting the key element in PON networks, the optical coupler needs to combine low loss with other characteristics as small dimensions, temperature range, and cost. In this sense, the 1x4 single fused-fiber appears as a simple option for a small number of users, whilst the technology of the Planar Lightwave Circuit (PLC) waveguides with ribbon fiber alignment emerges as the best cost-effective for the PON with the highest standardized splitting ratio (1×16, 1×32).

Optical multiplexers are key components in the near future access networks using WDM or CWDM as they offer an individual wavelength channel to each user with full transparency and privacy with other users. The most promising technology for access multiplexers for a relevant number of wavelengths (higher that 8) is based on integrated optical circuits (PLCs). The AWG design is the most advantageous. The wide recent commercialization is making it available at reasonable cost. In high port number designs, insertion loss in the range of 5–8 dB, channel crosstalk between 25 and 35 dB and, lately, athermal designs have been obtained with very low temperature sensitivity interesting for the outdoor access optical infrastructure.

As fiber is seen as a promising option for the access portion of the network, there has lately been a high interest in finding a standard for the optical connectors. This is perhaps the most important issue in order to achieve the generalization of FTTH. Historically, optical fiber has been perceived as difficult to work with, craft sensitive and expensive to implement. And while the higher cost for fiber has been due, in large part, to the expense of network electronics, it has also won a price because of the need for specialized connecting hardware, installation tooling and labor-intensive termination.

Interest in deploying fiber closer to the user has supposed an enormous interest in developing new connectors. Small Form Factor (SFF) connectors are promising to be key elements in the FTTH growth. All the SFF designs share some common characteristics: high integration and easy and fast installation. Anyway, their performance with single-mode fibers it is not yet clear. On the other hand, there are many different designs (LC, MT-RJ, MU, VF-45 . . .) and, perhaps, not all of them will finally survive. Once again the market will decide which of the designs will succeed, so we will have to wait for the massive deployment of optical access networks.

Mechanical (non fusion) splices could also be a good alternative to achieve the deployment of optical access networks. There are some places mainly in old European towns where it is difficult to achieve the desired environment to use a fusion splicer. In these situations, mechanical splices could be a solution (single-mode performance is not assured again). In conclusion, there is a renewed interest in mechanical splices **[FUJI]**.

REFERENCES

[Ackerman01] Ackerman, D.A.; Johnson, J.E.; Ketelsen, L.J.-P.; Geary, J.M.; Asous, W.A.; Walters, F.S.; Freund, J.M.; Hybertsen, M.S.; Glogovsky, K.G.; Lentz, C.W.; Reynolds, C.L.; Bylsma, R.B.; Dean, E.J.; Eng, L.E.; Qasiameh, O.; Gupta, R.; Ghanikanov, F.; Veliadis, 'A practical approach to wavelength selectable DWDM sources', Lasers and Electro-Optics Society, 2001. LEOS 2001. The 14th Annual Meeting of the IEEE, Volume 1, 2001

[Alcatel] http://www.alcatel.com

[Berger 01] Berger, J.D.; Zhang, Y.; Grade, J.D.; Lee, H.; Hrinya, S.; Jerman, H., 'Widely tunable external cavity diode laser based on a MEMS electrostatic rotary actuator', Optical Fiber Communication Conference and Exhibit, 2001. OFC 2001, Volume 2, 2001.

[BICC] http://www.biccgeneral.com

[Biernacki 99] Biernacki, P.D.; Frankel, M.Y.; Gingerich, M.E.; Matthews, P.J., 'A high-speed mixed digital-to-analog circuit board for accurate control of wavelength tunable lasers for fiber-optic communications', Lightwave Technology, Journal of, Volume 17, Issue 7, July 1999.

[Boos 99] Boos, N.; Beguin, A.; Beok, H.; Brun, M.; Fondeur, B.; Hiller, D.; Lerminiaux, C.; Lu, R.; Nicolas, C.; Rosala, F.; Ryszytiwskyj, W., 'Phasar devices for WDM networks', Nanostructures and Quantum Dots/WDM Components/VCSELs and Microcavaties/RF Photonics for CATV and HFC Systems, 1999 Digest of the LEOS Summer Topical Meetings, 1999, pp. II31–II32.

[BRIGG 93] P.R. Briggs et al., 'The Use of the High Air-Speed Blown-Cable System for Placing Optical-Fiber Cables in Ducts', NFOEC Conference, 1993.

[Boos 99] Boos, N.; Beguin, A.; Beok, H.; Brun, M.; Fondeur, B.; Hiller, D.; Lerminiaux, C.; Lu, R.; Nicolas, C.; Rosala, F.; Ryszytiwskyj, W., 'Phasar devices for WDM networks', Nanostructures and Quantum Dots/WDM Components/VCSELs and Microcavaties/RF Photonics for CATV and HFC Systems, 1999 Digest of the LEOS Summer Topical Meetings, 1999, pp. II31–II32.

[Borella 97] Borella, M.S.; Jue, J.P.; Banerjee, D.; Ramamurthy, B.; Mukherjee, B., Proceedings of the IEEE, Volume 85, Issue 8, Aug. 1997.

[Buldawoo97] Buldawoo, N; et al., 'A Semiconductor Laser Amplifier-Reflector for the future FTTH applications', European Conference on Optical Communications – ECOC 97 –, no. 448, September 1997.

[CAB 02] http://www.ca-botic.com/how.html

[Cableunet00] www.cableunet/fibrtech/pof.htm

[Cahill 98] Cahill, M.J.L.; Pendock, G.J.; Summerfield, M.A.; Lowery, A.J.; Sampson, D.D., 'Optimum optical amplifier location in spectrum-sliced WDM passive optical networks for customer access', Optical Fiber Commuunication Conference and Exhibit, 1998. OFC '98.

[Chang] 'VCSELs: Advances and Future Prospects', C.J.Chang-Hasnain, Bandwidth9

[CIR] 'Integrated Optics and the Rebirth of the Broadband Dream', Lawrence Gasman, Communications Industry Researchers.

[CITYNET02] http://www.citynettelecom.com/index.swf

[Cypress01] 'Components' www.fiberopticproducts.com

[Corning] 'Corning Infinicor CL 1000 Optical Fiber', www.corningfiber.com/.

[DIG] 'Digital Optical Link Developments', CERN, 1999.

[ECAST] 'VCSEL Transceiver Global Market & Technology Forecast 1999–2009', www.electronicast.com

[FEE 95] J.M. Fee et al., 'Analysis and Measurement of Friction in High Speed Air Blowing Installation of Fiber Optic Cable', NFOEC Conference, 1995.

[FEE 97] J.M. Fee et al., 'Analysis of Duct Displacement Factor in Fiber Optic Pulling', 1997.

[FUJI] Mechanical splices Datasheet from Fujikura.

[Gan 00] Chaoqin Gan; Mingde Zhang; Xiaohan Sun, 'A novel large-scale star coupler with only two EDFAs', IEEE Photonics Technology Letters, Volume 12, Issue 10, Oct. 2000, pp. 1391–1393.

[GRIF 93] Griffioen, W.; Plumettaz, S. A., 'Installation of Fiber Optic Cables in Duct', 1993.

[HAM] 'Optical Components: Emerging Trends and Vendors', WR Hambrecht Co.

[Harari 97] Joseph Harari, Guanghai Jin, Jean P. Vilcot and Didier Decoster (1997) Theoretical Study of p-i-n Photodetectors' Power Limitations from 2.5 to 60 GHz: IEEE Transactions on Microwave Theory and Techniques, Vol. 45, No. 8, August 1997, pp. 1332–1336.

[Hentschel95] Hentschel, C.; Leckel, E. 'Complete analysis of erbium-doped fiber amplifiers', Lightwave symposium, Hewlett-Packard Company, 1995.

[Hill 90] Hill, A.M.; Payne, D.B.; Blyth, K.J.; Forrester, D.S.; Silvertown, A.; Arkwright, J.W.; Faulkner, D.W.; Ballance, J.W., 'An experimental broadband and telephony passive optical network', Global Telecommunications Conference, 1990, and Exhibition, 'Communications: Connecting the Future', GLOBECOM '90, IEEE, vol. 3, 1990, pp. 1856–1860.

[Holloway 97] Holloway, W.T.; Keating, A.J.; Sampson, D.D., 'Multiwavelength source for spectrum-sliced WDM access networks and LAN's', IEEE Photonics Technology Letters, Volume 9, Issue 7, July 1997.

[Hsieh 99] Hsieh, C.S.; Cheng, W.H., 'Direct correlation between device and technological parameters in fusion fiber coupler', Lasers and Electro-Optics, 1999. CLEO '99. Summaries of Papers Presented at the Conference on, 1999, pp. 158–159.

[Hunziker 98] Hunziker, W., 'Ways to cost effective packaging: waveguide tapers and passive self-alignment', Lasers and Electro-Optics Society Annual Meeting, 1998. LEOS '98. IEEE, Volume 1, 1998, pp. 36–37.

[Hsieh 99] Hsieh, C.S.; Cheng, W.H., 'Direct correlation between device and technological parameters in fusion fiber coupler', Lasers and Electro-Optics, 1999. CLEO '99. Summaries of Papers Presented at the Conference on, 1999. pp. 158–159.

[IBM] 'SFF-1063/1250N-SW PTH Serial Optical Transceiver Design & Layout Guide', IBM Data Sheet.

[EEE] IEEE ISSCC 1999 Slide Supplement/Copyright IEEE.

[IEEE01] 'POF standard' www.ieee.org

[Iannone 00] Iannone, P.P.; Reichmann, K.C.; Smiljanic, A.; Frigo, N.J.; Gnauck, A.H.; Spiekman, L.H.; Derosier, R.M., 'A transparent WDM network featuring shared virtual rings', Lightwave Technology, Journal of, Volume 18, Issue 12, Dec 2000, pp. 1955–1963.

[Iwashita 97] Iwashita, K.; Watanabe, T.; Ota, T.; Shigematsu, T.; Watanabe, K., '1 × 8 star coupler waveguide device with multifiber-connector-compatible connections', Optical Fiber Communication. OFC 97, Conference on, 1997, pp. 110–111.

[Jacob 98] Jacob, D.; Prassas, M.; Willems, F.W.; Van Der Plaats, J.C.; Steenbergen, C.A.M., 'First prototype of bi-directional EDFA for single fiber tree-and-branch access networks carrying AM-VSB and digital multiwavelength signals', Optical Fiber Commuunication Conference and Exhibit, 1998. OFC '98, Technical Digest, 1998.

[Jinjuji 90] K. Jinguji et al., Electron. Lett., Vol. 26, 1990, p. 1326.

[Jung 98] Jung, D.K.; Shin, S.K.; Lee, C.-H.; Chung, Y.C., 'Wavelength-division-multiplexed passive optical network based on spectrum-slicing techniques', IEEE Photonics Technology Letters, Volume 10, Issue 9, Sept. 1998.

[Kim 01] Kim, H.; Chung, Y.C., 'Passive optical network for CDMA-based microcellular communication systems', Lightwave Technology, Journal of, Volume 19, Issue 3, March 2001.

[Lascomm02] http://www.lascomm.com

[Lin 97] L.Y. Lin, M.C. Wu, T. Itoh, T.A. Vang, R.E. Muller, D.L. Sivco and A.Y. Cho (1997) High-Power High-Speed Photodectors – Design, Analysis, and Experimental Demonstration: IEEE Transactions on Microwave Theory and Techniques, Vol 45, No 8, August 1997, pp. 1320–1331.

[Liou 97] Liou, K.-Y.; Stark, J.B.; Koren, U.; Burrows, E.C.; Zyskind, J.L.; Dreyer, K., 'System performance of an eight-channel WDM local access network employing a spectrum-sliced and delay-line-multiplexed LED source', IEEE Photonics Technology Letters, Volume 9, Issue 5, May 1997.

[Liou98] Liou, K.-Y.; Dreyer, K.K.; Burrows, E.C.; Zyskind, J.L.; Sulhoff, J.W.,

'A 24-channel WDM transmitter for the access systems using a loop-back spectrally sliced light-emitting diode', IEEE Photonics Technology Letters, Volume 10, Issue 2, Feb. 1998.

[LR 00] http://www.lightreading.com/document.asp?doc_id=1618

[METHOD] 'Methode Electronics 2000 Annual Report', www.methode.com

[MIT] 'Devices for ATM Fiber Optics Communication', Mitsubishi R&D Progress Reports, March 2001.

[Modiano 99] Modiano, E.; Barry R., 'Architectural considerations in the design of WDM-based optical access networks', Computer Networks 31 (1999) pp. 327–341.

[Murtaza 96] S. S. Murtaza et al. (1996), 'High-Finesse Resonant-Cavity Photodetectors with an Adjustable Resonance Frequency', IEEE J. of Lightwave Technology, Vol. 14, No. 6, June 1996. pp 1081–1089.

[Nadler 99] Nadler, C.K.; Wildermuth, E.K.; Lanker, M.; Hunziker, W.; Melchior, H., 'Polarization insensitive, low-loss, low-crosstalk wavelength multiplexer modules', Selected Topics in Quantum Electronics, IEEE Journal on, Volume 5, Issue 5, Sept.–Oct. 1999, pp. 1407–1412.

[Nolan 98] D.A. Nolan; W.J. Miller; R. Iron, 'Fiber-based band splitter', Optical Fiber Communication Conference and Exhibit, 1998. OFC '98, Technical Digest, 1998, pp. 339–340.

[OCC] http://www.occfiber.com.

[OFC 97] 'Plastic Optical Fibers', R. Brad Kummer, Tutorial, OFC '97, ThQ1.

[OIF] List of OIF documents. UPC file: contingut_boana.xls (see Section 5, 'Standards').

[O'Mahony88] O'Mahony M.J., 'Semiconductor Laser Optical Amplifiers for use in future fiber systems', Lightwave Technology, Journal of, Volume 6, no. 4, April 1988, pp. 531–544.

[OXF 99] Oxford Electronics Ltd. October 99

[PAL] 'Advances In Automation', Palomar Technologies.

[Pan98] Pan, J.J.; Huang, Y., 'Ultrabroadband monolithically-fused 1/spl times/4 fiber-optic couplers', Optical Fiber Communication Conference and Exhibit, 1998, OFC '98, Technical Digest, 1998, pp. 338–339.

[Pan 99] Pan, J.J.; Tian Zhu, 'New 1/spl times/N fiber coupler employing diffractive optical element', Lasers and Electro-Optics, 1999. CLEO '99. Summaries of Papers Presented at the Conference on, 1999, pp. 159–160.

[Park 01] Hee Su Park; Seok Hyun Yun; In Kag Hwang; Sang Bae Lee; Byoung Yoon Kim, 'All-fiber add-drop wavelength-division multiplexer based on intermodal coupling', IEEE Photonics Technology Letters, Volume 13, Issue 5, May 2001, pp. 460–462.

[Patel 99] Patel, R.R.; Bond, S.W.; Larson, M.C.; Pocha, M.D.; Garrett, H.E.; Lowry, M.E.; Deri, R.J., 'Multi-mode fiber coarse WDM grating router using broadband add/drop filters for wavelength re-use', LEOS '99. IEEE Lasers and Electro-Optics Society 1999 12th Annual Meeting, Volume 2, 1999, pp. 826–827.

[Pirelli] 'Fiber to the home Technical specifications', http://www.Pirelli.com

[POFESKA] http://www.pofeska.com/pofeskae/tece/whatspofe/whatspofe.htm

[POLY] American Polywater Corporation Nov 99.

[Prat 00] Prat, J.; Ruiz, S., 'Transparent supervision of optically amplified fiber links with received signal and ASE monitoring', Optics Communications 183, September 2000.

[R. Vacca 00] Jhon R. Vacca, 'Tha Cabling handbook', Prentice-Hall, Inc.

[Rigole 95] Rigole, P.-J.; Nilsson, S.; Backbom, L.; Klinga, T.; Wallin, J.; Stalnacke, B.; Berglind, E.; Stoltz, B. '114-nm wavelength tuning range of a vertical grating assisted codirectional coupler laser with a super structure grating distributed Bragg reflector', IEEE Photonics Technology Letters, Vol. 7, July 1995.

[Sarlet 00] Sarlet, G.; Morthier, G.; Baets, R., 'Control of widely tunable SSG-DBR lasers for dense wavelength division multiplexing', Lightwave Technology, Journal of, Volume 18, Issue 8, Aug. 2000.

[Sasaki 00] Sasaki, Y., 'Optical fiber devices', Transparent Optical Networks, 2000 2nd International Conference on, 2000, pp. 9–12.

[Seung-Tak01] Seung-Tak Lee; Chae, C.-J. 'Low-cost bidirectional optical amplifier using a unidirectional Er-doped fiber amplifier and a fiber Mach-Zehnder interferometer', IEEE Photonics Technology Letters, Volume 13, Issue 1, Jan. 2001.

[Siemens 00] 'Plastic fiber components', www.siemens.com

[Souza] A.B. de Souza, J. Prat, 'Reduction of the tuning time of 1.5 um 3-section DBR lasers in controlling the transient currents', Lasers and Electro-Optics Society, 2000. IEEE/LEOS '2000. pp. 589–590.

[SFF 00] 'Performance Comparison of Small Form Factor Fiber Optic Connectors', IEEE Trans on Advanced Packaging, Vol. 23, No. 2, p. 188, May 2000.

[SFF2] 'SFF Connectors. What's the best choice?', from Gemflex.com, different authors from AMP and Lucent.

[Takesue 00] Takesue, H.; Yamamoto, F.; Sugie, T, 'Novel node configuration for DWDM photonic access ring using CMLS', IEEE Photonics Technology Letters, Volume 12, Issue 12, Dec. 2000, pp. 1698–1700.

[TIMES] Times Fiber Communication INC(r) TFC 1-800-677

[Uetsuka99] Uetsuka, H., "WDM optical components using PLCs", Lasers and

Electro-Optics, 1999. CLEO/Pacific Rim '99. The Pacific Rim Conference on, Volume 3, 1999, pp. 585–586.

[VCSE] 'The ideal light source for datanets', K.S. Giboney et al., IEEE Spectrum, Feb. 1998.

[Witham 00] Witham, C.R.; Beranek, M.W.; Carlisle, B.R.; Chan, E.Y.; Koshinz, D.G., "Fiber-optic pigtail assembly and attachment alignment shift using a low-cost robotic platform", Electronic Components & Technology Conference, 2000. 2000 Proceedings. 50th, 2000, pp. 21–25.

[Wood 99] Wood, T.H.; Wilson, G.C.; Feldman, R.D.; Stiles, J.A., '(FTTH) system providing broad-band data over cable modems along with analog and digital video', IEEE Photonics Technology Letters, Volume 11, Issue 4, April 1999.

[Woodward98] Woodward, S.L.; Iannone, P.P.; Reichmann, K.C.; Frigo, N.J., 'A spectrally sliced PON employing Fabry-Perot lasers', IEEE Photonics Technology Letters, Volume 10, Issue 9, Sept. 1998.

[Yoshida 92] Yoshida, J.; Yamada, Y.; Kobayashi, M.; Tsuzuki, N., 'Optoelectronic and photonic integrated devices for optical modules in fiber optic subscriber systems', Communications, 1992. ICC '92, Conference record, SUPERCOMM/ICC '92, Discovering a New World of Communications, IEEE International Conference on, 1992, pp. 1784–1788 vol. 4.

[Zeng 94] Jen-Yu Zeng; Wann-Yih Guo; Yung-Kuang Chen, 'Lossless wide-band reflective star coupler using erbium-doped fiber amplifiers', IEEE Photonics Technology Letters, Volume 6, Issue 12, Dec. 1994, pp. 1485–1487.

Other web pages:

http://www.agere.com/
http://www.fcsi.fujitsu.com/
http://www.jdsuniphase.com/
http://www.luminentinc.com/
http://www.nortelnetworks.com/
http://www.semiconductors.philips.com

Chapter 6

TRANSMISSION IMPAIRMENTS AND SUPERVISION

6.1. INTRODUCTION

Access networks can have a large number of subscribers, any failure in the service represents significant loss for the service providers. Therefore, fault surveillance and performance monitoring are important issues to ensure the correct transfer of information to the users, and a real time, non-intrusive fault surveillance system is essential to reduce construction and maintenance costs and improving service reliability. In this line, maintenance involves the whole operations required to set up and maintain, within prescribed limits, any element entering into the setting-up of a connection to guarantee the performances of the optical carrier, and locate any fault in the access network.

Accurate and cost-effective signal monitoring technologies are being developed to replace the signal monitoring that is currently done in the electronic domain at many optical-to-electronic-to-optical (OEO) conversion locations in the network. Analog optical signal monitoring that measures optical signal parameters such as channel power, channel center wavelength, and channel optical signal to noise (OSNR) is useful and can provide important information with regard to the health and quality of an optical signal. Moreover, some digital techniques are also used to obtain information about the transmission performance. The cyclic redundancy check (CRC) uses parity bits generated by polynomial encoding of digital signals. Then, decoding algorithms

J. Prat et al., Fiber-to-the-Home Technologies, 303–331.

detect errors in the received digital signal. Error correction, if required, may be accomplished through the use of an automatic repeat-request (ARQ) system. Others digital monitoring tests make evaluation of block errors in synchronous networks by means of the bit interleaved parity (BIP-n) mechanism. The network elements monitor the block errors and send information about them to the transmission source end in the backward frames. The detection of errors leads to operation and maintenance signals (OAM) being generated and carried in the frames to the information source. These are basically remote error indication signals (REI). The quality of transmission may also be estimate whether is taking into account the measured BER as a function of received optical power level.

However, the above mentioned monitoring techniques are not sufficient to fully understand the signal quality due to the fact that several failures or impairments can take place in the optical transmission. These are principally: The fiber non-linear effects, power losses, crosstalk, frequency fluctuations, dispersion, noise, failures of amplifiers, transmitter laser power fault and fiber cable break. To have an estimation of all these degradation is an aspect of great importance for design parameters and configuration of the optical networks. Some of these degradations can be measured more easily at optical level. Therefore, in the last years many efforts have been concentrated in the development of transparent supervision systems which carry out monitoring of optical signal in the optical domain without realizing the OEO conversion **[Diaz 01]**.

However, there has not yet been sufficient discussion of the optimum testing method and wavelength, since optical fiber cables only began to be installed in subscriber networks about few years ago. In spite of that, several ingenious strategies have been developed to supervise the quality of the optical signal and fiber-fault identification with certain degrees of transparency. A two-step test procedure is advisable to assure good network performance:

- Network qualification /acceptance.
- Network monitoring.

Both acceptance and monitoring of in-service fibers are possible with the equipment available on the market. This is very important because the optical infrastructure carries aggregate traffic and is shared by different users. There are different techniques for non-intrusive monitoring/supervision and most of them are based on OTDR devices.

Another important aspect that must be taken into account is about the cost of the supervision strategies in FTTH systems. Since the quantity of installed optical fiber will increase rapidly, installation and maintenance costs must be reduced as much as possible. In this chapter, the main characteristics of surveillance systems are defined. The different impairments produced in an access network and several systems built to detect and locate faults are described. The optical signal supervision is not very developed in access. However, several methods that can be applied are also reviewed here, and the main equipments used in monitoring and in the supervision systems are described as well.

6.2. FAULTS AND IMPAIRMENTS

Faults and impairments produced in an optical transmission system determinate the quality of optical signal. Several parameters are used to make an estimation of Quality of Signal (QoS). Some of these parameters are defined in the following items.

- *Eye pattern:* It is the diagram that an oscilloscope display in which a pseudo-random digital data signal from a receiver is repetitively sampled and applied to the vertical input, while the data rate is used to trigger the horizontal sweep. System performance information can be derived by analyzing the display. An open eye pattern corresponds to minimal signal distortion. Distortion of the signal waveform due to intersymbol interference and noise appears as closure of the eye pattern.
- *Signal-to-noise ratio (SNR):* It is the ratio of the amplitude of the desired signal to the amplitude of noise signals at a given point in time. SNR is usually expressed in dB and in terms of peak values for impulse noise and root-mean-square values for random noise.
- *Bit error ratio (BER):* It is the number of erroneous bits divided by the total number of bits transmitted, received, or processed over some stipulated period. The BER is usually expressed as a coefficient and a power of 10; for example, 2.5 erroneous bits out of 100,000 bits transmitted would be 2.5 out of 105 or 2.5×10^{-5}.
- *Sensitivity:* In an electronic device, e.g., a communications system receiver, or detection device, e.g., PIN diode, the sensitivity is the minimum input signal required to produce a specified output signal having a specified signal-to-noise ratio, or other specified criteria.
- *Power penalty:* it refers to the increase in the average received power required (sensitivity) to achieve a specified SNR because of nonideal conditions.

6.2.1. Elements Failures

An important issue for the resilience, protection and Quality of Signal (QoS) is to detect failure event or degradation of performance of any element of the optical network. Hence, if a failure is produced in any network element, it is detected and transmitted to a management system. Moreover, environmental conditions and aging of the network elements produce degenerations that also must be monitored. These degenerations may be progressive over time and led to the data degradation and eventual loss of service. Other causes of signal loss are the bends in the cable or when the cable is forced, folded or maltreated.

The factors that influence in the light transmission are managed in a remote fiber test system. This system includes Remote Test Units (RTU) positioned at strategic points in the network. The failure events mentioned above can be measured periodically. The RTU include an optical test head, an optical test access unit, and a data network interface.

When loss of signal is produced in the downstream direction, it is detected by ONU that switches off the laser. In case of loss of signal in the upstream direction, it is detected by OLT, and a signal is sent to ONU in order to deactivate it for protection purposes.

6.2.2. Fiber cut

The main cause of failure in an access network is the fiber cut. Several strategies have been developed to detect, locate and solve failures in the fibers:

- *Dark fiber monitoring:* A fiber of a multi-fiber cable is monitored using an OTDR. It is assumed that the remaining fiber cables are in working order if this fiber is also in working order.
- *Active fiber monitoring:* In this method, an OTDR is connected to a fiber. When this is carrying important information, in non-intrusive monitoring scheme, the OTDR laser operates at a different wavelength from the in-service signals. For instance, when the network capacity is being leased to service providers. This is an expensive method which is not much used in the practice.
- *Monitoring of large numbers of fibers:* In this method an OTDR is shared among all fibers and one optical switch is used to select the test fiber. This way, monitoring cost is reduced.

6.2.3. Attenuation

It is due to absorption and scattering phenomena that are intrinsic to the glass compound used in the fiber, impurities absorption and absorption and scattering centers that are introduced during the drawing process. The main optical power loss in doped silica fiber is known as Rayleigh scattering and depends on the dopant.

There are three conventional methods to measure the fiber attenuation:

- *Cut back method:* A tungsten halogen lamp and monochromator can be used as light source to apply the same power in two fibers, one of length 1–2 m for reference and the under test. The attenuation is given by:

$$\alpha = \frac{10}{L}\log(P_0/P_L)$$

where P_0 is the output power from the short cutback length and P_L is the output power from a full length L.
- *Insertion loss method:* This is a nondestructive method where attenuation is measured by substitution or insertion loss, coupling the light through connectors. The accuracy of the measurement hangs on reproducibility of these couplings.
- *OTDR:* Optical Time Domain Reflectometer (OTDR) is also a nondestructive method that, in addition to attenuation, this technique can locate faults, measure splice loss and the attenuation coefficient. It needs to access to only one end and is very useful for field measurement.

Conventional methods are not suitable to test the individual branch loss in a branched fiber system because Fresnel reflections and Rayleigh backscattering from each branched fiber sum up. Thus, to measure the loss in each branch, it is necessary to test directly each fiber after branching or to use a special OTDR with wavelength tunable laser **[Tanaka 96]**. Conventional attenuation measurement techniques neither are suitable for very low attenuation measurement. In this case, a substitution technique using subcarrier modulation can be used. In this way, a 0.01 dB change in optical attenuation produces a 0.04 dB change in electrical attenuation by using double square-law detection **[Singh 00]**.

6.2.4. Optical beat interference (OBI)

Optical Beat Interference (OBI) occurs when light from two or more sources coherently arrive simultaneously on the same receiver photodiode creating beat signal at differential frequency, if the optical frequencies are close. OBI can be the mainly source of noise in subcarrier-multiplexed (SCM) passive-optical networks (PON) causing important disturbing products in the electrical frequency range. If the different frequency equals the RF of a subcarrier, then, the performance of channel will be impaired by the quadratic conversion of optical to electrical power of this frequency.

Subcarriers can be added to each PON because the demand for data service increases each time. Due to this, OBI may degrade system performance yet more, but operating the upstream laser in burst mode can reduce it significantly.

OBI depends on the distribution of optical power in the laser modes, hence, its effect decreases by using lasers with different mode spacing. However, in systems that use uncooled FP laser, the wavelength varies with temperature, then, the OBI effect will be intermittent. This noise can also be reduced by using sources with broad linewidth in order to spread the OBI noise power over a large bandwidth. Therefore, LEDs and FP laser are more suitable source for the transmitters since they have a broader spectrum than DFB laser. Employing slicing-spectrum techniques that broaden the interfering modes by driving the lasers below threshold lowered the noise by an additional 10 dB **[Woodward 98]**.

OBI will also impair the performance of a bidirectionaloptical link since Rayleigh-backscattering will cause the laser light to be back reflected, and it will interfere with the other signal beam at the receiver.

In the following, several factors that influence in the appearance of OBI affect in a system are enunciated:

- Number, type and, power of the optical sources
- Spectral distribution of the optical sources
- Wavelength separation of the optical spectra
- Number and linewidth of longitudinal modes for laser
- Polarization of the individual optical modes
- Data bit rate and modulation format
- Subcarrier frequency and modulation index
- Direction of the signals

6.2.5. Connector reflections

It is known that optical beat induced by interference of reflected signal beam from connectors, splices and fiber ends, because of refractive-index discontinuities, degrades the receiver sensitivity. Figure 6.2.1 shows how this impairment is produced in optical link with two connectors. In case that the maximum reflectivity of the connectors C1 and C2 is about –10 dB, the 10% of the optical beam is reflected in connector C2 toward connector C1 and the 10% of –10 dB is reflected in C1which goes back to the receiver interfering with the original signal. Therefore, the maximum interfered reflected beam amount is 1% of the signal beam (–20 dB).

The maximum beat intensity is $P_b = 2I_sI_r$, where I_s is signal beam photocurrent and I_r reflected beam photocurrent. In the worst of circumstances, when the signal beam and reflected beam interfere in phase, the induced interference reaches a maximum value, i.e. by using the experimental set-up shown in Figure 6.2.1 under the worst possible conditions, reflective interference level of –120 dB/Hz is obtained at 2.4 Gb/s NRZ which may causes 0.4 dB power

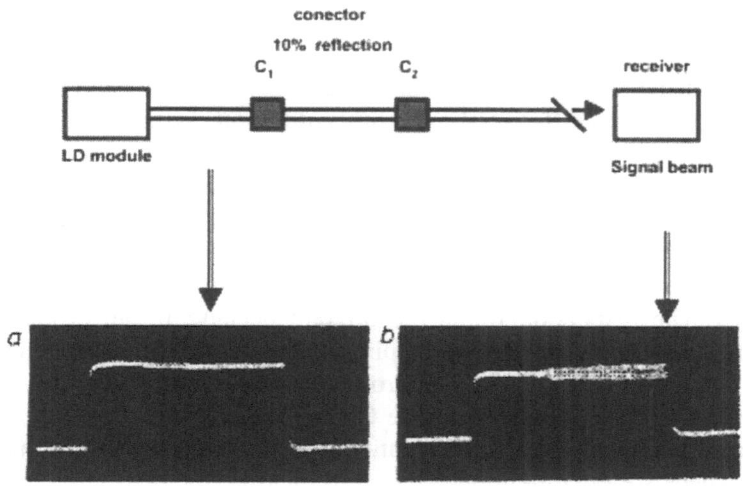

Figure 6.2.1. Effect of connector reflections.

penalty. Therefore, to maintain the power penalty less than 0.1 dB, RIN value should be less than –135 dB/Hz. Hence, optical connectors with less than –20 dB reflection requirements can be needed.

In a bi-directional optical link, the reflected light can lead to excess noise in the transmitter output and a degradation of receiver sensitivity due to the effect of the reflection-induced noise. In a lightwave system, a design parameter is BER = 10^{-9}. this parameter is given by:

$$\text{BER} = \frac{1}{2}\, ercf(Q/\sqrt{2})$$

BER varies with Q parameter and it is approximately 10^{-9} for $Q = 6$. Therefore, the receiver sensitivity corresponds to the average optical power (P_S) for which $Q = 6$. When thermal noise dominate for such a receiver, the current variances $\sigma_0^2 = \sigma_1^2 = \sigma_T^2$, $SNR \approx Q^2$ and $P_S = Q\sigma_T/R$, where R is the responsivity of the photodiode. When the reflected optical power in the connector closest to the node transceiver goes back to the receiver, an induced noise is produced and added to the total current variance, so that

$$\sigma^2 = 2\sigma_T^2 + \sigma_R^2$$

where $\sigma_R = 2rRP_S$ and r is the induced noise level measurement. i.e. to obtain a BER of 10^{-9} in an optical receiver with a sensitivity of –15 dB, the SNR should be about 15 dB. However, if reflection feedback arises to the receiver, a power penalty is produced. For $r = 0.1$ the receiver sensitivity must be –13 dB to achieve a SNR of about 15 dB ($Q = 6$). In this case the power penalty is 2 dB.

6.2.6. Dispersion

In access distances, dispersion mainly affects the signal in multi-mode fibers. This is attributed to differential paths of the different rays in the fiber that have different group velocities or delays related to the different mode. A single-mode fiber transports the energy in a single ray and the intermodal dispersion is not produced. However, intramodal dispersion D(λ) occurs due to the fact that different optical spectral components of the pulses present slight differences in their group velocities. This produces a pulse broadening along the distance. The dispersion limits the bit rate because it can produce inter-symbol interference by the overlapping of the adjacent pulses. This is also called chromatic dispersion and is zero at around 1.3 μm for step-index single-mode fiber and it only affects above Gbit/s rates.

Factors that contributes to chromatic dispersion are the material and waveguide. There are three basic techniques to measure the dispersion:

- *Time Domain:* the delays of the optical pulses at different wavelengths are measured, fitted to an analytical polynomial function and then differentiated directly to yield $D(\lambda)$ as a function of λ.
- *Frequency domain:* The phase shift $\phi(\lambda)$, equivalent to time delay in the time domain, is measured in sinewave modulated carriers at different optical wavelengths.
- *Interferometry:* The fringe visibility in a Michelson or Mach-Zehnder interferometer gives the delay of the different components of wavelength in the fiber.

6.2.7. Frequency instability

Because of both external and internal aspects in WDM PONs such as component aging, high drift of temperature and other possible factors, the channel frequency may runaway from its corresponding locking region and shift from its desirable value. Frequency instability constitutes a serious trouble when multiplexing various optical carriers. Therefore, periodical measurement of channel frequencies is indispensable for the system high performance maintenance, especially when no frequency stabilization system is included. The significant parameters that cause frequency drift in semiconductor laser diodes are:

- Ambient temperature fluctuations.
- Current injection variations.
- Aging effect.
- Spontaneous emission effect.
- Spurious mode hopping

The performance level of a frequency measurement system can be defined by several parameters as absoluteness, reproducibility, uncertainty, resolution, etc. Monochromator-based optical spectrum analyzer, interferometer-based optical spectrum analyzer or spectral analysis using a heterodyne spectroscopy method can provide the frequency measurements.

Recently, in **[Youn 01]** has been proposed another technique to monitor the optical frequencies of multiple WDM channels using pilot tones and an Arrayed-Waveguide Grating (AWG). The WDM channels, which include a pilot tone as shown in Figure 6.2.2, pass through an AWG in the monitoring module. Each channel is detected by a photodiode and processed with fast Fourier Transform to measure the amplitude of pilot tones. The frequency accuracy is given by the ratio of pilot tones.

6.2.8. Crosstalk

The use of wavelength channels in a WDM access network allows flexibility with regard to user demand for bandwidth, and helps integrate different protocols such as analog, ATM, IP and Gigabit Ethernet. However, due to the multiple wavelengths, intrachannel and interchannel crosstalk are produced. It occurs when a channel interferes at the optical multiplexer output with crosstalk components at the same wavelength. These components can be originated either by the same channel that traversed the cross-connect along a spurious path, or by other channels. In transparent WDM networks, intrachannel crosstalk is generally more severe than interchannel crosstalk and limits the number of cascadable nodes in a network. In a ring network with a centralized multiwavelength light source, the intrachanel crosstalk produces greater

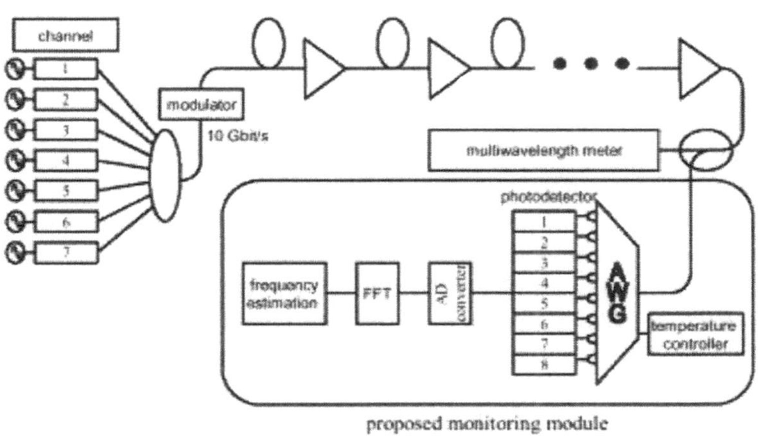

Figure 6.2.2. Scheme of optical frequency monitoring with AWG and pilot tones.

degradation due to the fact that the spectrum of such crosstalk is the same as of the upstream signal and is within the electrical bandwidth of the receiver.

Crosstalk also can arise from the spectra tails of the other channels entering the bandwidth of the selected channel or due to the imperfection of the components used in the network, and also due to the signal filtering characteristics of the optical devices used in the network.

6.2.9. Noise

There may be different sources of noise present in an optical fiber system. These are principally:

Amplified Spontaneous Emission noise (ASE)
In fiber optical communications the spontaneous emitted photons generated in optical amplifiers constitute an important noise source (ASE) to be considered. Moreover this noise benefits from the same mechanism of amplification along the doped fiber amplifier. Because of pump laser excitation, ASE noise exists even when no optical channel is input into the fiber amplifier.

Consequently, the determination of noise level in an EDFA without inserted optical channels into the fiber will serve for determining a quality path (without signal). In amplified WDM channels, the ASE noise level depends on the number of amplified channels and that their power decrease when an optical carrier is added and viceversa.

Unlike the case of single channel transmission, when using optical amplifiers to amplify or compensate signal loss of WDM channels, ASE noise, gain flatness, and gain competition must be well characterised so as to assess the requirements of the network healthy behaviour and keep its performances high.

Relative Intensity Noise (RIN)
This kind of noise is referred to the random intensity, phase and frequency fluctuations in the output of a semiconductor diode optical source even when it is biased at a constant current with negligible current fluctuations. Spontaneous emission, shot noise and Temperature variations are the fundamental mechanism to the RIN. The optical receiver converts the fluctuations of the RIN in current fluctuations that are added to the thermal and shot noise in the receptor. This noise improves as the injection-current level increases. Back reflection can increase the RIN by 10 to 20 dB. Usually, feedback ratio must be less than –60 dB in order to maintain an RIN of less than –140 dB/Hz. For frequencies near the relaxation-oscillation frequency, the RIN is considerably enhanced. However, for much higher frequencies, the laser

is not able to follow the fluctuation. Vertical-Cavity Surface-Emitting lasers (VCSELs) is able to emit several modes. The RIN spectrum exhibits several peaks corresponding to the relaxation-oscillation frequencies of the traverse modes. This can cause degradation of the noise performance.

Shot noise

The detected current by the photodiode, which can be expressed by $I_p = RP_{in}$, is not completely constant even when the incident optical signal has a constant power. Shot noise is produced as consequence of random electron-hole recombination. Thus, the output current of a photodiode in response to a optical signal is given by $I(t) = RP_{in} + i_s(t)$, where $i_s(t)$ corresponds to the shot noise that can be represented by Poisson statistics (Usually can be approximated by Gaussian statistics).

Thermal noise

This kind of noise is the dominant one in conventional systems. It is due to the random movement of electrons being at a finite temperature in any conductor even when a voltage is not applied. This movement arises in the resistors and semiconductors of an optical receiver and is added as noise to the current of the photodiode. Thus, the output current of the photodiode will be $I(t) = RP_{in} + i_s(t) + i_T(t)$ where $i_T(t)$, represent the current produced by thermal noise. The thermal noise follows Gaussian statistics.

APD noise

Since the responsivity of an Avalanche Photodiode (APD) is increased M times than the responsivity of a P-I-N photodiode, APD could improve the SNR by a factor of M^2. However, avalanche noise is added to the shot noise due to the internal gain mechanism of the APD and the SNR improvement is considerably reduced. This occurs due to the fact that the process of impact ionisation to achieve the typical APD gain generates secondary electron-hole pairs in the avalanche process. This pairs randomly generated are added to the shot noise corresponding to the primary electron-hole pairs. In **[Li 97]** is demonstrated that the excess noise factor decreases as avalanche region width decreases under both pure electron ionisation and pure hole ionisation.

6.2.10. Extinction Ratio

In an optical transmission system, the source of transmitter emits an optical pulse of constant energy for on-state, and no energy is emitted for off-state in an ideal situation, usually in NRZ modulation format. However, this is not true in real terms. A semiconductor laser emits a power P_1 for on-state and P_0 for off-state. The power P_0 depends on the bias current and the threshold current and produces a power penalty that is reflected in an increase of the minimum average received power required by the receiver to operate at a BER (i.e. of 10^{-9}). The relation P_1/P_0 represents the extinction ratio that can become a severe cause of signal degradation due to the energy accumulation in the optical receiver especially in a PON up-stream whether P_0 is a significant fraction of P_1 (when the laser is biased close to but above threshold). Typical extinction ratio is about 10 dB and the corresponding power penalty is negligible.

6.2.11. Differential optical path loss

In a star configuration of an optical passive network each user usually has a different path length. This circumstance makes that signal with different power arrive to the node in the up-stream link, hence a decision threshold for each user bit stream. This can become a critical trouble for optical access networks because of the amplification and decision circuit of the of the receivers have difficulties to adapt to the wide range of input power sufficiently fast. On the other hand, when the power is divided up among a given number of subscribers to broadcast, it is necessary to ensure that each receives at least the minimum amount of optical power required to detect a quality signal. When each user has a different path length, this can generate serious degradations for farther away users determined by the desired Signal-to-Noise Ratio (SNR), which might depend on the received optical power itself.

6.2.12. Timing jitter

This impairment arises when the decision instant to sample a signal is continually fluctuating bit to bit. Normally, signal is sampled at the peak of the pulse but there is a random variation (t in the sampling time due to the noise, instability and design of the clock-recovery circuits. Sampling time fluctuations produce a reduction of the sampled signal and hence SNR and the receiver performance are degraded. SNR can be improved increasing the received optical power inducing to a power penalty. In general, the RMS value of the timing jitter should be below 10% of the bit period for a negligible power penalty. These impairment measurements will be the main parameter to determine the OSNR in the different path, and crucial to establish the quality of the optical signal.

6.3. MAINTENANCE PROCEDURES

Optical Maintenance (OM) covers all the means to guarantee the performances of the optical carrier (fibers, splices, connectors, and passive components), and specifically to detect and locate any fault in the OAN. Today acceptance tests are done in two phases: acceptance of the components (fiber. splitters, filters, etc.) and acceptance of the infrastructure once the components are installed.

The first phase has as objective to verify if the components characteristics match the specification presented by the supplier.

In the second phase, after the acceptance tests for components have been approved, two basic tasks are necessary to implement an effective preventive maintenance scheme:

- Network qualification/acceptance: this establishes that the actual plan matches the specifications given by the network planner. Some procedures could be performed before any service runs over it.
- Network monitoring: It involves periodic measurement of the physical carrier and comparison to a reference measurement or to the network specifications given by the network planner

Among functions of operation, administration and maintenance (OAM) for the passive optical network (PON), it is stated that the optical line termination (OLT) receives information about the status of the optical network unit (ONU) through physical layer operation and maintenance (PLOAM) cells. Alarms are referred to events regarding link interruptions and bit error rate (BER) degradation, both upstream and downstream.

On line monitoring could be the most efficient solution also considering the performed cost analysis. However, it is limited to the use of an out of band wavelength between 1625 and 1650 nm. Although, the 1625 nm wavelength is not yet standardised for measurement maintenance purposes, several manufacturers have instruments that work with this value. On the market now available are several complete remote fiber test system including OTDR, passive components, and management system in modular solution chosen by the user.

In a remote fiber test system, there is a central management system positioned in the central office. The central management system is in charge of receiving the measures realized by the different remote test units (RTU) which are located in the remote nodes. These units monitor the optical fibers connected to their optical test access unit to detect faults, impairments and signal loss. The objectives of a remote fiber test system are:

- Reduction of the average repair time.
- Introduction of a predictive maintenance policy.

- Introduction of computerised documentation of the optical network.
- Standardisation and network construction tools.
- Improved efficiency of maintenance teams.
- Reduction of portable maintenance equipment.
- Provide automatic detection and precise location of faults.
- Minimise the problems encountered with traditional maintenance techniques.

6.4. MONITORING SYSTEMS

For fiber fault location and fiber identification, each fiber can be accessed independently in the case of single-star networks and PONs with splitters in the office, but not in the case of PONs with splitters outside the office. Figure 6.4.1 shows the structure of a typical PON and access methods for testing **[Nakao 96]**.

Optical Time Domain Reflectometer (OTDR) is the most popular tool for measuring the fault location **[Selvan 97]**, but it is impossible to measure the fault location remotely when the splitters are outside the office, because OTDRs cannot interpret superimposed reflections from multiple paths. To identify a

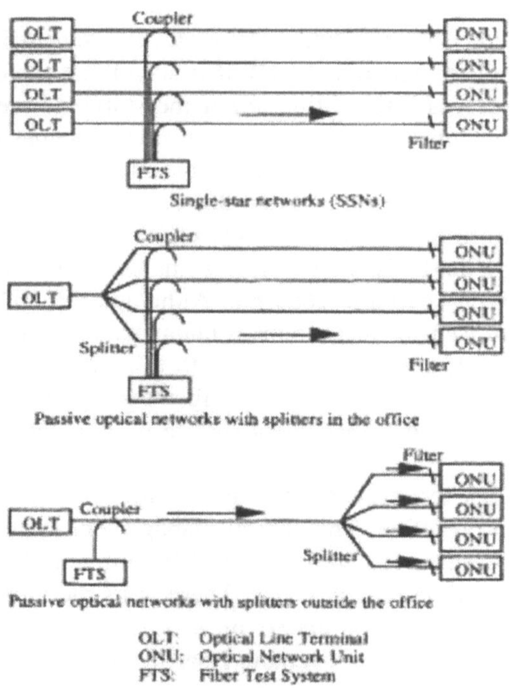

Figure 6.4.1. Testing structure of access networks.

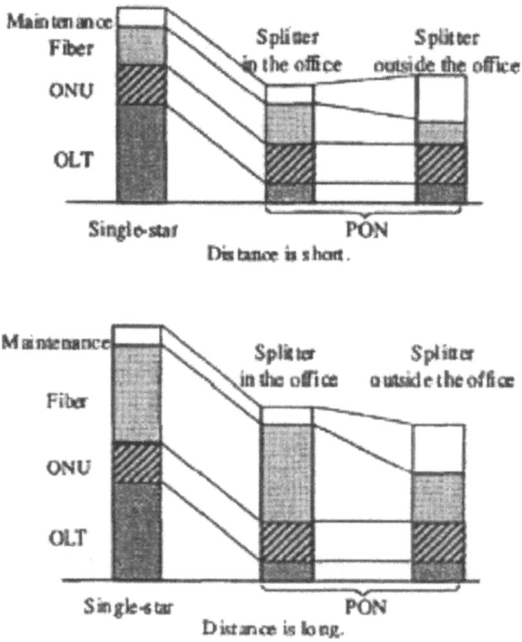

Figure 6.4.2. Network costs breakdown.

fiber, one must inject the light into the fiber at the splitter. Thus, when splitters are installed outside the office, the maintenance costs are high.

Figure 6.4.2 shows a breakdown of network cost, including those of equipment, cables, installation, and maintenance. The dividing line for choosing the splitter location should be calculated based on factors such as cable cost, labor cost, and frequency of fault occurrence.

Different remote fiber-testing systems have been developed according to the configurations of optical access networks and requirement of the subscribers and the service provider. These systems are described in the following.

6.4.1. Fiber monitoring between buildings

The NTT Access Networks Service Systems Laboratories, Japan, have implemented an automatic optical fiber operation support system (AURORA) proper to fiber faults monitoring in access rig networks **[Enomoto 99]**. It is a test system based on test light bypass modules. The system configuration, Figure 6.4.3, consists of a control terminal, test equipment modules (TEM) each of which contains an optical time domain reflectometer (OTDR) and test equipment controller (TC), optical fiber selectors (FS) which select test fibers, test

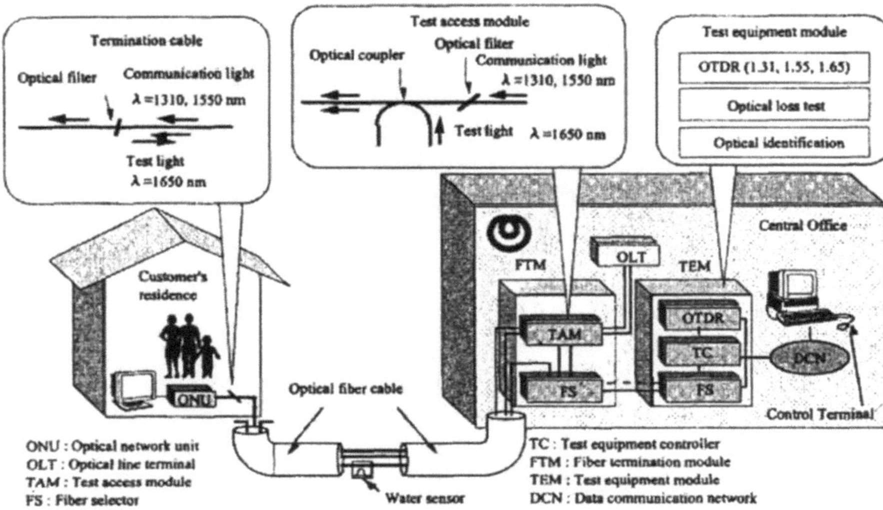

Figure 6.4.3. System configuration AURORA.

access modules (TAM) which introduce a test light into fiber lines, and termination cables with an optical filter that allows a communication light to pass but that cuts off the test light. The control terminal orders various optical fiber tests and administers fiber information. The various commands between the control terminal and TEMs are transmitted through a data communication network (DCN).

AURORA carries out automatic OTDR measurements and reveals fiber characteristics and fault locations with no degradation in transmission quality. Couplers were installed to pass the next cable to monitor the optical fiber cables between buildings, and also, three kinds of filter, F1, F2, and F3, were installed as show in Figure 6.4.4. The F2 filter allows the communication light whose wavelength is λ1 to pass and cut off the test light whose wavelength is λ2. They are arranged in front of the transmission equipment. The F1 and F3 filters allow the test light to pass and cut off the communication light. They are installed between the couplers and in front of the OTDR, respectively.

Figure 6.4.5 shows the influence of the communication light from TE0 on the transmission quality between TE1 and TE2. A required cutoff characteristic of the F1 filter is that there should be negligible degradation in the signal to noise ratio (SNR) as a result of the mutual communication light interference caused by TE0.

The test light bypass module configuration, as show in Fig. 6.4.6, use 1650 nm wavelength for maintenance tests and 1310 nm wavelength for communication light. Moreover, the module also use an optical filter which can pass 1310 nm, but cut off 1650 nm wavelength light.

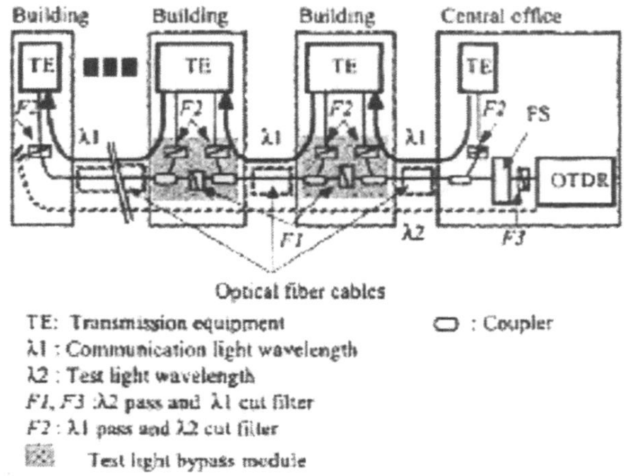

Figure 6.4.4. Basic configuration of testing system.

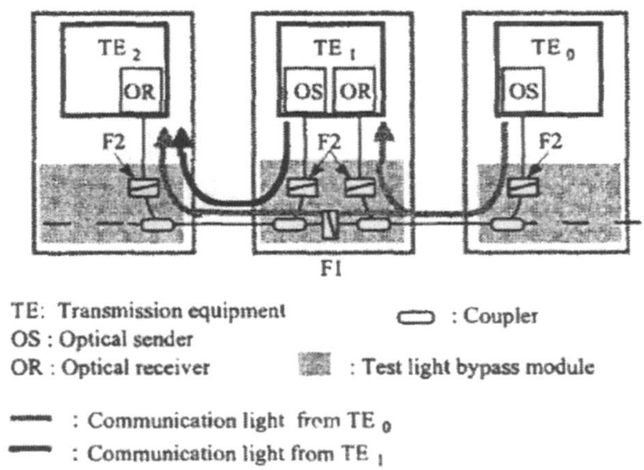

Figure 6.4.5. Influence of communication light from TE0 on TE1 and TE2.

An OTDR trace obtained using the test light bypass module is shown in Figure 6.4.7 where the testing system monitors cable between buildings.

6.4.2. Branches access network monitor

In passive optically amplified branched networks, erbium-doped fiber amplifiers (EDFAs) are often placed before the optical tree-coupler to support more fiber branches and to improve the transmitted signal quality. Figure 6.4.8 shows a new fiber-fault identification scheme for a $1 \times N$ optically ampli-

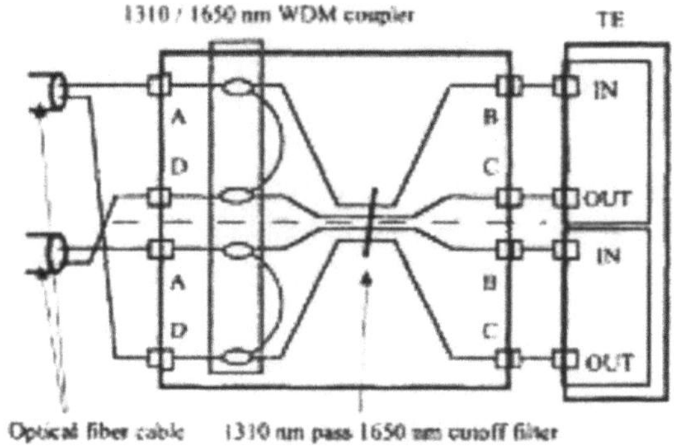

Figure 6.4.6. Configuration of test light bypass module.

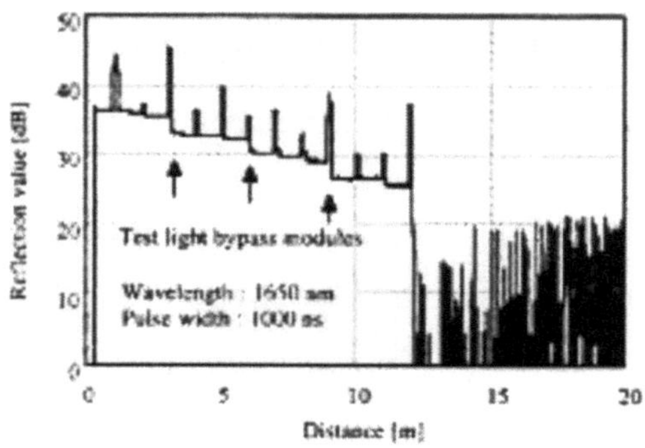

Figure 6.4.7. OTDR trace of optical fiber cables with test light bypass modules.

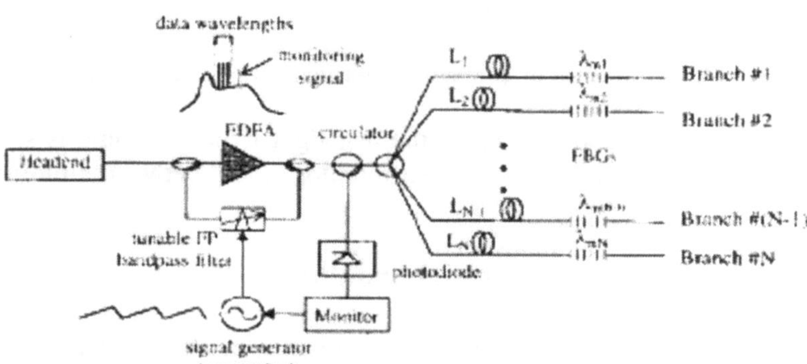

Figure 6.4.8. Fiber-fault identification scheme for branched access networks.

fied branched network that is simple to implement. It makes use of a FBG with a wavelength-sweeping monitoring laser source formed by feeding the ASE back into the EDFA through a wavelength tunable filter. The fiber branches can continuously be supervised by monitoring the periodic reflected optical pulses in time-domain **[Chan 99]**.

For a single fiber feeder as shown in Figure 6.4.9, several FBGs of different center wavelengths are used to locate the fault along the feeder **[Chan 97, Giles 98]**.

If the received ASE power at any designated monitoring wavelength deteriorates after certain time or is below the detection limit, this indicates that there might be a fault or a progressive degradation at the corresponding fiber branch.

The data channels are transmitted in the EDFAs flat-gain region of 1540–1555 nm while the ASE power in the unused gain spectrum, 1556–1562 nm, is used as the monitoring light source. The light source is constructed by looping back some of the EDFA emission to its input through a tunable Fabry-Perot etalon filter (FPF) generating a gain-clamped saturated laser emission with wavelength-tuning capability. By applying a periodic saw-toothed voltage at a few kilohertz to the FPF, the gain-clamped signal is made to sweep through the unused EDFA gain spectrum periodically and thus serves as a wavelength-sweeping monitoring light source. An FBG of distinct center reflection wavelength is placed at the end of each fiber branch as the branch identifier. For surveillance application, the FBGs must have stable thermal response, and can be achieved by means of temperature-compensated package.

In each sweeping cycle, the emission wavelength of the monitoring signal is swept continuously through the unused EDFA gain spectrum. As illustrated in Figure 6.4.10, when the FBG center wavelength matches with that of wavelength sweeping monitoring source, an optical pulse will be generated and reflected upstream to a monitoring photodiode via a circulator. Thus,

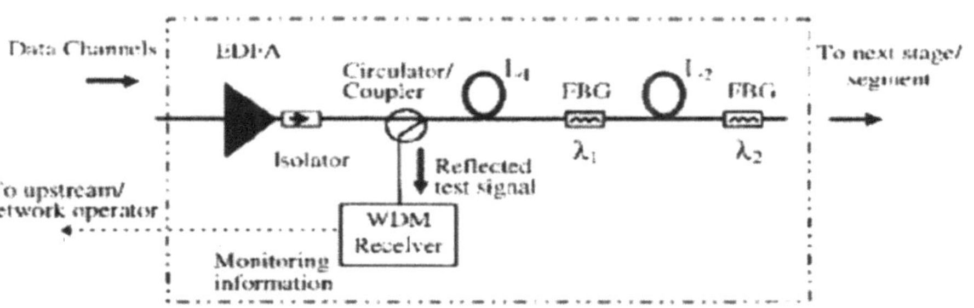

Figure 6.4.9. Fiber-fault identification scheme a single feeder networks.

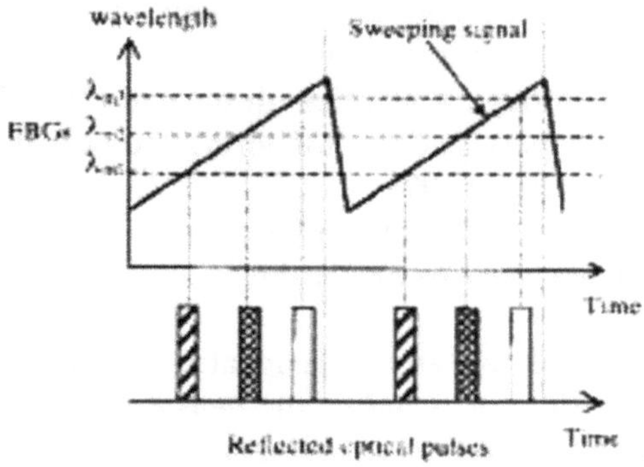

Figure 6.4.10. Operation principle.

the health status of a certain fiber branch is signified by the presence of the respective optical pulse in each sweeping period.

The detected optical pulse trains measured at the photodiode for different scenarios in a three branches access network are shown in Figure 6.4.11.

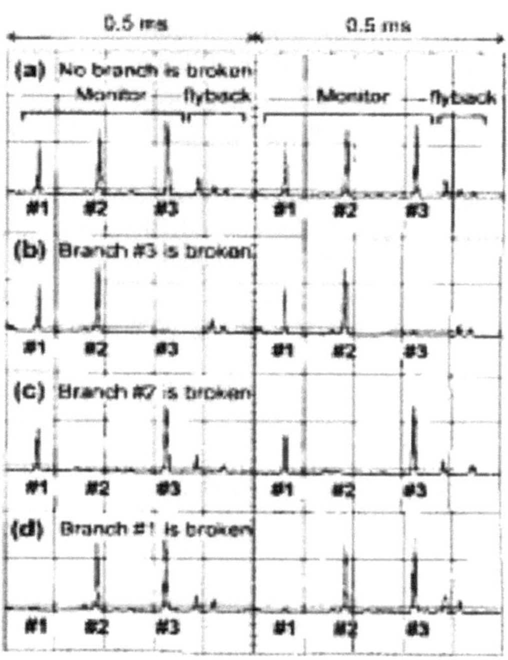

Figure 6.4.11. Reflected optical pulse train measured at the monitoring module for four different scenarios.

Any fiber cut is indicated by the absence of the corresponding optical pulse in each sweeping period. Figure 6.4.11(b–d) shows the detected optical pulse train for fiber cut in fiber branches # 3, # 2 and # 1, respectively.

6.5. TEST INSTRUMENTS

Test and monitoring devices carry out important task of measurement and verification of the operational characteristics of the passive optical components such as splitters, connectors, couplers, optical fiber, and those of active components such as sources, photodetectors and optical amplifiers. In the optical fibers, several parameters, such as core and cladding diameters, numerical aperture, refractive-index profile, fiber attenuation, dispersion, mode-field diameter for single mode fibers are measured with accuracy and precision by the manufacturer since this component is not easily replaceable after it has been installed.

On the other hand, the optical signal through the link is degraded due to noise, crosstalk, and the remaining of the above mentioned impairments. Therefore, it is necessary to realise maintenance and monitoring functions to know the effect of these impairments by means of measuring parameters as the bit error rate (BER) and optical signal to noise ratio (OSNR). This allows determining the signal quality and the status of the optical path in any link in or out of operation. Several devices are used to accomplish these functions that include, optical power meters, optical spectrum analyser, and optical time domain reflectometer (OTDR).

6.5.1. Optical power meter

The optical power detection is the most basic measurement in optical communications and is included in the most optical test equipments. A commercial optical power meter uses photodetectors to measure the optical power level which is usually given in dBm. This measurement instrument can also be used to measure the fiber attenuation by means of the cutback and insertion-loss methods. A typical power meter, which uses a Ge photodetector, allows a measuring range of –60 to +18 dBm in the 780 to 1600 nm band, whereas another with an InGaAs detector can measure signals in a –73 to 3 dBm range within the 840 to 1650 nm band.

More sophisticated instruments are build with an optical source that offers the possibility of automatic measure of optical loss in a fiber, an optical return-loss tester, a fault indicator, etc.

6.5.2. Optical spectrum analyser

At the expense of elevated sophistication, high performance levels could be reached. Commercially available optical spectrum analysers (OSA) may also provide the measurement of optical power as a function of wavelength, so optical power per channel and channel power equality measurement in WDM systems could be performed. The OSA is also used to measure crosstalk, ASE noise (hence the OSNR), frequency-domain intermodal dispersion, and channel faults due to any flawed element in a node or in the central office.

OSNR measurement
Once knowing the OSNR and ASE level before the WDM filtering, it is possible to derive at the degradation of OSNR from the enhancement of the ASE between the channels. In a small wavelength band around a channel, ASE will nearly increase. One only has to measure the additional ASE between two adjancent channels with an OSA and calculate the OSNR. In the MOON project, Polarization extinction method is also proposed to measure the OSNR. Since ASE is unpolarized and the optical signal remains almost polarized, the ASE noise and the optical signal power may be measured separately including a polarizer before the OSA as shown in Figure 6.5.1. Thus, and the OSNR can be calculated.

The OSA can be in general classified as: Monochromator-based optical spectrum analyser, Michelson interferometer-based optical spectrum analyser and spectral analysis using a heterodyne spectroscopy method.

Monochromator-based optical spectrum analyser
Figure 6.5.2 shows a basic configuration of a monochromator-based optical spectrum analyser. Acting as a rotating prism, the rotating diffraction grating separates the different wavelengths of light. The aperture width is variable and used to determine the wavelength resolution of the instrument (typically about 0.1 nm). The monochromator performances depend on the dispersive power of the grating, the length, and the width of the exit split.

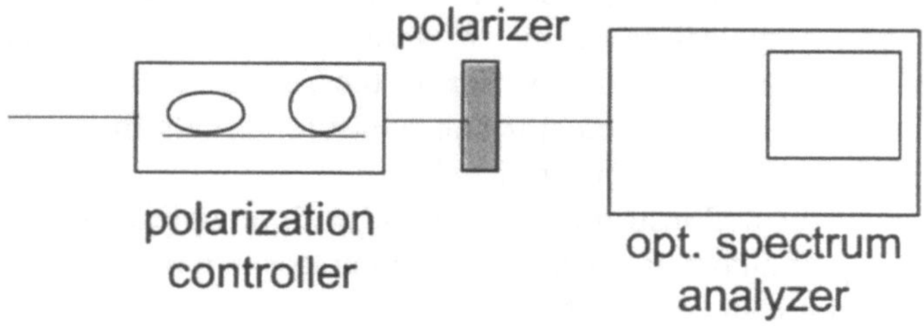

Figure 6.5.1. Proposed OSNR measurement set-up for WDM networks.

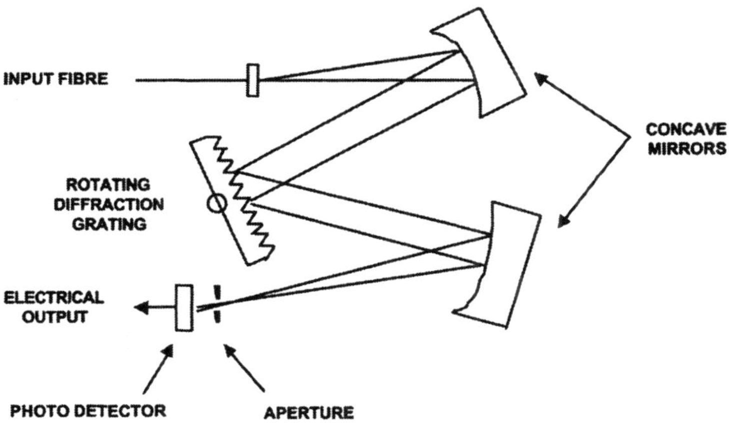

Figure 6.5.2. A monochromator-based optical spectrum analyser.

Michelson Interferometer-based optical spectrum analyser
By moving a Michelson interferometer arm, see Figure 6.5.3, and measuring the power in the interference pattern created between the signal and a delayed version, the Michelson-based optical spectrum analyzer makes direct measurement of the coherence length as well as a precise wavelength measurement. Therefore, calibrating these instruments with an absolute frequency reference may be needed when improved performances are required.

The environment perturbations and the imperfection of displacement of the interferometer mirror or the rotating grating decrease the performances of both the monochromator and interferometer-based optical spectrum analyser

Spectral analysis with heterodyne spectroscopy method
It is based on a frequency-scanning device (laser, or a scanning FP interferometer) that is heterodyned with an unknown frequency and with a reference frequency. This frequency measuring technique falls into the category of

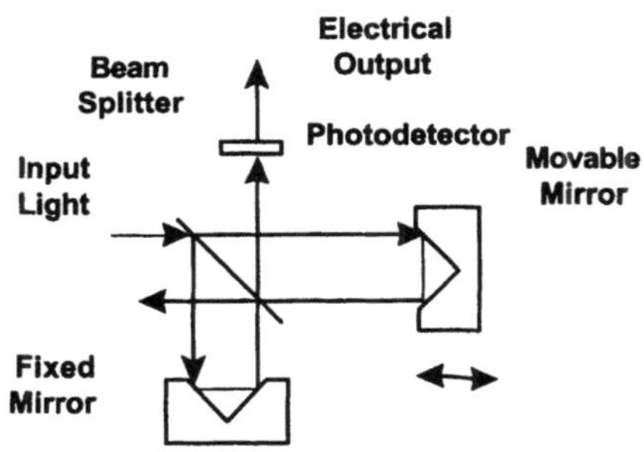

Figure 6.5.3. Michelson interferometer.

spectrum analysis. A prototype has been implemented in the Department of Signal Theory and Communications of Politechnical University of Catalonia **[Amrani 00]**.

The basic configuration of the measuring system is depicted in Figure 6.5.4. During each period of the scanning frequency, temporal pulses occur indicating coincidence between the heterodyned frequencies. The time difference between the pulses is converted into a frequency difference using the linear slope of the scanning frequency. Thus, the knowledge of this frequency difference and of the reference frequency will give an accurate measurement of the unknown frequency. Both, the frequency sweeping range and the detected pulses are digitised and processed by the computer. When the reference frequency is locked to an absolute frequency standard, the present system will allow an absolute measurement of the carrier frequencies **[Amrani 99]**.

6.5.3. Optical Time-Domain Reflectometer

Together with OSA, Optical Time-Domain Reflectometer (OTDR) is the more useful measuring device in optical fiber systems. For detecting the location of fiber breaks, fiber discontinuities, and fiber loss OTDR is a proper device. OTDRs with measurement ranges of typically 200 km are commercially available. OTDR consists of an optical source, a receiver, a data acquisition module, a central processing unit (CPU), an information-storage unit, an internal memory and a display.

One of the fibers loss mechanism is scattering caused by small variations in the material such as discontinuities. The OTDR instrument sends out light pulses which will be scattered along the whole fiber. Additionally a part of the pulses will also be reflected in locations where abrupt change in material density exists (e.g. connectors with air gap). The reflected and backscattered

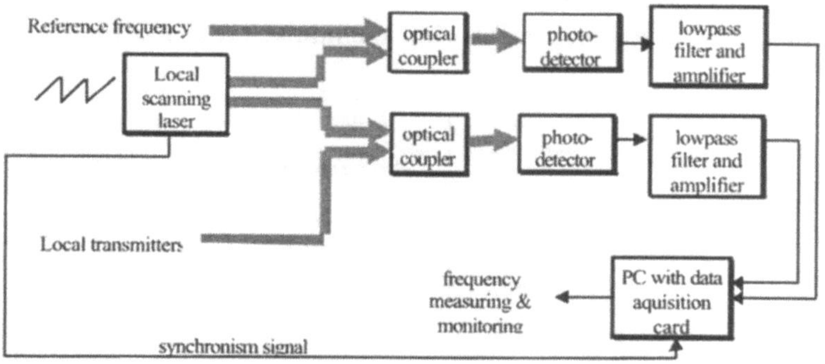

Figure 6.5.4. Schematic configuration of an optical spectrum analyser using a frequency-scanning laser.

light will be detected at the input of the fiber. From the time delay of the returned pulses, the location of discontinuities or fiber breaks can be calculated. A simple method for in-service OTDR monitoring a WDM link is the usage of an out-of band wavelength for the OTDR (e.g. 1300 nm) and bypass EDFAs by means of standard 1300 nm/1550 nm WDM couplers.

An OTDR with tunable source has been proposed in the NTT laboratories, Japan, to monitor faults in a branched optical network with a wavelength dependent optical splitter/router module. The OTDR is used to receive a returned light only from a specified branched fiber, by changing the test light wavelength across the whole transmission band of the module **[Tanaka 96]**.

6.5.4. BER Estimator

More comprehensive measures of signal quality are the bit error ratio (BER) which is conventionally measured at optical-electrical-optical (OEO) conversion locations through bit interleaved parity (BIP). The usual technique for BER measurement relies on transmitting of standardized pseudo-random test patterns using commercial bit error test sets. Another technique may rely on parity checks. While the first mentioned technique might be used for out-of service measurements by the maintenance staff, the second technique may be employed in in-service. However, not all optical network elements will contain the resources for bit-error detection of an arbitrary payload signal and very low bit-error rates cannot be investigated in that way. Therefore, faster approaches which are transparent to signal data format and/or bit rate are desired to enable the continued development of transparent optical networks.

Currently, most BER estimator employs the histogram analysis technique that consists of sampled signal values analysis as a means of estimating the signal Q factor and BER. The signal may be sampled randomly throughout the bit stream without regard to the decision time, this is to say an asynchronous fashions, or only at the decision time, synchronous sampling. Asynchronous sampling technique is completely transparent to signal bit rate and data format, while, the synchronous sampling approach is not transparent to bit rate, but is transparent to data format **[Downie 01]**, **[Hanik 99]**, **[Shake 98]**.

Figure 6.5.5 shows the inner architecture of a synchronous BER estimator. The intention is to estimate the BER only by knowing the signal histogram, the OSNR, the optical power, and the bandwidth of a received signal. The signal shape can be displayed as an eye diagram and the different sampled values of marks and spaces in the center of the eye can be converted in a histogram.

Even though the estimation algorithm does not allow an exact estimation of the Q-factor, the BER-estimator is a valuable and powerful tool for super-

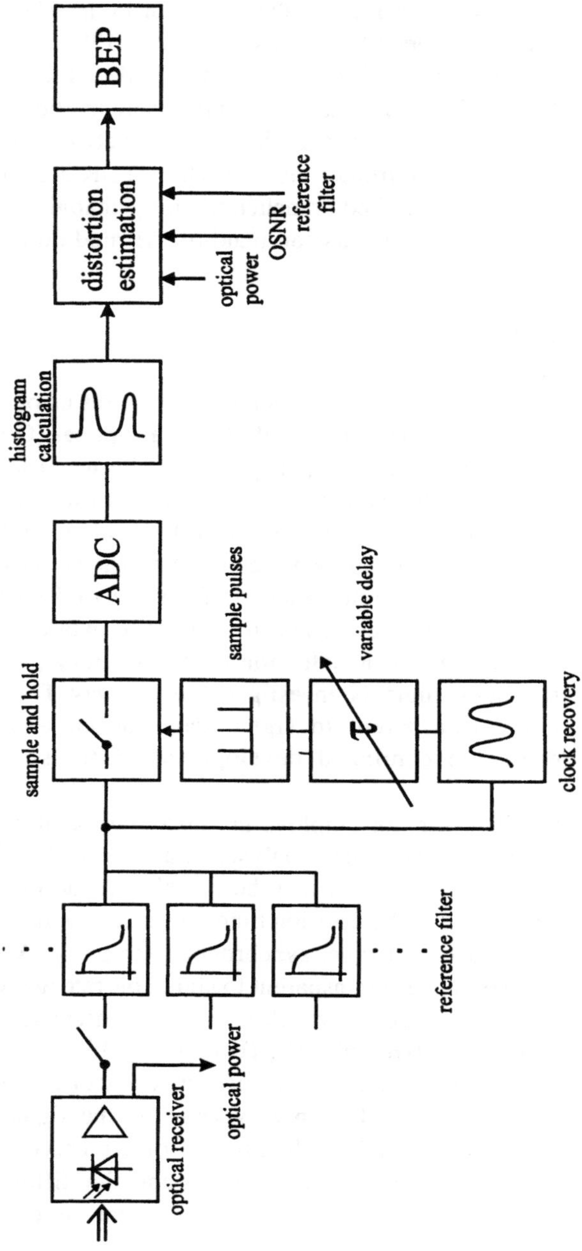

Figure 6.5.5. Architecture of BER estimator.

vision of optical signals because it is able to recognize slow signal degradations long before the client application.

SUMMARY

The access network consists of two parts, the collection and distribution network, and the feeder network. Monitoring of whole elements that compose of the optical access network must be made periodically to achieve an efficient network performance and to avoid lose of information. The maintenance costs may vary due to the splitter location, if it is located outside of central office, the maintenance costs will be higher but the costs by fiber cable and devices will lower

Initially, the supervision systems are centralized in the access node or central office because one supervision system for each user is an unaffordable cost due to high number of users in an access network. Also, it is possible to utilize portable supervision systems for fault detection at the user premises

The most of fiber supervision systems utilize the OTDR as measurement equipment. Fiber testing between buildings uses optical filters to cut the test light to the building and allow passing the data signal with no degradation in transmission quality.

Branches access network and single feeder network monitoring use fiber Bragg grating FBG with different center wavelengths in order to identify the faulty branch or to locate the fault along the feeder, using only one OTDR or one test equipment module for all fibers. In this monitoring scheme, it is not necessary to access each fiber independently; therefore, test light can be injected before the splitters in order that maintenance costs are reduced. In this way, by using the splitters outside the central office also it is possible to reduce equipments and optical fiber cable costs.

On the other side, channel and optical signal supervision strategies in optical transport networks can be utilize in order to supervise the optical signal quality in an optical node. The monitoring methods in photonic transport systems such as pilot tones can be apply in access networks topologies by using a pilot tone of different frequency with each optical signal. These frequencies must be out of the bandwidth of the data signal and can be inserting by FSK and ASK modulations.

The use of optical spectrum analyzer to monitor ASE, OSNR and wavelength variation of optical signal is very useful in wavelength-division multiple access techniques used in FTTH systems. An optical spectrum analyzer of low cost with a Fabry-Perot interferometer has been developed into an enhanced optical supervision system (EOSS). It is based on a frequency-scanning device using a heterodyne spectroscopy method.

BER estimation by means of histogram analysis is also widely used in optical networks supervision due to its low cost and accuracy of the technique. This method is based on sampled signal values that can be performed in either asynchronous or synchronous manner. The asynchronous method is transparent to signal bit rate and data format but a great amount unwanted samples could affect the accuracy of the measure. In contrast, synchronous method samples the signal only at the decision time. This technique is not transparent to bit rate, but is transparent to data format.

REFERENCES

[Amrani 00] Amrani, A.; Junyent, G.; Prat, J.; Comellas, J.; Ramdani, I.; Sales, V.; Roldan, J.; Rafel, A., 'Performance monitor for all-optical networks based on homodyne spectroscopy', IEEE Photonics Technology Letters, Volume 12, Issue 11, Nov. 2000.

[Amrani 99] Amrani, A.; Roldan, J.; Prat, J.; Rafel, A.; Junyent, G., 'Optical spectrum explorer for maintenance of optical networks', LEOS '99. IEEE Lasers and Electro-Optics Society 1999 12th Annual Meeting, Volume 2, 1999.

[Chan 99] Chun-Kit Chan; Tong, F.; Lian-Kuan Chen; Keang-Po Ho; Lam, D., 'Fiber-fault identification for branched access networks using a wavelength-sweeping monitoring source', IEEE Photonics Technology Letters, Volume, 11, Issue 5, May 1999.

[Chan 97] Chun-Kit Chan; Tong, F.; Lian-Kuan Chen; Jian Song; Lam, D., 'A practical passive surveillance scheme for optically amplified passive branched optical networks', IEEE Photonics Technology Letters, Volume 9, Issue 4, April 1997.

[Diaz 01] Diaz, O.; Rafel, A.; Prat, J., 'Functionality of an enhanced optical supervision system (EOSS) in optical transport networks', Networks and Optical Communications, NOC2001, June 2001.

[Downie 01] Downie, J.D.; Tebben, D.J., 'Performance monitoring of optical networks with synchronous and asynchronous sampling', Optical Fiber Communication Conference and Exhibit, 2001. OFC 2001, Volume 3, 2001.

[Enomoto 99] Enomoto, Y.; Daido, A.; Araki, N.; Honda, N.; Tomita, N., 'Design and system configuration of optical fiber line testing system using test light bypass modules', Communications, 1999. APCC/OECC '99. Fifth Asia-Pacific Conference on . . . and Fourth Optoelectronics and Communications Conference, 1999.

[Giles 98] Giles, C.R.; Strasser, T.; Dryer, K.; Doerr, C., 'Concatenated fiber grating optical monitor', IEEE Photonics Technology Letters, Volume 10, Issue 10, Oct. 1998.

[Hamazumi99] Hamazumi, Y.; Kawai, T.; Koga, M.; Sato, K., 'Compact 8 × 8 optical switching board implemented with connection supervision and optical power control function', IEEE Photonics Technology Letters, Volume 11, Issue 3, March 1999.

[Hanik 99] Hanik, N.; Gladisch, A.; Caspar, C.; Strebel, B., 'Application of amplitude histograms to monitor performance of optical channels', Electronics Letters, Volume 35, Issue 5, 4 March 1999.

[Heismann 96] Heismann, F.; Fatehi, M.T.; Korotky, S.K.; Veselka, J.J., 'Signal Tracking And Performance Monitoring In Multi-wavelength Optical Networks', Optical Communication, 1996. ECOC '96. 22nd European Conference on, Volume 3.

[Hill 94] Hill, Goff, 'Management and Functionality of Optical Networks', ECOC '94.

[Li 97] Li, K.F.; Ong, D.S.; David, J.P.R.; Robson, P.N.; Tozer, R.C.; Rees, G.J.; Grey, R., 'Avalanche photodiode (APD) noise dependence on avalanche region width', Device Research Conference Digest, 1997. 5th, 1997

[Nakao 96] Nakao, N.; Enomoto, T.; Kuroiwa, M., "Splitter Positions And Testing Wavelength For Optical Fiber Access Networks', Optical Communication, 1996. ECOC '96. 22nd European Conference on, Volume 2

[Selvan 97] Selvan, B.; Green, R.J.; Haigh, N.R., 'Network monitoring for passively split optical fiber networks', Optical Fiber Video, Audio and Data Systems (Digest No. 1997/038), IEE Colloquium on, 1997.

[Shake 98] Shake, I.; Takara, W.; Kawanishi, S.; Yamabayashi, Y., 'Optical signal quality monitoring method based on optical sampling', Electronics Letters, Volume 34, Issue 22, 29 Oct. 1998.

[Singh 00] Singh, R.; Kawakami, T., 'New technique for measuring low fiber optic attenuation using double square-law detection', Instrumentation and Measurement, IEEE Transactions on, Volume 49, Issue 1, Feb. 2000.

[Tanaka 96] Tanaka, K.; Tateda, M.; Inoue, Y., 'Measuring the individual attenuation distribution of passive branched optical networks', IEEE Photonics Technology Letters, Volume 8, Issue 7, July 1996.

[Woodward98] Woodward, S.L.; Stayt, J.W., Jr; Romero, D.M.; Freund, J.M.; Przybylek, G.J., 'A study of optical beat interference between Fabry-Perot lasers', IEEE Photonics Technology Letters, Volume 10, Issue 5, May 1998.

[Youn 01] Youn, C.J.; Shin, S.K.; Park, K.J.; Chung, Y.C., 'Optical frequency monitoring technique using arrayed-waveguide grating and pilot tones', Electronics Letters, Volume 37, Issue 16, 2 Aug. 2001.

Chapter 7

ECONOMIC CASE STUDY

INTRODUCTION

The main economical aspects related to the deployment of an optical access network will be dealt with. After a brief description of the current technology trends, two solutions are studied in order to point out the main advantages and drawbacks of each one. This analysis is focused on companies that want to deploy their own networks.

The case under study is focused on to determine which the cost-effective topology solution for PON is in order to reduce the CapEx (Capital Expenditures per user). Market costs have been taken into account to evaluate the deployment costs of different scenarios studied. The presented costs by the analysis will refer to civil work, cable, manholes, splices and couplers.

7.1. TECHNOLOGY TRENDS

Currently, the critical economic situation of most Telecommunications companies has changed with respect to investment mentality. A long time ago, new operators, who wanted to deploy their own networks, developed new cost-effective equipment without considering so much the deployment costs. But now, with the difficulty to make great investments, all of them are focused on reducing costs. Some studies comparing different topologies and strategic deployments have emerged **[M.Corner 00]**.

Actually, one of the most important costs in a network deployment is the civil work. Due to the high costs of infrastructures with ducts, companies

J. Prat et al., Fiber-to-the-Home Technologies, 333–347.

are looking for other cabling systems. Nowadays, many vendors offer the possibility of putting the cable in the sewer with mechanical robots to reduce the infrastructure costs. Other solutions could be the blowing cable or the use of dark fiber.[1]

An optical access network that uses dark fiber is cheaper, that is because the most representative cost of a deployment is the canalization infrastructure. In order to consider the reduction of costs derived of using dark fiber instead developing, a new infrastructure of ducts, Figure 7.1.1 shows a comparison between these two deployments costs.

However, it is not easy to establish the entire access network with dark fiber. Dark fiber does not arrive everywhere. We have to consider that it was deployed with other criteria different from the company that wanted to rent it, so little deployment with ducts to arrive everywhere is always needed.

Another important aspect in a network is the digital technology used. A variety of technologies are available for the access network, nevertheless, due to the fact that an access network connects a lot of users, the technology that offers a reduced cost per interface port with great bandwidth is the desired one. Figure 7.1.2 shows the currently relative costs per user interface port for different technologies (such as XDSL, Ethernet, ATM . . .) in the access tract. In conclusion we can determine that the Ethernet technology

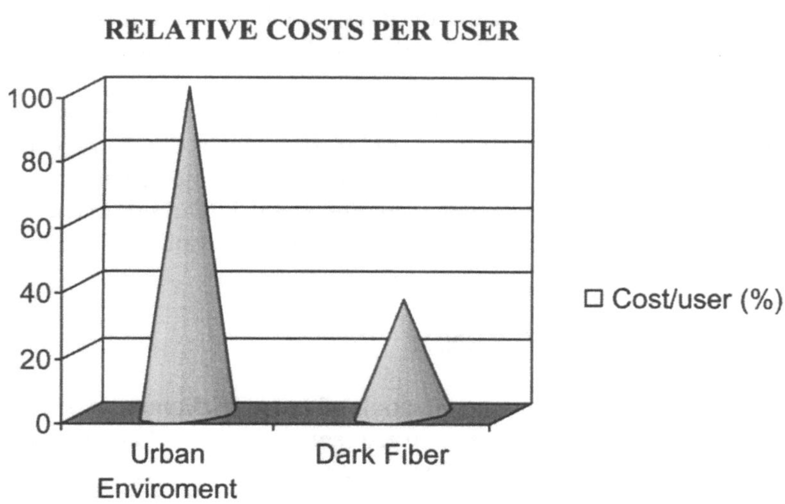

Figure 7.1.1. Relative costs per user.

[1] Dark Fiber is the fiber owned by companies (not necessarily telecommunications companies or operators) that do not use it. They rent it.

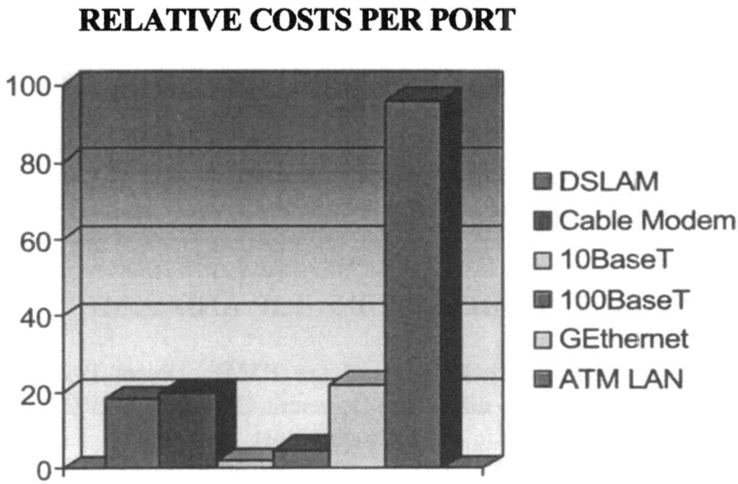

Figure 7.1.2. Relative costs per port.

has the lowest cost by user/port. In fact, Ethernet is the simplest technology and the one that has less hardware requirements.

On the other hand, if we consider the cost per Gigabit of Bandwidth (Figure 7.1.3) the 10 Gigabit Ethernet will be the technological solution with best cost-effective relation. Networks that can offer 10 Gigabit Ethernet have all the factors to have a great success.

Nowadays, Ethernet is growing faster than its competitors. Although in PON, ATM has a standard supported by the FSAN. The Ethernet group "EFM"

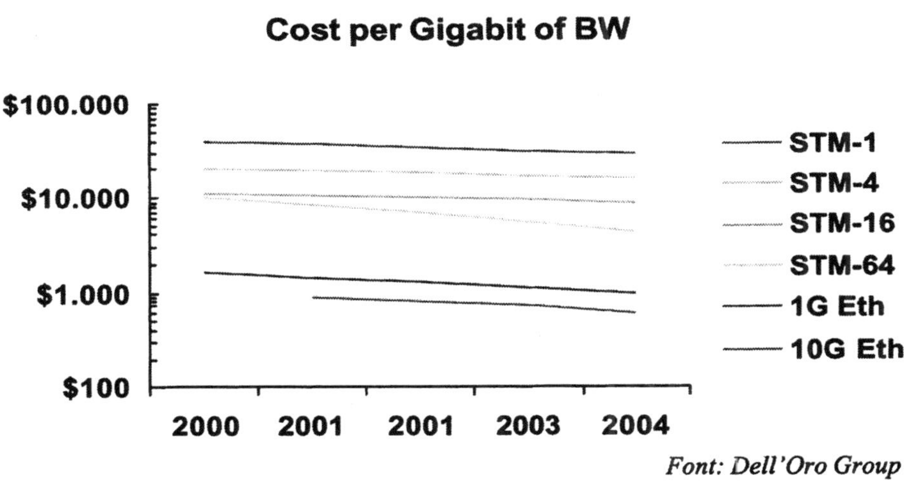

Figure 7.1.3. Cost per Gigabit of bandwidth.

is working hard to provide PON with a MAC protocol to accept Ethernet as a standard protocol for passive optical access networks. In fact, Ethernet is the most used technology around the world.

Some people are currently studying the impact costs of different topologies. Actually, one of the most important conclusions obtained through different analysis is that the star topology reduces the costs up to 20% percent in comparison with a ring topology **[Felipe 01]**.

7.2. SCENARIOS UNDER STUDY: P2P AND P2MP

Nowadays, P2P[2] (Point-To-Point) and the P2MP[3] (Point-To-Multi-Point) topologies are two issues under development. The P2P and P2MP are the two scenarios under study in the following costs example.

Our challenge is to obtain which is the cost-effective topology, and what is the difference in terms of costs. Although these topologies are currently used by the APON, they are also investigated by EFM.

As cost is an important factor, the studies try to identify the benefits of deploying a P2MP vs. P2P topology in Passive Optical Access Networks (PON) **[K.Shaneman 01]** in the network lay out design. With the cost of fiber being brought down, new deployment strategies could be considered to reduce implementation cost.

Before starting the analysis, a brief description of each model is given in Figure 7.2.1. The picture's scenario assumes N = 32 users and two fibers for each one. The P2P Ethernet Topology with Local Convergence points along the network is being deeply studied by different vendors just to provide Ethernet in high bandwidth access networks.

All the elements that take part in a network deployment of a P2P or P2MP topology are: the Central Office and all the active equipment used, the cable, the cabling system, the civil work, the splices, the couplers, the transceivers, the hand holes, the manholes and the user equipment: ONU (Optical Network Unit).

---

[2] P2P: Point-to-Point. Optical networks with dedicated link between the Central Office and the Customer.

[3] P2MP: Point-to-Multi-Point Optical networks with couplers between the Central Office and the Customer.

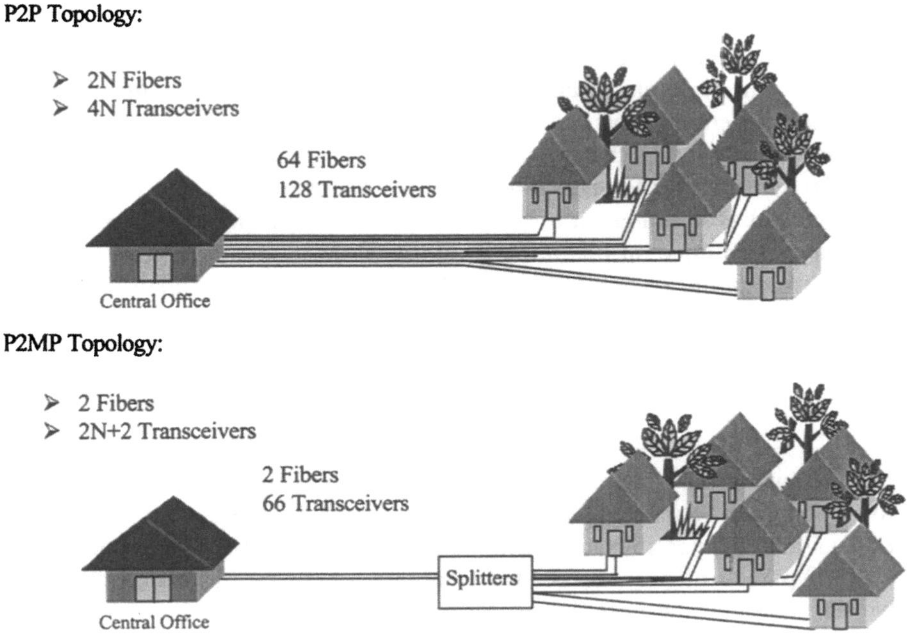

Figure 7.2.1. Scenario of a P2P and P2MP topologies.

7.3. COST ANALYSIS

In continuation, the difference in terms of cost of deploying a P2P or P2MP topology will be analyzed. Due to the fact that most new operators prefer to deploy their own networks instead of using dark fiber, and considering that a scenario with a duct's infrastructure is the expensive solution, the next analysis will study a scenario with a duct's infrastructure. Nevertheless, costs that have no effect with changing the topology of the external plant will not be considered in the analysis. Those elements are the Central Office, the active equipment of the ONU and the network tract that take part in the access loop.[4]

The representative costs taken into account in the analysis are: the canalization (civil work), cabling system (manually and mechanically), transceivers, spread, optical splitters, hand holes, and the tract between the Central Office and the splitter (in P2MP scenario).

---

[4] The fiber used to link the terminal box and the user. This wall-box can provide service to 16 users.

As it has been said before, the case under study uses an infrastructure with a duct's canalization. The types of ducts used are those with 2 × 125 mm + 3 × 63 mm of diameter, and the number of possible users assumed for the analysis is referred to the maximum capacity of fiber allowed by those ducts.

The study supposes that we have a uniform distribution of users along the canalization area deployment, the principal characteristics and the economical figures have been taken from different vendors and some research studies that have been done on this issue **[S.Figuerola]**.

The maximum capacity of fibers in the canalizations under analysis is 3.750 fibers, to obtain the total number of homes[5] that the network can connect a consideration of two fibers for user is assumed.

$$2 \times 125 \text{ mm} + 3 \times 63 \text{ mm} \Rightarrow 3.750 \text{ fibers (2 fibers for User)}$$
$$\Rightarrow 1.875 \text{ possible users.}$$

After using the maximum capacity of the ducts, we can provide service to 1.875 homes (possible users), but considering that a free space in ducts must be available for possible scalability and that some statistic studies shows that 80% of passed homes[6] are needed to obtain a penetration of users of 60%, we will consider the next figures for the example under study.

- 1.500 homes passed (80% of the potential users)
- 1.125 Users (60% of penetration)

General characteristics for the scenario under study are:

- 2 optical fibers per user.
- The modularity of the cable fiber in the access tract is 32. One cable can provide services to 16 users.
- Number of homes passed: 1.500.
- Number of Users: 1.125.
- The network Topology is divided into 3 tracts: Feed, Distribution and Access.
- The Access tract is not significant. It is the same for different topologies. This tract goes from the terminal box[7] to the user.
- Each terminal box has a capacity of 16 users.

To obtain the total length of cable and canalization needed for 1.500 homes passed the figure of density of 0,625 users/m has been used. This figure and others come from different analysis realized in other studies.

---

[5] Home: Houses with 1 user.

[6] Home passed: To obtain the number of users, a number of homes passed with fiber is needed although some of these homes will never be connected as users.

[7] Terminal box: It is the wall-box that distributes the fiber into 16 users.

Scenario 1(P2P):

- Deployment of P2P Topology needs 4.750 m of canalization and 14.400 m of optical fiber cable.

Scenario 2(P2MP):

- The Local Convergence Point is 600 m far from the node.
- Deployment of P2MP Topology needs 2.400 m of canalization, 7.275 m of optical fiber cable and 600 m of canalization for the tract from the central office to the LCP point.
- This topology allows free space inside the canalization ducts for possible scalability (using the same ducts as P2P).

7.3.1. Unitary costs

All the costs figures used for the analysis are in average from 5 market prices given by vendors and constructors companies. The precedence of some results is not shown because they are not transcendent for the example.

Cable:

- A = cable of 144 optical fibers = 11,4 €/m
- Z = cable of 96 fo[8] = 7,2 €/m
- B = cable of 32 fo = 3,36 €/m

Canalizations:

- C = ducts of 2 × 125 mm + 3 × 63 mm = 69 €/m
- D = duct of 3 × 40 mm = 30 €/m
- L = Cable spread = 1,2 €/m
- N = splitter (1:16) = 240 €/splitter
- M = hand hole of 40 × 40 × 65 mm = 630 €
- E = Transceivers = 60 €

7.4. SCENARIO 1: CASE P2P. DEPLOYMENT COSTS

Basic characteristics and the deployment costs for the P2P topology are shown here.

- Canalization length = 4.750 m
- Total cable length = 14.400 m
- 1.125 users

---

[8] fo = optical fiber.

- 1.500 homes passed
- 2.250 optical transceivers (2 transceiver per user)

Total Deployment Costs

The below equation obtains the total costs for an infrastructures deployment with the assumptions taken from scenario 1.

$$14.400 \times \left(0{,}34 \times \frac{432}{144} \times A + 0{,}06 \times \frac{144}{32} \times B\right) +$$

$$14.400 \times \left(0{,}34 \times \frac{432}{144} + 0{,}06 \times \frac{144}{32}\right) \times L +$$

$$4.750 \times (0{,}23 \times C + 0{,}77 \times D) + 2.250 \times E = \mathbf{522.905\ €}$$

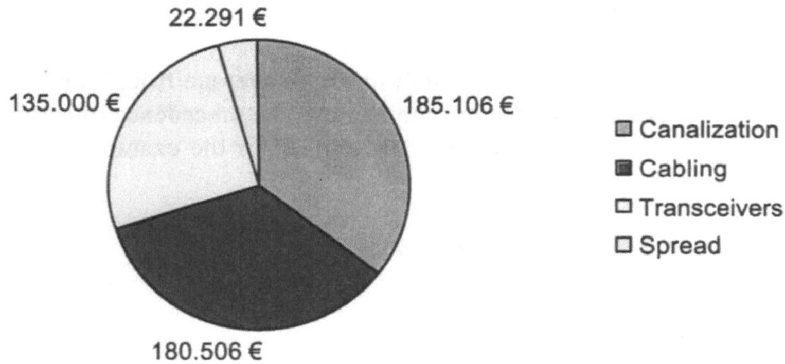

Assuming a total cost of 522.905 € in terms of infrastructure, a CapEx of 465 € per user is obtained.

7.5. SCENARIO 2: CASE P2MP. DEPLOYMENT COSTS

The P2MP topology has the following basic characteristics and deployment costs. Two fibers per user are used, one for upstream and the other for downstream traffic.

- 1.125 users.
- Canalization length = 2.400 m + 600 m.
- Cabling length= 7.725 m.
- The canalization length is reduced about 44% in comparison with the P2P topology. The reason is that only two fibers link the node and the optical splitter.
- The splitter is 600m far from the node.
- Number of optical splitters = 70.

- Two hand holes are needed for the installation of the splitters.
- Number of transceivers = 1.125 + 70 = 1.195.
- 70 fibers from the node (central Office) to the splitter + 10 dark fibers = 1 cable of 96 fibers.

Total Deployment Cost

The next equation obtains the total costs for an infrastructures deployment with the assumptions taken from scenario 2.

$$600 \times Z + 600 \times D + 7.725 \times \left(0{,}25 \times \frac{432}{144} \times A + 0{,}11 \times \frac{144}{32} \times B\right)$$

$$+ 7.725 \times \left(0{,}25 \times \frac{432}{144} + 0{,}11 \times \frac{144}{32}\right) \times L +$$

$$2.400 \times (0{,}23 \times C + 0{,}77 \times D) + 5 \times M +$$

$$70 \times N + (1.125 + 70) \times \mathrm{E} = \mathbf{293.460\ €}$$

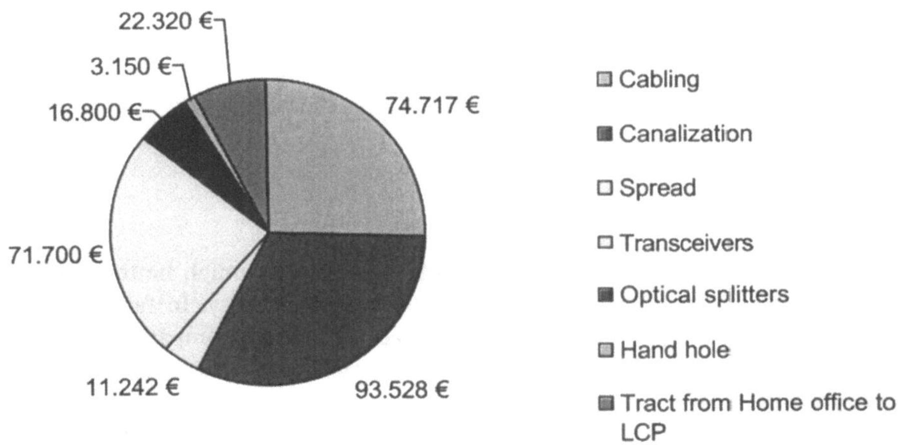

Assuming a total cost of 293.460€ in terms of infrastructure, a CapEx of 260€ per user is obtained. Due to the fact that this is the cost-effective solution, we should calculate the value of a general CapEx adding the costs of active user equipment, the switches from the Home Office and the repercussions of the access loop.

In order to determine how these new costs affect the analysis, the following assumption of equipment is needed:

- 1.125 users;
- 1.125 ONU's (O/E converter);
- 24 Switches with 48 ports GEthernet;

- 46 terminal wall-box;
- 10 meters of cable of 2 fibers per user ⇒ 2.250m of cable 2 fo.

After having determined the necessary equipment, the unitary prices of those elements are (some prices are approximated):

- ONU = 400€
- Termination Wall-Box for 32 fibers = 140€
- Cable of 2 fo = 0,8€/m
- Switch with 48 ports Geth.+ 2GBIC + SW = 7500€

Once we have all the premises defined, a global cost must be defined by:

$$293.460€ + 1.125 \times 400€ + 24 \times 7500€ + 10 \times 2 \times 1.125 \times 0{,}8€ + 32 \times 140€ = \mathbf{936.940€}$$

Finally a cost for the deployment and the equipment of the access network is obtained, and then in order to obtain a CapEx per user, a simply division must be done. This result shows the cost by user connected for a new operator who wants to deploy its own network.

CapEx = 832 €

7.6. RESULTS

In conclusion, the scenario with a P2MP topology is much better than the P2P in terms of cost, scalability, feasibility and flexibility. In fact, the total cost is reduced about a 40%. In order to establish a visual comparison between the two topologies, Figure 7.6.1 presents graphically the results obtained analytically before.

Referencing the above results, one consideration must be taken into account:

- In the distribution tract, it has been assumed that the two topologies use the same number of hand hole, so we only compare the hand hole used to put the splitters.

Once we have arrived at this point, only a visual global costs perspective of the two technologies is left, Figure 7.6.2 shows this case.

In conclusion, P2MP, as the best cost-effective topology, is a good consideration. The benefits are reflected in the infrastructures, cabling, and optical transceivers costs reduction although, with the P2MP topology, some costs such as splitters and hand hole are added (which are not important any longer). Furthermore, the management is easier.

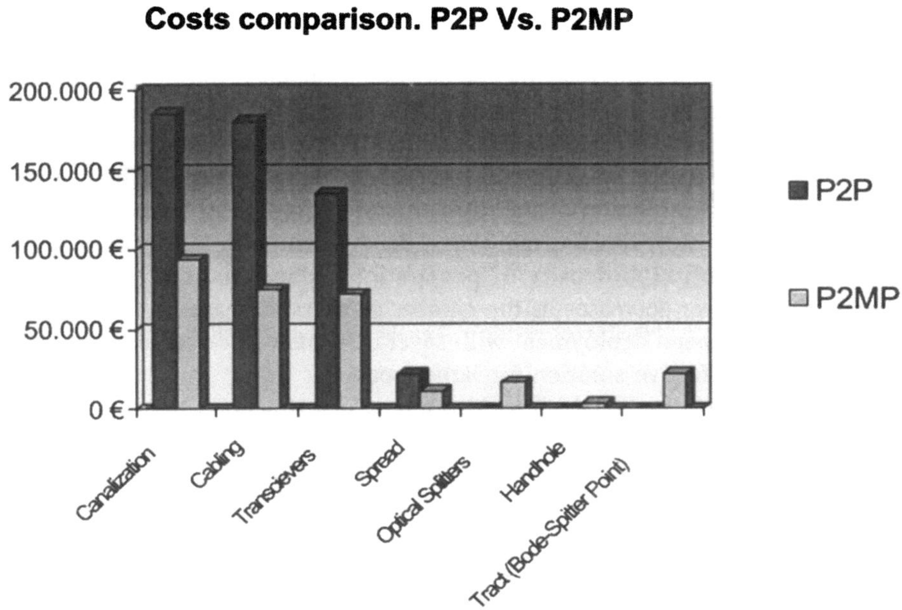

Figure 7.6.1. Infrastructures deployment costs comparison of P2P versus P2MP.

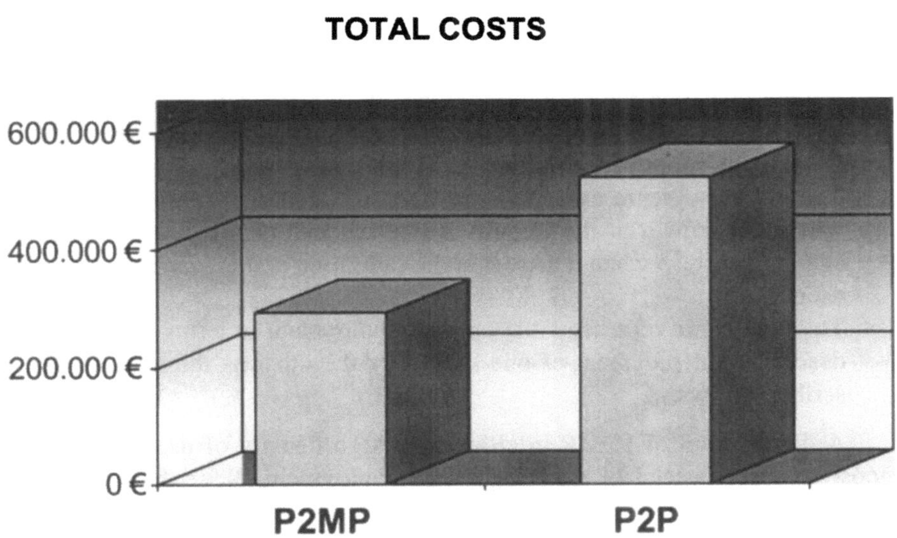

Figure 7.6.2. Total costs of infrastructures deployment.

7.6.1. Simulations

In order to conclude the example, some results obtained through simulations of a real area deployment will be shown. The analysis is focused on a European city and the topology analyzed is the P2MP, the new scenario presented assumes 60.000 meters of canalization and 37.000 homes passed.[9] This scenario will only reflect a relation of the infrastructure deployment costs.

Simulation results demonstrate that a deployment with P2MP topology in a large area with high density of penetration reduces the cost per user. In fact, cost per user decreases as the number of users increase. As a result, we can determine that a deployment with several Points of Convergence is really the best cost-effective solution for large areas.

Another aspect analyzed is the economical investment. A network deployment can be gradual or all at the initial, in other words, we can invest as we grow instead of realizing all the deployment investment at the beginning.

Simulation A

The simulation A shown in Figure 7.6.3 represents a scenario with several LCP's along the area deployment), in order to reduce the total costs and implement the P2MP topology. A penetration index of users of 60% (more or less) is estimated. The simulation has the following characteristics:

- Each curve starting from the upper left side represents the relative cost per user in scenarios with 1, 2, 3 . . . LCP's respectively.
- Each Local Convergence Point is designed to cover approximately 6.000 m of canalization.
- The deployment of the different Local Convergence Points (LCP) is realized at the same time for each curve. Once during the deployment, the LCP's arrives at the saturation point the cost/client grows, it is due to the difficulty in realizing more canalization.
- The red curve represents the costs due to a scenario with a P2P topology where the growth of the Central Office (node) is limited, it can grow but at a high cost, so when it arrives at the saturation point it can not serve more users.
- The blue curve represents the optimal deployment in terms of cost/user. It describes a deployment of one LCP after the other as the number of subscribers grows.

A deployment of P2MP in cities with high density of users reduces the cost per user instead of realizing a P2P deployment. Nevertheless, the cost

---

[9] Home passed: A home passed is the number of homes that have to be cabled in order to obtain a percentage (%) of clients.

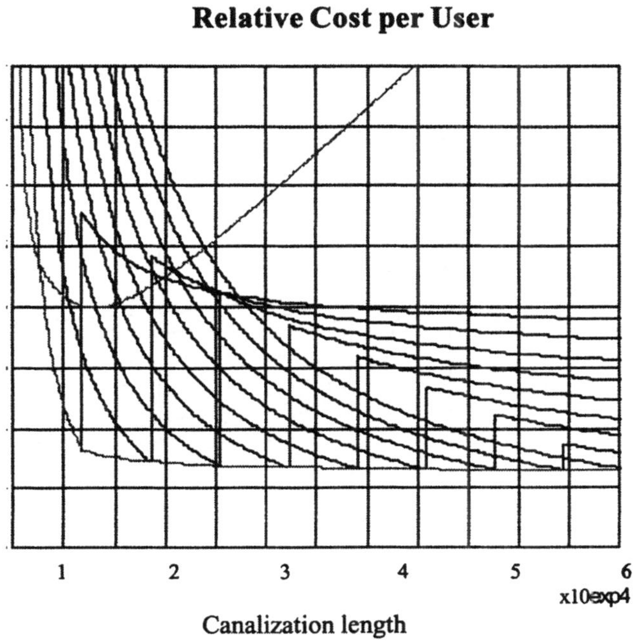

Figure 7.6.3.

among topologies in cities with a low density of users is high, so is then is better a P2P deployment. Then a deployment of P2P (always if the penetration factor is low) is better **[K.Shaneman 01]**.

Simulation B

The simulation B represented by the Figure 7.6.4 shows a scenario with several LCP's along the area deployment in order to reduce the total costs and implement the P2MP topology. A penetration index of users of 60% (more or less) is estimated. The simulation has the following characteristics:

- Each curve starting from the upper left side represents the relative cost per user in scenarios with 1, 2, 3 . . . LCP's respectively.
- Each Local Convergence Point and the respective canalization are dimensioned to support all the possible users that there are under the area they have to supply. For example, the third curve has 3 LCP's end each one and its canalization is designed to provide service to the existing users under the area it covers (23.500 m of canalization).
- The deployment of the different Local Convergence Points (LCP) is realized at the same time for each curve. They do not have a saturation point, because each LCP has been previously designed to attend the maximum number of users that can accept.

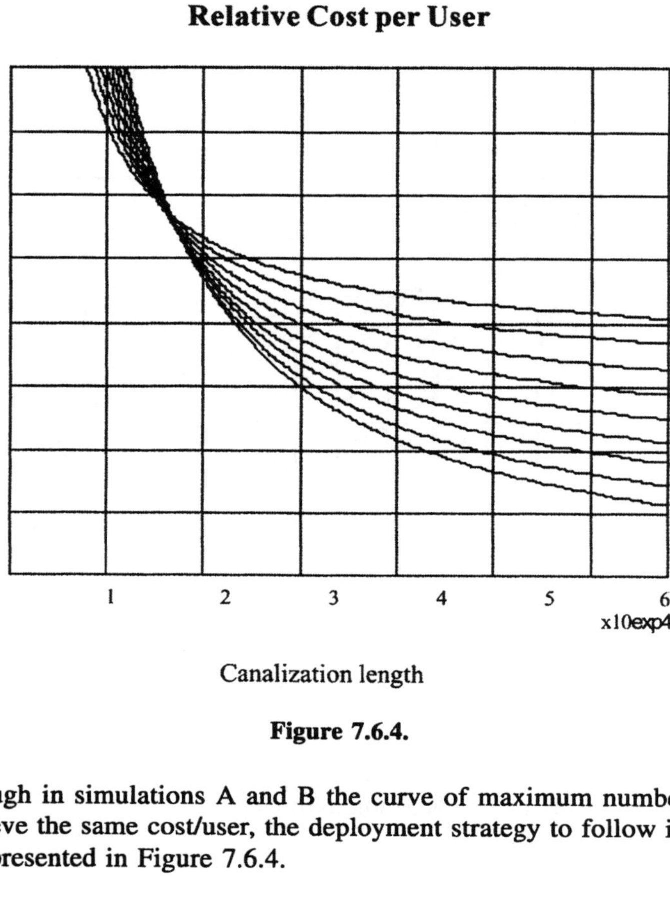

Figure 7.6.4.

Although in simulations A and B the curve of maximum number of LCP can achieve the same cost/user, the deployment strategy to follow is the blue curve represented in Figure 7.6.4.

7.7. CONCLUSIONS

After this analysis, the reader will note that not only the technology is an important issue, but the cost as well is an important factor in network deployments.

New operators, who want to deploy their own access network, could drastically reduce the CapEx by using P2MP vs. P2P topologies. Currently a total CapEx per user in P2MP is about 832€ (approximately).

In order to reduce the deployment costs of cabling and canalization, dark fiber is a good solution. Nevertheless, it is not easy to arrive everywhere and supply all the users with dark fiber.

Other advantages are the implementation facility, for example, dark fiber can be used along the first tract from the Central Office (node) to the splitter or LCP. Then the rest of the canalization could be deployed using ducts. The reduction in the fiber costs should be significant.

A network deployment has to be developed gradually in order to reduce the current cost per user, and the area that a LCP has to supply must be delimitated and designed previously before the deployment attending to some users density figures.

REFERENCES

[BOANA 01] UPC (Optical Communications Group), 'BOANA (Nortel project)', Networks Dec. 2001.

[EFM 01] Ethernet in the Firs Mail. Tutorial. IEEE 802 LMSC. Portland, OR. July 2001.

[Felipe01] J. Felipe Lobos, 'Red de acceso FTTx/xETHERNET', XI Telecom I+D 2001.

[G.Powell] Guy Powell, 'Implementing Passive Optical Network (PON) in the Acces Network', Telecoms & Transmisión Networks Conference, Nov. 2001.

[K.Shaneman01] Keith Shaneman, 'Deploying All-Optical Access Networks', Corning Access Solutions, EFM 2001.

[LightReading] LightReading, 'PON: The Dream Is Alive', Nov. 2001 'PON: The Dream Is Alive', Nov. 2001.

[M.Corner00] M. Corner, 'Residential Access Networks', Corning Cables System. August 2000.

[Pirelli 01] Pirelli Telecom, 'Developing Fiber To the Home', Nov. 2001.

[S.Figuerola02] S. Figuerola, J. Prat, 'Especificaciones Funcionales y de diseño para la planta exterior de una red óptica con tecnología FTTH', UPC January 2002.

[T.Volotinen00] 'An integrable and cost effective FTTH- network structure', OFCON. Jan. 2000.

MIX
Papier aus verantwortungsvollen Quellen
Paper from responsible sources
FSC® C105338

If you have any concerns about our products,
you can contact us on
ProductSafety@springernature.com

In case Publisher is established outside the EU,
the EU authorized representative is:
Springer Nature Customer Service Center GmbH
Europaplatz 3, 69115 Heidelberg, Germany

Printed by Libri Plureos GmbH
in Hamburg, Germany